Lecture Notes in Electrical Engineering

Volume 258

For further volumes:
http://www.springer.com/series/7818

Veena S. Chakravarthi
Yasha Jyothi M. Shirur
Rekha P.
Editors

Proceedings of International Conference on VLSI, Communication, Advanced Devices, Signals & Systems and Networking (VCASAN-2013)

Springer

Editors
Veena S. Chakravarthi
Yasha Jyothi M. Shirur
Electronics and Communication Engineering
B.N.M. Institute of Technology
Bangalore, Karnataka
India

Rekha P.
Telecommunication Engineering
B.N.M. Institute of Technology
Bangalore, Karnataka
India

ISSN 1876-1100 ISSN 1876-1119 (electronic)
ISBN 978-81-322-1523-3 ISBN 978-81-322-1524-0 (eBook)
DOI 10.1007/978-81-322-1524-0
Springer New Delhi Heidelberg New York Dordrecht London

Library of Congress Control Number: 2013941343

Printed on acid-free paper

Springer is part of Springer Science+Business Media (www.springer.com)

I.S.N. PRASAD, I.A.S.
Principal Secretary to Government

BANGALORE
THE IT CAPITAL OF INDIA

Department of Information Technology, Biotechnology and Science & Technology, Government of Karnataka

6th Floor, 5th Stage, M.S. Buildings,
Dr. Ambedkar Veedhi, Bangalore - 560 001. INDIA
Ph : 22280562, 22374314, 22032434
Fax : 22288340

Message

Bangalore has emerged as the knowledge hub of Asia owing to the abundant pool of technical manpower and research institutes both in the public and private sectors. To maintain this pride of place, the state Government has brought out its own ESDM Policy and accepted the recommendation of the Karnataka ICT Group 2020, with a focus on the electronics sector, as the future growth potential lies there.

In this context, It is hearty to note that BNM Institute of Technology, Bangalore is organizing an International Conference on VLSI, Communication, Advanced devices, Signals & Systems and Networking VCASAN-2013 during 17th -19th July'2013. The focus areas of the conference are subjects of immense relevance and are receiving global attention as they are integrated to our day to day living in many ways.

I am sure the conference will provide a great opportunity to the participants to meet with thought leaders and eminent speakers in the various sessions.

I wish the Event all success.

(I.S.N. Prasad)
Principal Secretary
Department of IT, Biotechnology
and Science and Technology
Government of Karnataka

URL: www.bangaloreitbt.in

email : itbtsec@gmail.com

Visvesvaraya Technological University

"Jnana Sangama", Belgaum - 590 018

Prof. H. Maheshappa
Vice Chancellor

Ref. No. VTU/VCS/2013/2116
16.04.13

MESSAGE

I am extremely delighted to know that BNM Institute of Technology, Bengaluru is conducting an International Conference VCASAN 2013 from 17th of July to 19th of July 2013.

The themes of the Conference VLSI, Communication, Signals & systems and Networking with smart electronics are indeed brilliant ideas. I am confident that this conference would offer a revitalizing podium for an extensive outlook to scholastic and to the working professionals to intermingle and swap their thoughts.

It would be a great opportunity for the academic circles, research and industry to move together with the familiar idea to signify their latest achievements and possibilities of collaborations can be explored.

I congratulate the management and the staffs on this momentous occasion. I wish VCASAN-2013 a grand success.

(Prof. H.Maheshappa)
Vice Chancellor

The Principal,
BNM Institute of Technology,
12th Main Road, Banashankari 2nd Stage,
Bengaluru-560 070

Phone : (0831) 2405454, Fax : (0831) 2405456, email: vc@vtu.ac.in, website : www.vtu.ac.in

Mindtree

April 29, 2013

Message :

I wish to congratulate BNMIT for conducting the International Conference VCASAN – 2013 in a grand scale. The innovation in Information Technology gets triggered by innovations in the VLSI and Networking space and the ripple effect is being seen be it through wave of products based on Internet or Cloud. The eminent speakers in the conference and multitude of relevant tracks will add colors and tremendously benefit the students, faculties and attendees in knowing the latest trends and gearing up for future. I wish the conference all the success and look forward to more such from BNMIT.

Welcome to Possible!

Best Regards,

Janakiraman
President & CTO

✆ : 91-80-26711780, 81, 82
Telefax : 91-80-26710881
Email : bnmitprincipal@gmail.com
Website : www.bnmit.org

Vidyaya Amrutham Ashnuthe

B.N.M. Institute of Technology

Administrative Office & Campus : P.O. No. 7087, 27th Cross,
12th Main Road, Banashankari II Stage, Bangalore - 560 070, INDIA

Narayan Rao R. Maanay
Secretary

Message

I am extremely happy that an International Conference, VCASAN 2013 is being organized at BNM Institute of Technology between 17th – 19th July, 2013. I am equally happy that there is a very good response from both India and abroad. This Conference on VLSI Communication, Advanced devices, Signals & Systems And Networking is definitely going to be very comprehensive and is expected to bring together some of the eminent personalities, researchers and thinkers from around the Globe. I am sure that this will provide a platform which will be research rich, refreshing and at the same time converge the knowledge of great minds of the academia with the industry. The varied agenda for the conference, I hope will not only bring out current technologies but also provide healthy exchanges, encourage and create a platform that will hold high the flag of Bangalore, the Electronics City of India.

I am delighted to congratulate Dr.Veena.S.Chakravarthi and her team, for organizing the conference working under the advisory committee comprising of Director Prof. T.J. Ramamurthy, Dean Dr. K. Ranga and Principal Dr.M.S.Suresh and eminent personalities in the field.

Springer Publications will be publishing this conference which would inspire and motivate the budding researchers, to come up with the best possible data and information.

I wish that, this three day International Conference achieves all the goals that has been set and accomplish its mission.

With every good wish,

Narayan Rao Maanay

Secretary,
BNM Institute of Technology

Vidyaya Amrutham Aahnuthe

B.N.M. Institute of Technology

(Affiliated to Visvesvaraya Technological University & approved by AICTE, New Delhi)
Administrative Office & Campus 12th Main Road, Banashankari II Stage, Bengaluru 560 070
☏: 91-80-26711780 - 81 - 82 Fax 91-80-26710881
Email . bnmitdirector@gmail.com Website : www.bnmit.org

Prof. T. J. Rama Murthy
B.E., M.Sc(Engg) M.I.E., M.I.S.T.E. M.I.E.E.E
Director

Message

I am pleased to welcome all the guests and participants to the International conference on VLSI, Communication, Advanced devices, Signals & Systems And Networking VCASAN-2013 in Bangalore, the silicon city of India!

Over the years the BNM Institute of Technology has surely evolved to become surely one of the most preferred T schools of Karnataka. The management and faculty believe that true learning can happen in a research rich environment. BNMIT constantly endeavors activities ensuring knowledge acquisition and sharing. VCASAN-2013 is a step forward in this direction. VCASAN 2013 provides an excellent forum for exchange of information and discussion on a wide variety of advanced electronics fields of VLSI, Communication, Advanced Devices, Signals & Systems and Networking, current developments and future trends.

Over the years, dramatic improvements have been made in the field of electronics enabling many SMART Products impacting human lifestyle. Currently we are now in a very dynamic and challenging time for SMART Electronics development. Today's SMART Products are an indispensable part of human life.

With more than 400 technical experts, representing academia, Industry and research organizations coming together at such a conference, we expect to have great educational experience and exchange of ideas in emerging technologies.

The conference site is situated in the centre of the city and reachable to all popular sights and points of interest in Bangalore. Take your time to visit the city of Mysore, the cultural capital of Karnataka.

I am looking forward to meeting you all in Bangalore during VCASAN 2013 and sharing the most pleasant, interesting and fruitful conference.

Prof. T. J. Rama Murthy
Director

Vidyaya Amrutham Ashnuthe

B.N.M. Institute of Technology

(Affiliated to Visvesvaraya Technological University & Approved by AICTE, New Delhi)
Administrative Office & Campus : 12th Main Road, Banashankari II Stage, Bengaluru - 560 070
✆: 91-80-26711780 - 81 - 82 Fax : 91-80-26710881
Email : bnmitprincipal@gmail.com Website : www.bnmit.org

I am happy to note that our Institute is organizing an International conference on VLSI, Communication, Advanced Devices, signals & systems and Networking (VCASAN) from July 17, 2013 to July 19, 2013. In this conference the topics related to both hardware & software are included along with several series of lectures from renowned professors. This conference brings all the researchers, academicians and industrialists on a single platform to analyze the latest problems and their adoptability. The outcome of this conference will help in dissemination of knowledge in the area of **"SMART ELECTRONICS"**

I wish the conference all the success and the efforts of organizers are to be appreciated, in their endeavor to conduct the conference in befitting manner

Prof. Ranga. K
Dean, BNMIT

Dean
B.N.M. Institute of Technology
P B No 7087, 27th Cross, 12th Main
Banashankari - II Stage
Bangalore - 560 070

Vidyaya Amrutham Ashnuthe

B.N.M. Institute of Technology

(Affiliated to Visvesvaraya Technological University & Approved by AICTE, New Delhi)
Administrative Office & Campus : 12th Main Road, Banashankari II Stage, Bengaluru - 560 070
☏: 91-80-26711780 - 81 - 82 Fax : 91-80-26710881
Email : bnmitprincipal@gmail.com Website : www.bnmit.org

Message

Dissemination of knowledge is the duty of all technical people. Knowledge is most effectively exchanged between persons when they talk to each other than through the print media of journals. A conference serves this important purpose. That is the time when researchers exchange ideas, seek new ones and renew old ones. VCASAN-2013, an International conference on VLSI, Communication, Advanced devices, Signals and Systems and Networking organized by BNMIT, has set to do this as extensively as possible bringing together academicians, industrial scientists and students from around the globe. This conference also covers a wide range of topics that is fertile for interdisciplinary research. Today's research hinges on our ability for convergence of technologies and barrier less exploration. VCASAN presents the delegates an opportunity to listen to industry leaders, active researchers and students throwing open a plethora of opportunities for new ideas. The pre-conference tutorials provide a good refresher course on VLSI, wireless Sensor Networks, Signal Processing and Analog design. I assume the opportunity presented for interaction by the international conference VCASAN is very wide for the delegates.

The organizing committee and its team have managed this stupendous task of organizing a conference of this magnitude with meticulous planning. This should encourage delegates to participate actively. I wish the conference a grand success.

Dr. M.S.Suresh
Organizing chair
Principal, BNMIT

Vidyaya Amrutham Ashnuthe

✆ : 91-80-26711780, 81, 82
Telefax : 91-80-26710881
Email : bnmitprincipal@gmail.com
Website : www.bnmit.org

B. N. M. Institute of Technology

Administrative Office & Campus : P.O. No. 7087, 27th Cross,
12th Main Road, Banashankari II Stage, Bangalore - 560 070, INDIA

(Affiliated to Visvesvaraya Technological University & Approved by AICTE, New Delhi)

Message

On behalf of the program committee I would like to welcome you to the international conference in VLSI, Communication, Advance devices, Signals & Systems And Networking VCASAN-2013.
The overwhelming response to our call-for-papers indicates the common interest among researchers, industry participants and academicians. The response has been good in all the announced tracks of VLSI, Communication, Advanced devices, Signals & systems And Networking. To our pleasure several invited talks from session chairs has been organized with a theme of "SMART ELECTRONICS", which opens the mind of the researchers beyond one's own field by looking into complementary fields. A separate UG/PG student track based on their projects is planned to encourage students to share their work. The selection of the student track papers for oral presentation will be done by expert review committee looking at the innovative content in their contributions.

First day, the Preconference day on 17th July starts with the inaugural session, followed by keynote speeches by distinguished experts across the globe. Afternoon session has tutorials consisting of five parallel tracks. On the remaining two days, the four parallel oral presentations of selected papers will be held in separate halls. We have arranged the parallel sessions such that overlap of related research areas is kept to a minimum. On the Second day of the conference, the evening will be closed by a special cultural event, followed by a banquet dinner.

In all, we will have close to fifty technical sessions during three days of conference, including five invited keynotes speeches, half-day tutorials on five most relevant topics. An industrial exhibition showcasing the latest technological educational offerings and an interesting Panel discussion during the closing session at the last day of the conference is being planned.

I am grateful to all the sponsors for their participation and encouragement. My heartfelt thanks are due to all invited chairs for their careful preparation of the keynote sessions and special articles for publication. I am grateful to all dignitaries for their encouraging messages for the conference.

Last but not the least, I wish to express my appreciation to my co committee members for all the support in making this conference most memorable one.
I am looking forward to seeing you for a most educational experience in Bangalore.

Veena S. Chakravarthi

Program Chair VCASAN-2013

Dr. Veena S Chakravarthi
Professor, ECE Department, BNMIT
12th Main, 27th Cross
Banashankari - II Stage, Bangalore - 560 070
Mobile: 98459 80607

B.N.M. Institute of Technology

Vidyaya Amrutham Ashnuthe

(Affiliated to Visvesvaraya Technological University and approved by AICTE, New Delhi).
Administrative Office & Campus : 12th Main Road, BSK II Stage, Bangalore - 560 070.
✆: 91-80-26711781-82 Fax : 91-80-26710881.
E-mail : inmaanay@vsnl.com Website : www.bnmit.org

Dr.S.B. Bhanu Prashanth

Prof & Head, Dept of ECE

MESSAGE

In my capacity as the head of the organizing department and a technical chair, I deem it a privilege to write this message for the proceedings of VCASAN-2013. Looking back, it was conceptualized to encompass the major streams of electronics engineering as the conference themes, with an objective to encourage the varied interests of academicians and researchers. And the extent to which the needs have been gratified is evidential today. The research papers were critically reviewed by experts in the respective areas and the appraisals have added significant technical value to the original contributions I believe.

Reflecting on the records, I congratulate and compliment the organizing committee for their stupendous work, and sincerely wish the conference to create a long-lasting influence on its participants, particularly on the young researchers who have tried to make an exemplar shift in their approach from 'Answering the Questions' to 'Questioning the Answers'.

Date: 24th April 2013

Best Wishes and Regards

24/4/'13

Dr. S.B. Bhanu Prashanth
Professor & HOD
Electronics & Communication Engineering
BNM Institute of Technology
Bangalore - 560 070

Bhageerathi Bai Narayana Rao Maanay Charities

Preface

Vidyaya Amrutham Ashnuthe

Hearty welcome to you all for the VCASAN-2013 conference!

It gives us immense pleasure to present you *Proceedings of the International Conference on VLSI, Communication, Advanced Devices, Signals and Systems and Networking* (*VCASAN-2013*).

This book is a collection of technical papers presented by renowned researchers, keynote speakers and academicians in the International Conference VCASAN-2013, organized by B.N.M. Institute of Technology (BNMIT), Bangalore, India during July 17–19, 2013. We are happy to include some of the interesting keynotes/articles from experts, in the proceedings of VCASAN-2013. Thanks to all great minds for sharing their view points in defining future trends.

The content of the book is useful to engineers, researchers, academicians as well as industry professionals by providing global trends in cutting-edge technologies in electronics and communication engineering.

The overwhelming response to our call for papers indicates the common interest among researchers, industry participants and academicians. For VCASAN-2013, we have received more than 125 draft technical papers from different places within and outside India. After the review process, 60 papers were selected for publication. The response has been good in all the announced tracks and outstanding in VLSI and Signals and Systems.

All submitted research papers have undergone critical reviews by an expert review committee and only accepted papers are included in the proceedings of VCASAN-2013, published by Springer. We would like to thank Springer for publishing the proceedings in the prestigious series of Lecture Notes in Electrical Engineering.

Professor T.J. Rama Murthy has recaptured B.N.M. Institute of Technology's growing years.

We would like to express thanks to all the dignitaries for their encouraging messages and technical experts in the fields for their keynotes. Our heartfelt thanks to all authors for their outstanding contributions and in particular the members of the technical review committee for their competent evaluation of the large number of submissions.

We thank BNMIT Management and Advisory Committee for their continuous support during the entire process of bringing out this book. Furthermore, we would like to express our appreciation to the Technical Committee for review coordination.

We thank Organizing Committee, Program Committee, Publicity Committee, Registration Committee and Finance Committee for their great efforts and time in organizing such an event.

We are sure you would enjoy reading the proceedings of VCASAN-2013 as much as we enjoyed putting it together.

Happy Reading!

Dr. Veena S. Chakravarthi

Yasha Jyothi M. Shirur

Rekha P.

Patrons and Committees

Patrons

Prof. B. C. Rajanna, Chairman, BNMEI, Bengaluru

Prof. Sunanda P. Jadhav, Founder Secretary, BNMEI, Bengaluru

Mr. I. S. N. Prasad, Principal Secretary, Department of IT, BT and S&T, Govt. of Karnataka, Bengaluru

Dr. H. Maheshappa, Vice Chancellor, VTU, Belgaum, Bengaluru

Sri Naryan Rao R. Maanay, Secretary, BNMEI, Bengaluru

Prof. K. R. Iqbal Ahmed, Member, G. C., BNMIT, M. A. N. U. Central University, Hyderabad

Sri Ashok R. Maanay, Joint Secretary, BNMEI, Bengaluru

Dr. C. Naganna, Member, G. C., BNMIT, Bengaluru

Advisory Committee

Dr. S. Sadagopan, Founder Director, Indian Institute of Information Technology, Bengaluru

Dr. Santanu Das, Founder, Domanicom Corporation Inc., USA

Dr. B. S. Sonde, Retired Professor, IISc, Bengaluru

Mr. S. Janakiraman, President and CEO, Mindtree Ltd., Bengaluru

Mr. C. S. Mohan, Consultant, Aquantia Semiconductor India Pvt. Ltd., Bengaluru

Mr. Vijay Muktamath, Research Fellow, Bionic Vision, University of Melbourne, Australia

Dr. Devendra Jalihal, Professor, Department of ECE, IIT Madras, Chennai

Dr. K. E. Prakash, Registrar, VTU, Belgaum, Bengaluru

Mr. Vilas Bhade, Vice President, Engineering, Synapse India, Bengaluru

Dr. C. P. Ravikumar, Technical Director, University Relations, Texas Instruments, Bengaluru

Mr. M. N. Kumar, Vice President India Operation, Aquantia Semiconductor India Pvt. Ltd., Bengaluru

Prof. T. J. Rama Murthy, Director, BNMIT, Bengaluru

Dr. K. Ranga, Dean, BNMIT, Bengaluru

Dr. M. S. Suresh, Principal, BNMIT, Bengaluru

Program Committee

Program Chair Dr. Veena S. Chakravarthi, Professor, Department of ECE, BNMIT, Bengaluru

Dr. Devendra Jalihal, Professor, Department of ECE, IIT Madras, Chennai

Mr. Ajay G. Gupta, Director, Department of Computer Science, Old Dominion University, Norfolk, United States

Mr. Vijay Muktamath, Researcher, BioVision, Australia

Dr. Swaroop Ghosh, Assistant Professor, Department of CSE, University of South Florida, US

Dr. S. B. Bhanu Prashanth, Professor and Head, Department of ECE, BNMIT, Bengaluru

Dr. Vijaya C., Dean, R & D, SDMCET, Dharwad

Dr. Raghunath Govindachari, CTO, Mindtree Consulting Pvt. Ltd., Bengaluru

Dr. A. Srinivas, Professor, Department of ECE, PESIT, Bengaluru

Mr. M. N. Kumar, Vice President India Operation, Aquantia Semiconductor India Pvt. Ltd., Bengaluru

Dr. Siva Yellampalli, Professor, UTL Technologies, VTU Extension Center, Bengaluru

Prof. Rajashree Narendra, Department of Telecom, BNMIT, Bengaluru

Dr. S. Rajesh, Director, Cadence Design Systems, Bengaluru

Mrs. Yasha Jyothi Shirur, Assoc.Professor, Department of ECE, BNMIT, Bengaluru

Mrs. P. Rekha, Assoc.Professor, Department of Telecom, BNMIT, Bengaluru

Organising Committee

Organising Chair Dr. M. S. Suresh, Principal, BNMIT, Bengaluru

Dr. K. Ranga

Prof. Rajeashree Narendra

Dr. Veena S.Chakravarthi

Dr. G. N. Krishnamurthy

Dr. P. A. Vijaya

Dr. B. G. Prasad

Dr. M. N. Chandrashekaraiah

Dr. Nagalingappa

Dr. H. A. Vidya

Dr. S. Manjunath

Dr. Thipperudrappa

Dr. Jayanna

Mr. P. S. Ramesh

Mr. R. M. Puttaswamy

Mr. D. N. Krishna Kumar

Mr. N. Shesha Prasad

Mrs. Veena S. Murthy

Mrs. S. Bindu

Mr. Subodh Kumar Panda

Mrs. P. Prabhavathi

Mr. Shivananda

Mr. P. Venkat Rao

Mrs. Vrunda Kusanur

Mrs. A. Sumathi

Mrs. B. L. Sujaya

Mrs. J. P. Anuradha

Mrs. K. Mamatha

Mrs. Rashmi S. Bhaskar

Mrs. Anuradha V. Rao

Mr. P. Sudir

Mrs. Priya R. Sankpal

Mrs. Ashwini S. Savanth

Mrs. Padmaja Jain

Mrs. K. Chellamma

Mrs. M. Ayesha Khanum

Mrs. N. Chaitra

Mrs. Priyadarshini K. Desai

Mr. K. N. Kiran

Mrs. B. P. Aruna Rao

Mrs. Keerti Kulkarni

Mr. Shivasharanappa B. Biradar

Mrs. Reena Kulkarni

Mr. C. Chandrashekar

Mrs. N. Rashmi

Mrs. M. V. Sireesha

Mr. Rajesh Sudi

Mrs. N. Sindhu

Technical Review Committee

Chair Dr. S. B. Bhanu Prashanth, Professor & Head, Department of ECE, BNMIT, Bengaluru

Prof. Rajashree Narendra, Head, Department of Telecom, BNMIT, Bengaluru

Dr. N. Anand Mohan

Mr. Ananth Krishna

Mr. Vilas Bhade

Mr. C. S. Mohan

Mr. Dinesh Annayya

Dr. Bindu A. Thomas

Dr. Jayadeva Jayadeva

Mr. G. S. Girish

Dr. Gopinath Karmakar

Mr. K. N. Hosur

Mrs. C. N. Jayashree

Dr. K. S. Gurumurthy

Mrs. Kamala Yeresime

Dr. C. Kavitha

Mrs. Kavitha shankar

Dr. K. L. Sudha

Mrs. Koppad Deepali

Mr. Mayur Koulavkar

Mr. D. N. Krishnakumar

Mr. Krishnappa Suresh

Dr. KSRC Murthy

Dr. Kulkarni Meghana

Mr. R. Kumaraswamy

Mr. H. V. Kumarswamy

Mr. M. N. Kumar

Mr. Mahendra Dixit

Mr. Manjunath Nayak

Mr. Mayank Chakraverty

Mr. Mukhtamutt Vijay

Mr. Mundkur Visweswar

Mr. P. Nagaraju

Mr. Narasimha Murthy

Mr. R. Narendra

Mr. S. S. Navalagund

Mr. Nitin Pujari

Mrs. P. Prabhavathi

Mrs. Pramila Manohar

Dr. B. G. Prasad

Mrs. R. Jayagowri

Dr. S. Ravishankar

Dr. S. Rajesh

Dr. Veena S. Chakravarthi

Dr. Sadashivappa

Dr. Sahana Gowda

Mr. Sandeep Agarwal

Mrs. Saritha Chakrasali

Mr. Senthil kumar

Dr. Shashank Nagaraj

Mr. Shivakumar Sundaram

Mr. Shreedhar Joshi

Mrs. Sita Prakash

Dr. Siva Subbarao Patange

Mr. Somashekhar Bhat

Mr. R. Sreekantaswamy

Dr. A. Srinivas

Mr. H. C. Srinivasaiah

Dr. A. Srinivasan

Mrs. Sujatha Gopinath

Mr. Sujit K. Nayak

Dr. Sunil Manvi

Dr. M. S. Suresh

Mr.Tharakarama Raju

Mr. K. Thippeswamy

Dr. Uthara Kumari

Dr. L. Vibha

Dr. P. A. Vijaya

Dr. C. Vijaya

Dr. Siva Yellampalli

Contents

Part II VLSI

Part I
Keynotes

Chapter 1
Leveraging Today's Innovation for Tomorrow's Success Connecting the World from Cloud Data Centers to Consumer's Tablets

Faraj Aalaei

Abstract The creation and consumption of data is exponentially increasing, driven by everything from networked mobile devices to consumer apps to big data in enterprise organizations. The trends toward more cloud-based computing and centralized hosting are, in turn, triggering an unprecedented migration and consolidation of data into large-scale data centers.

1.1 Introduction

The creation and consumption of data is exponentially increasing, driven by everything from networked mobile devices to consumer apps to big data in enterprise organizations. The trends toward more cloud-based computing and centralized hosting are, in turn, triggering an unprecedented migration and consolidation of data into large-scale data centers. The same kind of processing demands are beginning to emerge in the enterprise space, with data consumption on the rise and trends like "bring your own device" are adding higher and higher requirements to those networks, both wired and wireless.

The question is how will all of this impact the future of network connectivity? What solutions will address the tsunami of data that is unfolding?

F. Aalaei (✉)
President and CEO, Aquantia Corporation, 700 Tasman Drive, Milpitas, CA 95035, USA
e-mail: Faraj.Aalaei@aquantia.com

V. Chakravarthi et al. (eds.), *Proceedings of International Conference on VLSI, Communication, Advanced Devices, Signals & Systems and Networking (VCASAN-2013)*, Lecture Notes in Electrical Engineering 258, DOI: 10.1007/978-81-322-1524-0_1,

1.1.1 The Data Center Dilemma

Industry analysts forecast that there will be more than 19 billion connected devices by 2016, from smart phones to home gateways to laptops and tablets. For both consumers and enterprise workers, those devices will stream everything from corporate files to games and video content. The majority of that data traffic will flow through data centers that form the cloud, where IP traffic is expected to reach 1.3 ZB by 2016.

More than forty million Ethernet ports shipped into data center servers in 2012. But with all of that data consolidation in the cloud, Gigabit Ethernet connectivity still represented more than 90 % of all ports, and as the trends accelerate, Gigabit Ethernet is quickly becoming insufficient. Combine that with virtualization and the step function in processor capabilities brought by the Romley generation of servers launched in 2012, and data center operators are struggling to keep up with the I/O bottleneck while also keeping costs and power consumption levels in check.

The clear solution for data centers is to migrate to 10Gigabit Ethernet. Indeed, 10GbE shipments are expected to ramp sharply in the coming years, particularly with 10GBASE-T native on Romley servers, easing adoption for data center operators. Research from Gartner and Bank of America Merrill Lynch predicts a 47 % compound annual growth rate in 10GbE switching and controllers from 2011 to 2015, resulting in more than 50 million units shipping in 2015. All the pieces are in place for the rapid adoption of 10GbE in data centers.

1.2 The 10GBASE-T Challenge

Widespread deployment of 10GbE in data centers requires a cost-effective connectivity solution that is backward compatible and has the capability to leverage existing cabling, a combination of attributes favoring 10GbE over copper (10GBASE-T). But 10GBASE-T is one of the toughest communication technologies devised to date by the IEEE. More than $500 million in venture capital and public company money has been spent so far to try and devise a solution. Multiple start-ups have tried and failed to address the challenges of 10GBASE-T. The stakes are enormous.

The challenges of 10GBASE-T lie in cabling characteristics and technical requirements. Over 100 m of CAT6A cable, for example, attenuation is close to 50 dB, leading to a margin to Shannon limit of less than 2 dB. Enabling BER < 10–15 operation under worst-case environments renders the design of a 10GBASE-T transceiver extremely challenging.

From its inception, Aquantia has been solely focused on addressing the challenges of 10GBASE-T. And where many companies took a brute force approach to solving this problem, leading to high power and large die size, Aquantia's team of experts addressed the challenges of 10GBASE-T by creating the Mixed-Mode Signal Processing (MMSP) architecture.

The MMSP's leading building block is a unique Analog Front-End (AFE) that partially cancels impairments on the incoming signal (such as cross talk and echo). As a result of this front-end signal processing, the ensuing Analog-to-Digital Converter (ADC) can be designed with lesser requirements, leading to a lower power and smaller size AFE. Post-ADC, the amount of signal processing needed to be performed on the signal is greatly reduced, which results in a less complex, lower power and smaller die size digital architecture.

The other domain where Aquantia's solution excels is in FEC (Forward Error Correction) using an LDPC (Low Density Parity Check), critical in achieving the target BER. LDPC requires several thousands of complex Hyperbolic ArcTangent calculations on each data frame performed by several hundred processor elements that need to communicate data between them at full speed. Aquantia has significantly simplified the Hyperbolic ArcTangent function through a mathematic approximation without any loss of performance for specific application of LDPC, while reducing complexity by many folds.

Also, to reduce switching power consumption on the processing nodes, Aquantia implemented a different communication method in which only those bits that are corrected in each processing steps are communicated, reducing the switching power between the nodes by at least an order of magnitude. Aquantia's target LDPC showed to have more than 3x power improvement compared to conventional implementations in similar process nodes.

The migration of data centers to 10GbE requires cost-sensitive solutions that address the diversity and reality of high-volume, multi-vendor environments. Aquantia is playing a critical role in the transformation of the data center to 10GbE, having delivered the world's most mature, lowest power and smallest footprint 40 nm 10GBASE-T PHY in production, for high-density switches, and the world's only 10GBASE-T solution for LAN-on-Motherboard (LOM) servers.

The growth of 10GBASE-T is set to accelerate in the coming quarters as a result of the world's only integrated MAC/PHY for LOM now being available on Romley servers. That solution addresses all the attributes of BASE-T that IT organizations have dealt with for the past 40 years: ease of use with existing cabling, backward compatibility, and continuing power and cost reduction following Moore's Law. As a result, Crehan Research projects that more than 5 million 10GBASE-T ports will ship in 2013, with that number reaching 40 million by 2016.

1.3 Beyond the Data Center

The next logical opportunity for 10GBASE-T is in enterprise and home network environments, where Ethernet already is ubiquitous and traffic demands are rapidly increasing, thanks again to the proliferation of mobile devices, video and multimedia traffic.

Many of the drivers for the migration to 10GbE in these markets are the same as they are in the data center world. The ever-increasing growth of traffic on mobile networks is creating a bottleneck where the wireless network meets the wired infrastructure. The pico and femto cells that extend mobile networks are deployed in indoor environments where 99 percent of all cabling is already Copper BASE-T—ideal for upgrading to higher bandwidth.

Standards for Wi-Fi also are advancing, such as with the 802.11c-based wireless LANs requiring a gigabit or more per antenna—again bolstering the argument for upgrading the wired infrastructure connecting the WiFi Access Points to the core network. And in SOHO and home network environments, a broad range of connected mobile and fixed devices are contributing vast amounts of traffic that increases the bandwidth requirements of those networks. Ethernet already is ubiquitous in those environments as a standard for connecting everything from routers to set-top boxes to game consoles to Internet-connected TVs. The vast amount of copper wiring that connects those devices presents another opportunity to upgrade the infrastructure, leveraging 10GBASE-T as a low-cost, low-power connectivity solution.

From the consolidation of traffic in data centers to the proliferation of connected devices to the growing importance of home and enterprise networks, the data tsunami that has started to unfold represents a considerable challenge at the connectivity level. In most cases, Gigabit Ethernet is still the dominant interconnect being deployed in both large-scale data centers and backhauls of cellular and WiFi networks. But, as we have discussed, Gigabit Ethernet is quickly becoming utterly insufficient and is creating an unsustainable situation, opening the opportunity for a 10Gigabit Ethernet connectivity solution to upgrade this massive set of networks. Among all 10GbE physical layer alternatives, 10GBASE-T represents the most cost- and power-effective solution for addressing the future of communications connectivity. Aquantia is at the forefront of this exciting transition taking place in large-scale data centers today, as it is also leading the way in the next phase about to unfold in Enterprise, SMB/SoHo and the home.

Chapter 2
Technology for SMART HOME

Santanu Das

Abstract The proliferation of internet of things beating human population, trending needs of aging population demanding service on demand, easy access to scalable servers and clouds of services anytime anywhere wide range of marketing is driving the home to SMART HOME which will soon be a reality. This paper presents the technologies responsible for this transformation.

2.1 Introduction

A SMART HOME is a HOME where in all man's comfort, entertainment, and esthetic needs are governed by technology. For example, ambience parameters like lighting/temperature needs are made intelligent to sense the need and provide. In a SMART HOME, a person can have all necessary services right from wake up alarms to personalized calendars with essential reminders to diet needs to medical support available to his disposal with an ease of a click of buttons or gestures. This is possible because of the advancement of wide variety of sensor technologies and high processors technologies.

2.2 Drivers for SMART HOME

SMART HOME is driven by the soon to be necessities like

- Energy conservation and sustainable design
- Home automation/security

S. Das (✉)
Domani Systems Inc, 2 Trap Falls Road, Suite 106, Shelton, CT 06484, USA
e-mail: santanu44@aol.com

V. Chakravarthi et al. (eds.), *Proceedings of International Conference on VLSI, Communication, Advanced Devices, Signals & Systems and Networking (VCASAN-2013)*, Lecture Notes in Electrical Engineering 258, DOI: 10.1007/978-81-322-1524-0_2,

- Home assistance (Tele-Assistance)
- e-Health (Telemedicine)
- e-Education
- Remote monitoring of appliances
- Communications and entertainment
- IT services for home and home office

A conceptual SMART HOME model is shown in Fig. 2.1.

Many contributes of this possibilities are electrical power distribution players, smart building control specialists, building application companies providing HVAC, lighting, security for homes, house hold appliances providers, SW, IT, communication equipment players and service providers like telecom operators (Fig. 2.2).

2.3 SMARTNET

The technical model of the SMART HOME is shown in Fig. 2.3.

In a SMART HOME, the SMART circuit constitutes a Master sever with cloud capability with the uplink interfaces on DSL, PON/WIFI on one side and other need-based pluggable service modules on the home side. The service modules can be variety of modules based on any short-range technologies like WIFI, ZIGBee, etc.

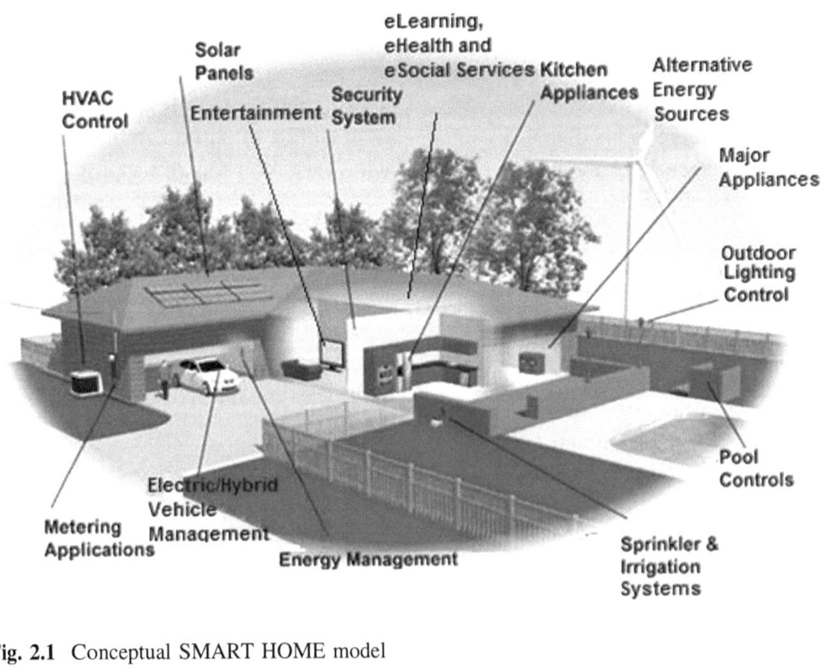

Fig. 2.1 Conceptual SMART HOME model

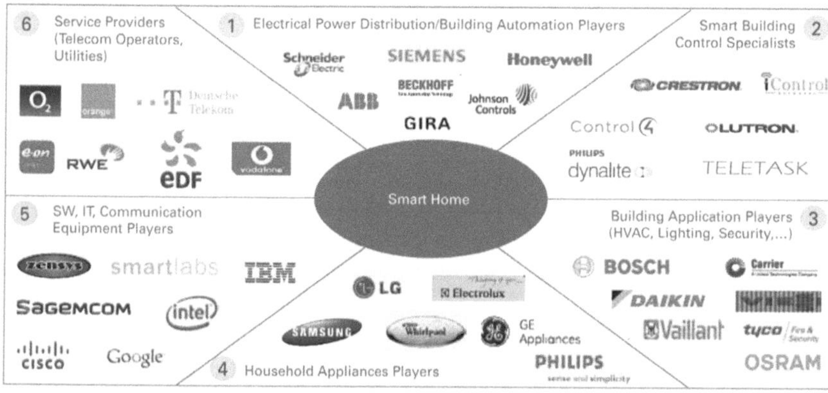

Fig. 2.2 Major architects of SMART HOME. Logos referred in this figure are respective properties of the companies

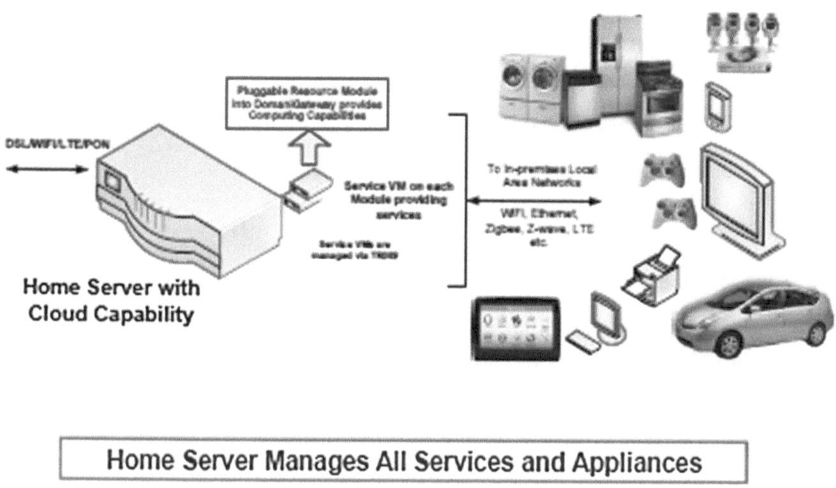

Fig. 2.3 HOME server manages all services and appliances

catering to internal automation/data transfer and controls for automated appliances for both utility and entertainment as shown in Fig. 2.3.

2.4 Requirements of SMART HOME Technologies

It is necessary to integrate entertainment and nonentertainment services like Video, Audio, Games to IT, Home Security, Skype, Browser on TV, etc. The technology should be shared by many service providers with security being par mount importance. Most of the services have to be in software so that they are flexible, upgradable and most importantly to avoid cluttering of multiple devices. The SMART HOME products should be on the fly provisionable and upgradable with easy user interface. It is essential for all the products on SMART HOME Network to be of low-power products, and they have to be of low cost affordable by larger mass.

The HOME Network has to support heterogeneous technologies to support multiple services. The HOME server is the heart of HOME Network, where most of the service applications run. One such Domani's HOME server architecture is shown in Fig. 2.4.

User or third party applications run on separate Virtual Machine (VM) with a standard interface to HOME Server.

The Domani's intelligent software architecture is shown in Fig. 2.5. The other important attributes of server software are in its ease of local and remote upgradability, maintenance, and security.

The multivendor service support frame work is achieved by TR-069-based management infrastructure as shown in Fig. 2.6.

The salient features of Domani's server are the following:

- Central Server, Home Server, and IP set-top box have a virtualized architecture
- Linux is the host operating system
  - In all three subsystems
- KVM is our preferred Hypervisor
  - On top of Linux
- Virtual Machines are specialized
  - Depending on the particular services it supports
- Each Virtual Machine has its built in Security Layer
- A common Security Layer isolates the Virtual Machines
  - Common Security Layer is based on Fuzzy Logic

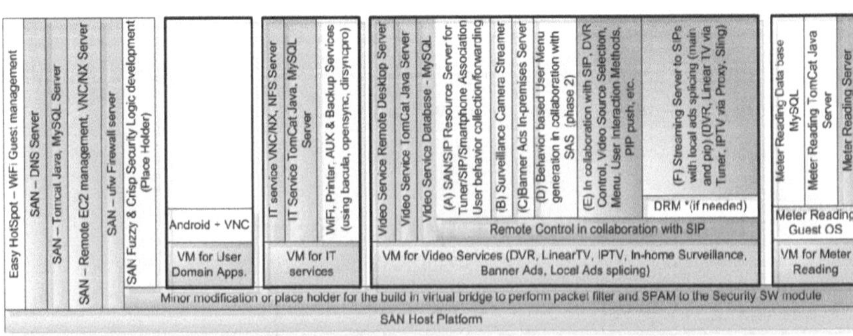

Fig. 2.4 Domani's HOME SERVER architecture

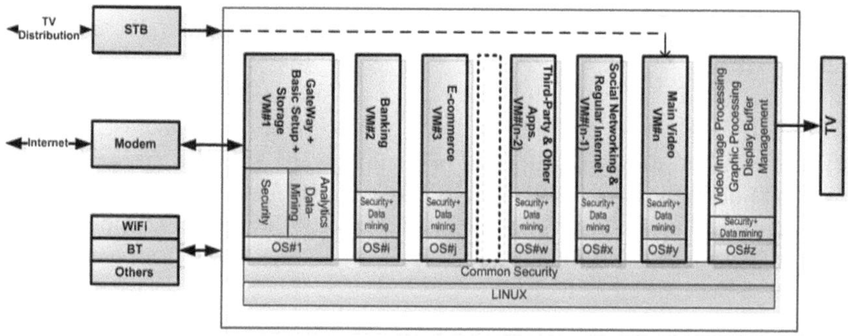

Fig. 2.5 Domani's software architecture

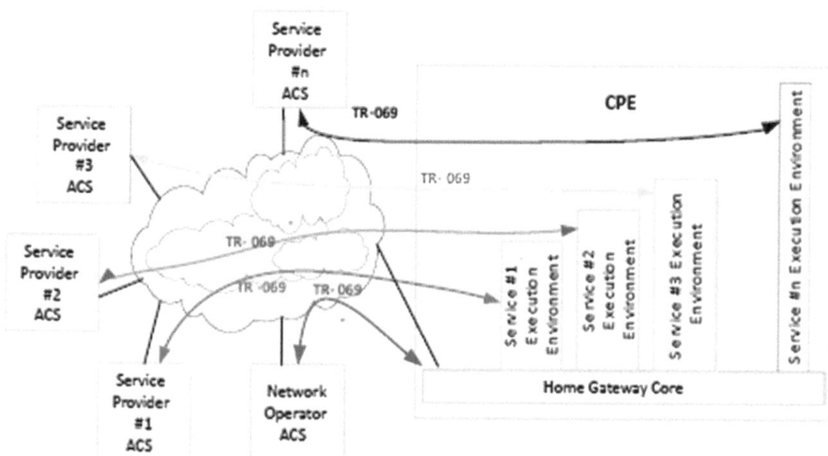

Fig. 2.6 Architecture compatible with OSGi and home gateway initiative

2.4.1 Supported Services

Variety of services like e-learning, telemedicine, VOIP phone, and video conferencing, environmental and appliance monitoring, and many more can be supported by Domani's platform.

2.4.2 SMART Sensor Network

Another essential building block of the SMART Home is the smart sensor network. The sensors are used to sense different parameters in a network configuration which are controlled automatically at home.

2.5 Conclusion

It is to be noted that major building blocks of SMART HOME technology are SMART HOME server and the sensor networks. Necessary features, functional requirements and macro architecture of the hardware, software are defined, and few of the differentiators are identified. The essential features of Domani's server which is suitable for SMART HOME is described. The need for supporting different VM devices is highlighted.

Chapter 3
Industry Trends and EDA Tool Vendor Readiness for New Inventions in Chip Design

Dennis Brophy

Abstract Today, there are two distinct branches for the design of System on Chips (SoC's) and Embedded Systems. Yet, at the advent of systems that will do everything, all the time, everywhere the notions of SoC and embedded systems design need to grow and expand to encompass all of a system's computational and physical elements. Often referred to as cyber-physical systems (CPS) when these two are combined, it is at the heart of design of the emerging smart energy grids, automotive domain, healthcare, entertainment, consumer applications, and more. The economic and societal potential is greater than that we have already witnessed with mobile consumer devices. The seeds of support for CPS will enable machine-to-machine systems, the Internet of Things, and more. What impact will this have on design? What might the EDA tools to design cyber-physical systems look like? And when will this happen?

D. Brophy (✉)
Director of Strategic Business Development, Mentor Graphic Corporation, 8005 SW Boeckman Road, Wilsonville, OR 97070, USA
e-mail: dennis_brophy@mentor.com

V. Chakravarthi et al. (eds.), *Proceedings of International Conference on VLSI, Communication, Advanced Devices, Signals & Systems and Networking (VCASAN-2013)*, Lecture Notes in Electrical Engineering 258, DOI: 10.1007/978-81-322-1524-0_3,

Chapter 4
Semiconductor Technology Enabling Smart Electronics

Lode Lauwers

Abstract This paper captures new trends in semiconductor technology, and how it influences and enables the design of future systems, and to the creation of "smart" devices. The paper also explains some of the smart electronics trends in mems, imagers, biomedical applications and illustrates the collaborative model of imec to achieve make this possible.

Keywords Through silicon vias · 3D System integration · Fabless and fablite companies · FinFETs · Aspect ratio trapping · SST-MRAM · Resistive RAMs

4.1 Engine for System Innovation

Future systems will be able to compute much faster and store many times more data than today's ICs. At the same time, they will use less energy. All this thanks to the relentless drive of the IC industry to keep on scaling chips. As the number of options keeps on increasing with higher degrees of complexity, resulting in an explosion of cost of new technologies generation after generation, pioneering collaboration will be vital to bring the required new processing technologies for this to market. Innovation will encompass new materials, transistor types, and integration and design methods.

In process technology R&D, new technologies and processes are developed for deep-submicron CMOS scaling, for both logic and memory technology, building on unique expertise in fundamental material understanding.

The logos used in the articles are properties of respective companies and photos are properties of IMEC.

L. Lauwers (✉)
Vice President Business Core CMOS, IMEC, Leuven, Belgium
e-mail: lauwersl@imec.be

V. Chakravarthi et al. (eds.), *Proceedings of International Conference on VLSI, Communication, Advanced Devices, Signals & Systems and Networking (VCASAN-2013)*, Lecture Notes in Electrical Engineering 258, DOI: 10.1007/978-81-322-1524-0_4,

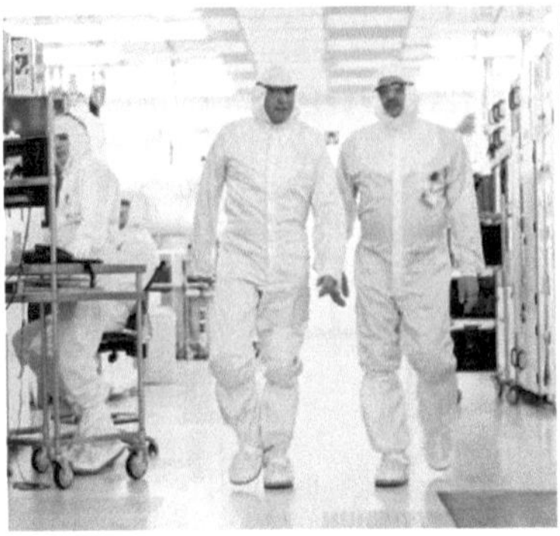

With a horizon of 10 years ahead, options need to be explored to create chips with dimensions smaller than 5 nm, an effort which, other than in "classical" geometrical scaling, is now aiming for the introduction, generation after generation, of alternative materials and new device and system architectures. It is the only (creative) way forward in order to bring the required improvement on the power/performance/area versus cost curve.

Beyond this dimensional-driven "roadmap" scaling of devices and circuits, total system scaling will be enabled by new routes of system-oriented technology innovation. One important route is stacking chips. Large programs have been launched in recent years on 3D system integration, using Through Silicon Via's, which allow to stack chips of various nature: heterogeneous integration of different substrate technologies, or optimized chips in memory and logic nodes. In complement of advanced stacking approaches, new programs have been launched on high-bandwidth optical input/output, creating optical solutions for high-bandwidth communication between the components of CMOS systems (Figs. 4.1, 4.2).

Fig. 4.1 Stacked chips, die to wafer

Fig. 4.2 Through silicon vias (TSV); exposed Cu nails

Finally, when designing future technologies, one witnesses an increasing need for advanced system studies, which give companies upfront knowledge about imminent and emerging technologies, so-called “design enablement.” This allows in a very early stage to tune the technology creation to a new application class, and an early technology design co-creation, bringing a first generation of a product on the market faster.

Above trends in semiconductor technology, are imposing new challenges on the future R and D, and new results in those areas will influence if not enable the design of future systems, or the creation of “smart” devices.

4.2 Enabling the Technology Roadmap

When striving for enablement of the technology roadmap and further scaling, significant efforts are required in fundamental understanding of materials, often building on in-depth and novel analysis and characterization techniques, modeling and device knowhow. Such efforts and the costs thereof can often not be born anymore by a single company, hence the imminent need for partnership structures, where results are generated in a state-of-the-art research infrastructure, allowing its stakeholders to mitigate risks, derive valuable insight for future development, transfer components for solutions, yet at a fraction of the total cost. Such stakeholders range from equipment to material suppliers, over IC manufacturers and foundries as well as design houses and fabless and fablite companies. All find their “angle of interest” and value in the platform results.

And indeed, due to the increased link between the technology components and the possibilities and/or limitations at circuit and system level, a technology can no longer be a black box, a “burden for others,” but mandates the right in-depth “insight” from its current and future users. Such insight enables design technology co-optimization through early access to such advanced technologies, long before these are implemented in the fabs. At the end of the day, each of the platform contributing partners will derive the value, essentially by being able to implement their roadmaps faster, with lower risk and within cost.

The recent trend of leading fabless design companies increasing their investments in process technology R&D, is illustrating that there is a new, disruptive paradigm setting the scene for future research and that system and architecture innovation at product level will be enabled through in-depth technology leadership.

4.3 Logic Scaling: Exploring for Future Nodes

In the development of advanced technology, the IC industry works to simultaneously improve the power, performance, area, and cost. That way, we can keep introducing new generations of chips with extra performance and a lower power use, but without significant cost increase. Hence, the needs to explore, develop, and implement solutions for a whole range of future technology nodes, from 14 nm down to 5 nm, and even beyond.

For the nodes that are nearest, technology solutions need to be engineered for insertion in production. And for those nodes that are farther out, a path finding pipeline needs to be developed, exploring options, and building fundamental understanding of materials and their interactions.

FinFETs (fin-shaped field-effect transistor) are the most promising transistor architecture for 14-nm technology and beyond, thanks to their excellent electrical characteristics and small variability. During the last 10 years, significant efforts have been spent in developing this technology. Currently, a 14-nm technology platform based on FinFETs is available and is the basis for further scaling toward 10 nm and possibly beyond. To scale FinFETs to 10 nm and below, one challenge is to improve the channel mobility and reduce the parasitic resistance. A solution to boost silicon mobility performance is using source/drain stressors (Figs. 4.3, 4.4).

Conventional FinFET technology using silicon channels will reach its limits around the 7-nm technology node. A solution to scale further is to introduce high-mobility channel materials. For the 10- and 7-nm technology node, focus is on the introduction of high-mobility channel materials, such as germanium and III/V compounds, and in particular, two main challenges namely how to implement nonsilicon materials into the device architecture and how to overcome some of the fundamental physics of the gate stack related to passivation. One future integrated technology could encompass germanium for pFET and III/V materials for nFET. In this context, various techniques on how to realize nonsilicon channels are compared, by decreasing defects in filled trenches through Aspect Ratio Trapping (ART), or building it on a strain-relaxed buffer layer. At the same time, fundamental work needs to be done in order to link material to electrical parameters, by

Fig. 4.3 FINFET made in IMEC 300 mm line

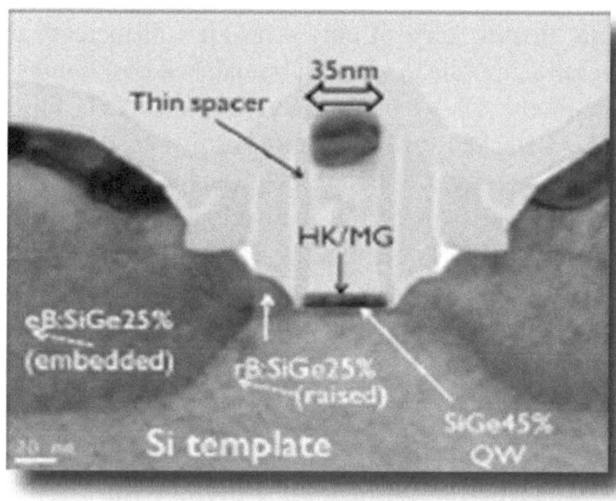

Fig. 4.4 Mobility boosting via S/D stressors, high-mobility channels

studying influence of oxide trapping on the gate stack mobility in high-mobility germanium and III–V channels.

Further down the road, for 5 nm and beyond, disruptive concepts are screened, such as vertical nano wire devices and spin-based devices. Explorations are ongoing how to build devices from new materials, such as graphene and 2D nano lattices.

In parallel, one will need to find the suitable solutions for scaling the interconnects, which on its turn will require the introduction of novel concepts and materials. On the radar screen are novel copper metallization methods, but also finding suitable alternatives for copper. This includes material screening to find candidates with k-values of 2 or below. Exploratory experiments investigate the use of carbon nanotubes as interconnect technology.

4.4 Extending Memory Scaling Beyond 20 nm

The increasing performance and functionality of smart mobile platforms requires more memory, at a lower cost per bit. So also here, further scaling is necessary.

But as with logic, it is proving more difficult to maintain the scaling pace and find solutions for the technical challenges. Exploration of new materials, cell, and array architectures and even new memory concepts is ongoing at high pace. Given the results of this R&D, one will probably see disruptive changes in memory appearing sooner than in logic chips.

SST-MRAM is a likely successor candidate for both DRAM and embedded SRAM applications. This is a disruptive technology in that it is based on magnetic rather than electronic storage.

For nonvolatile memories, NAND Flash has been the success of the past decade. The limits of conventional NAND flash are possibly extended toward 12–10-nm technology, by the introduction of hybrid-floating gate, a concept that allows to limit the cell-to-cell interference in high-density NAND with a scaled thickness of the hybrid floating gate of only 4 nm. It will increase the memory capacity of next-generation mobile devices, at unmatched cost figures.

A likely successor for conventional NAND Flash is 3D SONOS technology. This is a stacked vertical memory array consisting of 8, 16, 32…layers. In this area of 3D memory, research focus and expertise is in channel engineering and mobility, and the material innovation for vertical cells.

Farther out, a potential candidate for nonvolatile memory is resistive RAM (RRAM). Solid advances in the fundamental understanding of the switching mechanisms now allows to narrow down the selection of potential RRAM stacks and demonstrate functional memory cell arrays matching specs of novel systems.

4.5 Enabling Advanced Patterning

Cost has always been one of the main drivers of semiconductor R&D. In the future, though, aggressive scaling threatens to put quite a burden on the cost of new technology nodes.

That is why industry is increasingly investing in cost-effective patterning solutions, by combining the most advanced lithography tools and world-class expertise in critical process steps.

A key focus of in lithography is on the EUV production readiness in the areas of EUV tool, resists, masks, and defectivity, combined with learning regarding EUV insertion in advanced technology nodes. Such work is by excellence suited to be executed in a shared facility, where the burden of huge investments can be shared and combined with unique expertise at sufficient scale. First-generation EUV production tool is forecasted to be available around the beginning of 2014.

As a complementary patterning solution, recently, a lot of attention went into directed self-assembly. Though still quite a way from implementation, the technique is based on making two-component chemicals orient themselves based on guiding patterns. It is promising, because it can augment the performance of a less critical lithographic step into more refined patterns. Applications are possible in regular structures, contact holes, etc.

4.6 Smart Applications for a Better World

CMOS scaling has brought us systems with tremendous compute power. At the same time, we have engineered new fabrication techniques and new ways to use the characteristics of many materials. These possibilities have already led to many breakthrough applications that have changed the way we live and work.

Also in the future, semiconductor technology will continue to play its catalytic and fostering role toward new applications; deeply scaled technologies will allow for new breakthroughs in a variety of areas such as wireless communication, life science and healthcare, imaging and visualization, and energy. Only it will require more than before multidisciplinary innovation, blending nano electronics expertise with innovation ideas in the above-mentioned areas.

"Smart" devices will interact intelligently with their (our) environment, sensors and displays will be its multidimensional tentacles toward the world. Underlying technologies, the ultra-low-power radios (<100 uW power), harvesting power cells, CMOS sensors on MEMS- and NEMS-based platforms, new substrates (s.a.GaN for power devices and LED) integrated on silicon, and flexible electronics (low cost or large area solutions based on organic or oxide electronics) for lighting or displays.

In life science and pharmaceutics, we envision mass-produced nanoscale systems that will replace today's bulky and expensive equipment with onc-chip solutions. These systems will be the basis of a myriad of new possibilities, new applications, and new industries.

Development of high-quality imagers and sensors will create new enabling platforms, e.g., for cell inspection in the medical domain, with vast application possibilities and benefits for society. Advanced microchips will allow for highly multiplexed, hence disruptively cheaper, single-molecule genetic analysis. Of course, such developments require dedicated skill sets, and multiyear research collaboration, sustained efforts combining world-leading expertise in nanophotonics, CMOS sensors, technology integration and fabrication, as well as from the various medical domains.

Finally, organic and new generation silicon photovoltaic's will be vital to address our planet's pressing challenges in, e.g., sustainable energy generation (Figs. 4.5, 4.6).

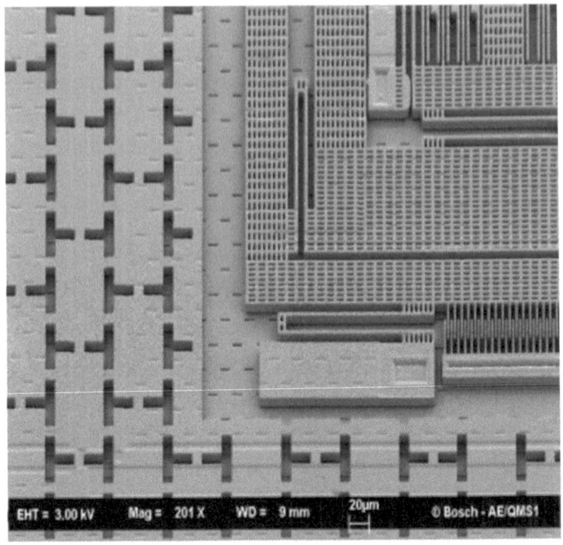

Fig. 4.5 SiGe MEMS processing (IMEC FAB)

Fig. 4.6 CMOS specialty imager (IMEC FAB)

4.7 Realizing Cost Efficiency Through Collaborative Research

Above challenges in CMOS process technology, energy, healthcare, or wireless technology are addressed in IMEC's nano electronics platform. Business models range from multipartner collaborative programs—proven to be a great way to mitigate the risk and share the cost of advanced research—up to bilateral or custom development or direct licensing in the various domains.

A staff of about 2,000 people, one-third of which are industrial residents and PhD students, are chartered with collaborative, open-innovation mode R&D, striving for excellence in fundamental understanding, process steps and

Fig. 4.7 IMEC's partnership ecosystem (March 2013)

integration, and application enablement, all of this eventually finding its path into the individual companies' private process and product development.

IMEC's core program in the CMOS scaling area, which started about 10 years ago, today is the largest collaborative effort in semiconductor R&D, involving the world's premier IC makers such as Global foundries, INTEL, Micron, Samsung, TSMC, SK Hynix, Toshiba/Sandisk, Qualcomm, Sony, Panasonic, Fujitsu, nVidia, Altera, Xilinx.

By blending this span of IDM's, foundries, fablite and fabless companies, with all leading material and equipment companies, more than 50 in total, IMEC has gathered in its Core Program the world's largest, yet unique ecosystem for semiconductor R&D (Figs. 4.7, 4.8).

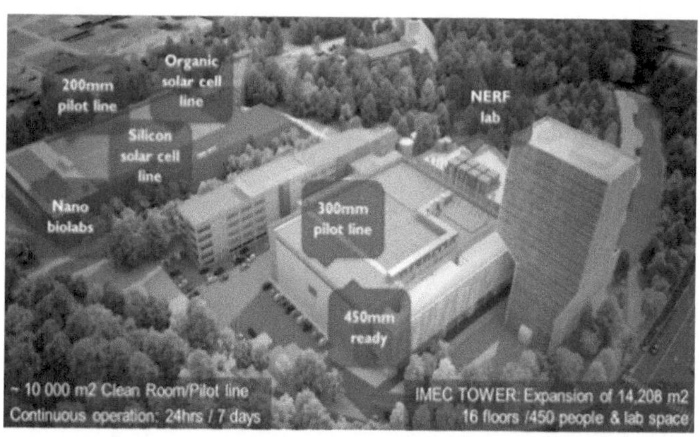

Fig. 4.8 IMEC's infrastructure and campus (animated, March 2013)

4.8 Creating Value for the Semiconductor Industry: The Next 10 Years

Looking forward into the scaling roadmap, a host of opportunities and challenges are ahead of us. Experts now see a path to scaling to 5 nm and possibly beyond. So we are ready for another 10 years of scaling.

But it would not be easy scaling. More than ever, in the following years, we will need the "R" in R&D to come up with solutions and innovations. More effort is expected to be required in exploratory programs.

4.8.1 Innovation in Devices, Materials, and Process Steps

System innovation also, because it will become key to assess correctly which applications will profit best from scaling, and how. Hand in Hand with product innovation, because only new generations of applications profiting from the performance, energy, and cost gains of tomorrow's technologies, will provide the payback for the upfront investments.

4.8.2 450 mm Wafers on the Horizon

Along the scaling roadmap, somewhere in the second half of the decade, the leading industry will make the transition toward 450 mm wafers. It is expected that at one point in time, the leading R&D will be done on the latest, 450 mm tools. On a worldwide scale, consortia are gearing up infrastructure and equipment level programs, a new infrastructure challenge is ahead and is expected to be another component of the overall set of instruments to keep mastering the cost of future systems. Leading edge 300 + 450 mm pilot infrastructure is expected to be an innovation driver for another decade. It will be a platform for cost-efficient high-quality R&D, with the knowledge and expertise of all participating world-class partners providing value and a great leverage.

Chapter 5
Advent of Internet 2.0 and Its Implications on the Semiconductor Industry

Arun Bellari

Abstract The talk will address how the computing and communications world has changed as the Internet migrates to its next phase. The new applications like social networking, advanced search engines, and PDA applications have integrated and evolved the telecom, video transmission, and data networks into a single entity. The need for searching any kind of information and quickly providing to customers is becoming the competitive advantage to service providers. This has made the networks migrate to storage, cloud, access, and delivery model. The new standards and SoCs based on that will continue to evolve and create opportunities to develop new technologies and products in the future electronics field.

A. Bellari (✉)
Managing Director, ASARVA Technologies Inc, Asarwa Road, Ahmedabad,
Gujarat 380016, India
e-mail: anbellary@gmail.com

V. Chakravarthi et al. (eds.), *Proceedings of International Conference on VLSI, Communication, Advanced Devices, Signals & Systems and Networking (VCASAN-2013)*,
Lecture Notes in Electrical Engineering 258, DOI: 10.1007/978-81-322-1524-0_5,

Chapter 6
Retracing a Humble Path: BNM Institute of Technology

T. J. Ramamurthy

Abstract As BNM Institute of Technology, Bangalore enters its thirteenth year, I feel privileged to be part of every stage of its growth since its inception. In this article, I have tried to recall my presence as a part of BNMIT's growing years.

When I joined BNM Institute of Technology as Founder Principal in 2001, I was responsible for the futures of a hundred and eighty students. I had great ambition to make this institute a house of learning that would stand out from the crowd.

With the backing of ten trusted faculty members and the BNMEI management, we had three Engineering branches namely Electronics & Communication Engineering, Computer Science Engineering and Information Science Engineering. In the subsequent year, two more branches, namely Telecommunication Engineering and Electrical & Electronics Engineering were added to meet the growing demand of engineers in these domains.

During initial years, our focus was to impart quality engineering education with a difference, and this paid exceedingly good dividends when we secured good results in the University examinations. Simultaneously, we also defined strong back-end processes for academic management (Fig. 6.1).

Growing years: In 2005, I moved on to become Director of the Institution, thereby handing over regular academic responsibilities to Dr. K Ranga, the new Principal, a very able and experienced academician. My responsibility since then has been to manage Administration and Finances of the Institute. During my time as Director, the popularity of BNM Institute of Technology has grown so much so that it has become the preferred T-School for most engineering aspirants.

T. J. Ramamurthy (✉)
Director, BNM Institute of Technology, Bangalore, India
e-mail: bnmitdirector@gmail.com

V. Chakravarthi et al. (eds.), *Proceedings of International Conference on VLSI, Communication, Advanced Devices, Signals & Systems and Networking (VCASAN-2013)*, Lecture Notes in Electrical Engineering 258, DOI: 10.1007/978-81-322-1524-0_6,

Fig. 6.1 BNMIT: The inception

This has lead to the student strength leaping from a mere 270 to a whooping 480 over the years.

My tenure as director has been ably supported by Principals Dr. K. Uday Kumar in the past and now Dr. M. S. Suresh.

I believe it is our strength to constantly attract excellent, experienced and dedicated faculty members and all hundred and fifty of them today are pillars of strength to our Institute.

Allow me to take you through the major milestones of our institute (Fig. 6.2, Table 6.1).

6.1 BNMIT Today

BNMIT produces five hundred excellent engineers in all major streams of engineering who are engaged in occupation spanning industries to research.

It boasts of a highly professional teaching fraternity who believe in constant knowledge acquisition and dissemination through holistic learning processes such as seminars, workshops, symposiums, and conferences. They also undertake consultancies in active cutting edge technologies such as VLSI, Wireless Sensor Networks, Optical Communication to name a few (Fig. 6.3).

Apart from providing sound academic teaching, the Institute also encourages Industry Interfacing Performance enhancing programs, Placement training and Career support programs, multiple skill acquisitions, and a research-rich learning atmosphere.

Fig. 6.2 BNMIT growing years

Table 6.1 Major milestones of our institute

Years	Milestones
2001	Inception of BNMIT with ECE, CSE and ISE courses
2002	Two more branches, TCE and EEE added
2005	Schools of Management studies providing MBA program
2008	Research Centers in ECE, EEE, and Mathematics established
2009	Research Center in CSE established
2011	Research center in Business Administration established
2010	Masters program in VLSI design and Embedded Systems in ECE and CSE courses introduced
2011	Masters program in CAID in EEE added
2012	Masters program CNE in CSE and Mechanical Engineering undergraduate course added

Fig. 6.3 Auditorium complex

I feel proud to see the young generation blossom with Associations for every discipline, be it Cultural Activities, Music, Nature, Adventure, outdoor or sports.

With this strong background, I am sure B. N. M. Institute of Technology is a temple of learning where a true aspirant will definitely achieve all that he has set out to achieve for a successful life professionally and personally. As an ambassador of this institution, I wholeheartedly welcome everyone to be a part of our endeavor to carve a roadmap to success.

Chapter 7
Platform SOC for SMART Home

Veena S. Chakravarthi

Abstract SMART homes are soon going to be the reality with all the necessary technology available and changing lifestyle demanding it. Today, internet is being used for range of applications from banking needs to entertainment to day planners to appliance control. With internet capabilities and superior processing power of VLSI technology coupled with availability of wide range of sensors, it is no more a dream to have a SMART Home. This paper discussed the hardware and software technologies which can be used to realize the SMART Home. SMART Home hardware is defined as a Platform SoC identifying the specific differentiators which makes it a preferred candidate for such applications. Proposed PSoC is required to support many interfaces, standards to cater to wider functions of a Home to name a few Entertainments, healthcare, appliance control, environmental control, etc. User-friendly GUI, low power, and faster response will be the major differentiators of such platform SoCs. This paper attempts to identify the functional requirements of a basic Platform SOC and gives insight into PSoC architecture for SMART Home solution.

Keywords Platform SoC · PSOC · SOCs · Z wave · Zigbee · X-10 · TCP/IP · Internet of Things IOT

7.1 Introduction

Availability of wide variety of sensors, faster processing power of these signals supported by internet drive the realization of SMART home technology. SMART Home technologies permits communication between different electronics gadgets

V. S. Chakravarthi (✉)
IEEE Member, Professor, Department of Electronics and Communication Engineering, BNM Institute of Technology, Bangalore, India
e-mail: scveena@bnmit.org

V. Chakravarthi et al. (eds.), *Proceedings of International Conference on VLSI, Communication, Advanced Devices, Signals & Systems and Networking (VCASAN-2013)*, Lecture Notes in Electrical Engineering 258, DOI: 10.1007/978-81-322-1524-0_7,

and appliances and intelligently monitors and self-control them automatically as per the needs of the persons living there. This can be achieved by having all the necessary electronics and appliances networked and connected on to internet. These unhindered growth of source and consumers of information is the driving force of speedy adoption of technologies for SMART Home. All the house hold electrical appliances, communication devices like phones, mobile phones, entertainment electronics like Television, Radio, and gaming devices are connected as *things* networked using internet. This is achieved by *internet of things* technology. In *Internet of things, the thing* is an object which is a constituent of World Wide Web which can be addressed by means of standard internet protocol of TCP/IP [1]. All these *things* irrespective of their functions, it is absolutely essential to have the interface technology which will support multiple standards and technologies. Hence, future *things* will not be application specific but have to be more generic. Hence, platform-based systems (PSOCs) on chip will be one of the promising solutions for such requirements. It is required to develop much generation of devices of different combinations, configurations, and interfaces with the base platform. This paper presents one such platform SoC which can be used as a thing for a SMART Home gateway on to internet. PSoC is required to support all necessary functions of communication, appliance controller, entertainment needs, and home security and many other Home needs. It is also necessary to be able to be networked on to internet through TCP/IP standard.

7.2 Functional Specification of PSOC for SMART Home

PSoC for SMART Home should support following essential features:

7.2.1 VOIP Call Processors

The PSoC should support multiple channels of Call processing, Call queuing, Caller ID services, ADSI, SIP, and H.323 as both client and gateway. It should support voice mail services with directory. The conceptual system part of PSoC is shown in Fig. 7.1.

7.2.2 IP Media Processors

The media processor block of PSoC should support 2D/3D Graphics, Transport Processor, Multistandard audio and video decoders and support worldwide Conditional Access and Digital Rights Management standards. The block should support functionality of Integrated H.264 video encoder to support recording of

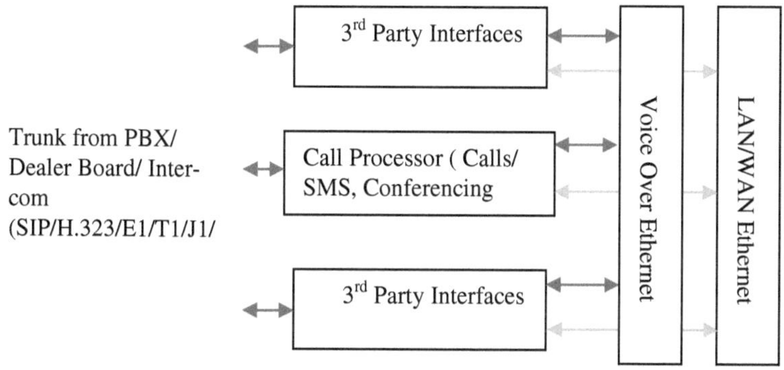

Fig. 7.1 Call processor block of PSoC

analog programming providing the consumer the ability to have seamless recording for legacy broadcasts and High Definition video content. It should also support features like video chat. It should also support wireless video distribution technology which enable to receive video between PCs and mobile devices to access and process personal content for internet streaming services. Support for conditional broadcasting and digital security required flexible security algorithm. In addition, it should support features listed below.

- Video codecs including H.264, VC-1, AVS, MPEG4, etc.
- Audio codecs including MP3, AAC, AC3, etc.
- Optimized latency profile for interactivity
- Error resiliency and concealment
- Crypto engine for HDCP 2.0 and DTCP-IP
- Graphics accelerator for GUI
- MII, USB, SPI, SDIO interfaces
- Integrated HDMI 1.3
- Display processor
- Video postprocessor

7.2.3 Real-Time Home Appliance Control

Control of electrical appliances like electric bulbs, fans, air coolers and others through Internet enabled mobile device in real time should be supported with a sensitive, highly responsive, and fast control circuitry. The flexibility is needed to easily add new devices and control functions. Remote monitoring and control is to be supported or easy monitoring and maintenance. Scalability is preferred. Support for X10 technology, Z wave and Zigbee technology can be used to implement the controllability of the electric appliances from anywhere in the world [2]. Houses and appliances and control functions are coded, and control message is transferred to the devices through X10 protocol-based controller. Z Wave or Zigbee core can be integrated for in side home communication. The Mobile controller can use GSM, thus making it a true remote controller on Internet.

7.2.4 *Health Monitor Block*

Health Monitor block is required to contain two subsystems like embedded health monitor system with advanced sensors which can be embedded into appliances, elderly person as wearable electronics or clothing and private health network to collect and maintain data base of health parameters for continuous monitoring.

7.2.5 *Home Security System*

This system part enables monitoring locally or by remote through high-resolution cameras fitted near the house and control the keyless automatic home lock system.

7.2.6 *Other Differentiators*

It is required to have PSoC designed for very low power, highly responsive and scalable with both local and remote control capability. It is preferable to operate PSoC by energy harvesting or by alternate energy sources. This will be the major enablers for SMART Home PSoC. Intelligence is another major differentiator of PSoC which enables smart applications like context awareness, environment control based on one's preferences, etc. Capacity of withstanding harsh conditions and rough operations and affordable security are other differentiators for this development.

7.3 PSoC Architecture

The Hardware platform is defined as a PSoC which enables the implementation of all the features mentioned in previous section. The PSoC block diagram is shown in Fig. 7.2.

The PSoC consists of modular bus–based architecture with flexibility to scale up the performance either by integrating processor core or by functional blocks.

All high-performance blocks like Processor cores, DMAC, and DSP core GEMAC blocks are the bus masters arbitrating for internal packet memory extendable to external SDRAM. Flexibility is added with many programmable Arbitration schemes supported in hardware.

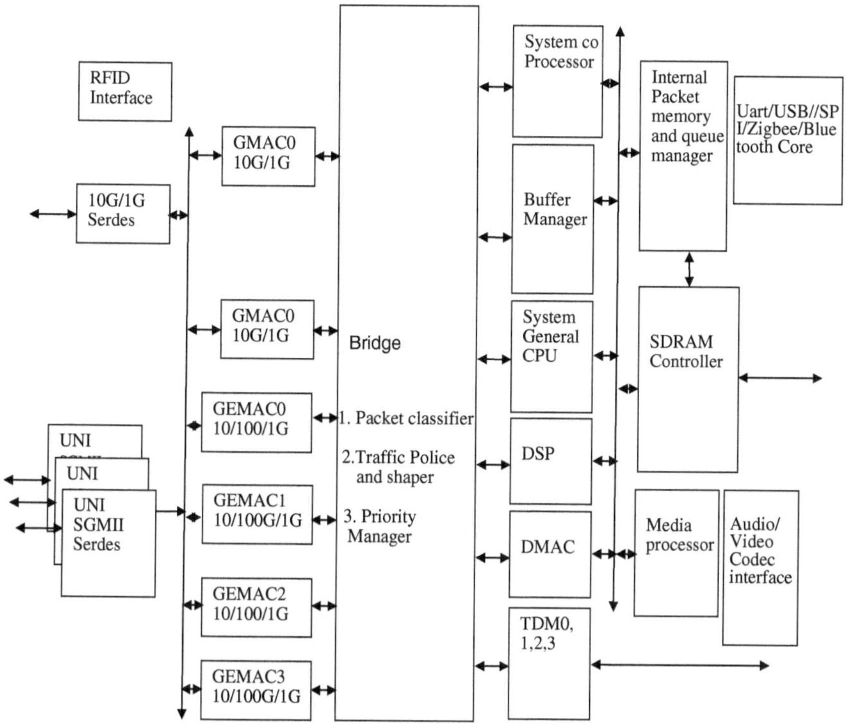

Fig. 7.2 Block diagram of PSoC for SMART Home

Slow interfaces like UART/USB/Zigbee and Bluetooth cores provide debug and other interfaces to required radios, analog front-ends, for implementing functions like health monitoring, home security, and appliance control interfaces.

Performance like throughput per master and delay in arbitration success for a master can be monitored by extra hardware which will help in programming right arbitration scheme.

As power consumption has to be curtailed to bare minimum and it is required to explore the possibility of on-the-fly clock switching and control, wake on active techniques in addition to normal low-power design techniques like clock/power gating, etc.

7.4 Internet of Things

Explosion of data generated and consumed in SMART Home environment can only be handled by technology like Internet of Things. It is required for PSoC to have necessary interfaces which will front end and communicate with range of sensors, process them to some extent in hardware and accelerate to a major extent before handing over to the software layers. It is required to have minimum 10 Gig back-end interface to support typical mid-size SMART Home. It is, however, essential to have internet always. It is highly essential for this PSoC to be self-monitoring, self-diagnosing, and self-healing. Integration of Multifrequency radios, different communication protocols, intelligent routing algorithms for latency optimization, and special low-power techniques will ensure the success of this IOT for SMART Homes.

7.5 Conclusion

As technology is penetrating into the home to make it SMART, it is necessary to keep in mind the flexibility, low cost, low power, and friendly user interface as a necessary feature of the *Things*. However, internet is most commonly used by most of the new generation; it becomes easy to build the solution around internet. Hence, the Internet of Things is posing to be the promising technology for SMART home. Because of the scalability needs and verity of the interfaces required by the SMART home technology consumers, products have to be built using PSoC. A typical PSoC architecture considering the needs of SMART Home is dealt in this paper. The highlights of the architecture are the configurability, possibility of different products based on PSoC architecture.

References

1. Internet of Things in 2020, a road map for the future, INFSO D.4 Networked enterprise and RFID Info G.2 Micro and Nano systems in co-operation with the RFID working group of the European Technology Platform on SMART Systems Integration (EPoSS), Sept 2008
2. Shahriyar R, Hoque E, Sohan SM, Naim I, Akbar MM, Khan MK (2008) Remote controlling of home appliances using mobile telephony. Int J Smart Home 2(3):37–54

Chapter 8
Electronics and Digital Computing Techniques for Images and Image Processing

P. A. Vijaya

Abstract Images are signals produced by the imaging devices and can be digitized and processed by using the modern digital computers. Image processing is a subject that deals with the various mathematical and computing techniques/algorithms and the devices that can be used for improvement of image quality, processing of images and analysis. Image processing has a wide variety of applications in day-to-day activities and has gained popularity commercially and also in academic. In this paper, an introduction to the importance of images and imaging devices and their history and applications before the advent of digital computers and digital camera are provided in brief. The advantages of digital computers for digital image processing, uses and applications of image processing and the common image processing operations/techniques that are in use and practice are mentioned. Finally, the future scope and the research work that can be carried out in this area are highlighted for the prospective academicians, industrialists, and researchers, for the welfare of the society and mankind.

Keywords Images · Imaging devices · Digital camera · Digital computer · Image processing · Image analysis and applications

8.1 Introduction

Images are signals produced by special devices called imaging devices [1]. Image is a two-dimensional rectilinear array of pixels. An image pixel may be represented in terms of gray level (Black and White image) or in terms of RGB (Red Green Blue) values [2]. Image processing is a subject that deals with the various mathematical and computing techniques/algorithms and the devices that can be used for

---

P. A. Vijaya (✉)
MIEEE, MISTE, Professor, Department of ECE, BNMIT, Bangalore, India
e-mail: pavmkv@gmail.com

V. Chakravarthi et al. (eds.), *Proceedings of International Conference on VLSI, Communication, Advanced Devices, Signals & Systems and Networking (VCASAN-2013)*, Lecture Notes in Electrical Engineering 258, DOI: 10.1007/978-81-322-1524-0_8,

processing of images. Because of wide variety of applications in commercial field, medical field, forensics, military and defense, satellite and remote signal processing etc., it is a very interesting and hot area of research for academicians, industrialists, scientists, engineers and doctors. It is an interdisciplinary topic for all the electrical science disciplines such as Electronics and Communication, Computer science, Information Science, Telecommunication, Electrical and Instrumentation. In this paper, some basic concepts of images and imaging devices, electronics for digital camera (a popular embedded system) and digital computers for image processing, various image processing operations, image analysis, image processing applications, and future research scope are discussed, in the various sections.

8.2 Images and Imaging Devices

The research work on images and invention of imaging devices has started around the year 1600 itself. The devices that were designed for capturing images can be broadly classified into 2 types. They are: (1) Direct image plane imaging devices and (2) Transform imaging devices [1].

8.2.1 Direct Image Plane Imaging Devices

The magnifying glasses were invented in the year 1600 for viewing images of small objects/characters more clearly, telescope was invented by Galileo in the year 1609 for viewing the images of very far off objects and microscopes were invented in the year 1660 for viewing and analyzing the images of objects that cannot be seen with human eyes [1]. In nineteenth century, photography became an important means of scientific research, and cameras were the commercial items that entered people everyday life. Photography played a decisive role in the invention of X-rays and later the radiography. Electronic cameras and electronic televisions were invented in the middle of the nineteenth century. Later, Radars and Sonars were also designed. Latest microscopes like atomic microscopes and tunnel microscopes were designed. Later, linear tomography and laminography techniques were used in capturing images. All these devices belong to direct image plane imaging devices. They produce images that can be directly viewed by eyes, and they need direct access to individual locations for accessing objects to be resolved.

8.2.2 Transform Imaging Devices

In transform imaging, image information retrieval and image formation/reconstruction are essentially separated so that the images can be processed and images can be reconstructed. X-ray crystallography, which uses indirect imaging was invented in 1912 by Max von Laue. This technique was used in 1953 by J. Watson and F. Crick in determining the spiral structure of DNA in the human cell, which is a remarkable scientific achievement. Holography was invented in 1948 by D. Gabor for improving the resolution power of the electronic microscopes. Invention of lasers and holography led for optical information processing. In 1970, digital holography was suggested by Langmann. In digital holography, digital computers are used for reconstruction of holograms, computer synthesis of holograms for 3-D visualization, image reconstruction, holography and Magnetic Resonance Tomography (MRI).

8.3 Digital Computers for Digital Imaging and Image Processing

The use of digital computers and processors for digital imaging and image processing have the following features/advantages.

- **Flexibility and adaptability**: Change/modification in the hardware is not necessary. Any analog signal can be digitized using ADC (Analog to Digital Converter) and can be processed by developing software techniques/algorithms.
- **Integration**: Optical information processing and Digital signal processing systems can be integrated. Digital signal is an ideal means for integrating imaging devices and image signals to computer networks and other information processing systems. The only limitation is the memory and processing speed in a digital computer.

8.4 Digital Camera as an Embedded System

High-resolution cameras have been designed as embedded systems for capturing color images. Digital camera usually has a lens, imager, and digitizer, codec, DSP or microcontroller, flash memory, image processing techniques or algorithms in ROM of the microcontroller, display units, and operating switches. Digital image sources may be commercial earth resource satellites, airborne scanner, airborne solid state camera, scanning microdensitometer, and high-resolution video camera. A typical block diagram of a digital camera, which is a famous commercial product designed and marketed by the embedded industries, is given in Fig. 8.1, [3]. It encompasses digital electronics, advanced VLSI special purpose chips for various operations, image processing algorithms, storage and transmission capabilities, which is an important design topic for electronics engineers and computer scientists. High-resolution cameras are used for capturing quality pictures/ photographs.

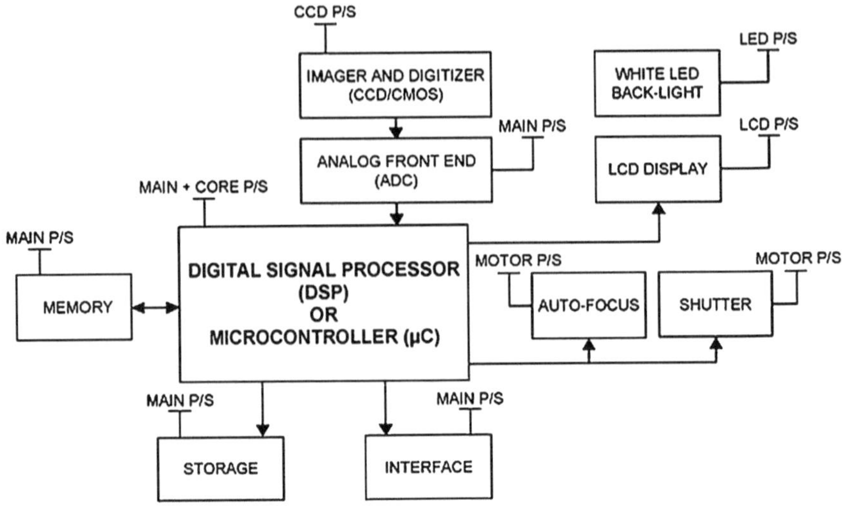

Fig. 8.1 Block diagram of a digital camera. (P/S: Power Supply)

8.5 Uses/Applications of Image Processing

Image processing is widely used or required in the following areas/fields.

- Commercial photography, film industry, cartoon simulation, animation, etc
- TV broadcasting, multimedia communication and applications
- In air and space photo reconnaissance and remote sensing applications
- Radiology, diagnostics, ophthalmology, etc. in medical field
- Automatic devices, industrial applications, pattern recognition [4], computer vision [5], forensics, security [6], military, defence, etc.

8.6 Image Processing Operations

The images captured using low-resolution camera, images taken by cameras in satellites, spacecrafts, remote-sensing satellites, X-ray images, CT scanner outputs, Forensic images, etc. may not be very clear and may have noise and changes in the orientation. Therefore, it is necessary to apply some preprocessing operations and analyze the good quality images for further operations/decisions. The preprocessing operations and also the other operations used on images for image quality and various image processing applications are discussed in brief [1–8].

Image formation and reconstruction: It has a fundamental impact on image quality. It is desirable to reconstruct images with the lowest possible noise without sacrificing image accuracy and spatial resolution. These are important for images taken from X-rays, CT scanners, etc. for diagnostics and many biologists are doing research in this topic. There are some analytical reconstruction techniques such as Filtered Back-Propagation (FBP) method and also iterative reconstruction techniques.

Image restoration: It is the operation of taking a corrupted/noisy image and estimating the clean original image. Corruption may come in many forms such as motion blur, noise, and focus mistake while using camera.

Image preparation/enhancement: Image enhancement is the process of adjusting digital images so that the results are more suitable for display or further analysis. It involves removal of noise and increase brightness in an image to identify key features. It can be done in spatial domain or frequency domain. Simplest methods are histogram equalization techniques. There is no generic image enhancement method which works in all case.

Image quantification/parameter estimation: Probabilistic approaches such as Bayesian technique and maximum likely hood approach, unsupervised learning techniques, Artificial Neural Networks, etc. are used for parameter estimation or image quantification to classify or group biomedical images for analysis, classification, and diagnostics.

Automated image analysis: Image analysis is the extraction of meaningful information from digital images by means digital image processing techniques. Image analysis tasks can be as simple as reading bar coded tags or as sophisticated as identifying a person using a standard biometric such as face [4], iris [6], finger print, DNA, palm print, etc [7, 9]. Computers are indispensable for the analysis of large amounts of data, for tasks that require complex computation, or for the extraction of quantitative information. On the other hand, the human visual cortex is an excellent image analysis apparatus, especially for extracting high-level information, and for many applications including medicine, security and remote sensing. Human analysts still cannot be replaced by computers. For this reason, many important image analysis tools such as edge detectors and neural networks are inspired by human visual perception models [8]. Computer image analysis largely contains the fields of computer or machine vision, and medical imaging and makes heavy use of pattern recognition [9], digital geometry [7], and signal processing [10]. Digital Image Analysis is when a computer or electrical device automatically studies an image to obtain useful information from it. Note that the device is often a computer but may also be an electrical circuit, a digital camera or a mobile phone. It involves image segmentation, motion detection, video tracking, optical flow, medical scan analysis, 3D Pose Estimation, etc.

Virtual imaging/image modeling: Virtual imaging is defined as images created by an optical display system that creates a larger screen than is physically presented to the user. These optical display systems are used in Head-Mounted Displays (HMD) or goggles. HMDs can also be used to provide Augmented Vision (AV), Augmented Reality (AR), Virtual Reality (VR), or Mixed Reality (MR), through differing optical design forms and the information provided. MR is very similar to AR, but in this case, the real world images are provided by a camera or cameras mounted on the HMD. These images are combined with images created by a computer using fully occluded optics to create an immersive visual experience. VR is usually a fully stereoscopic 3D environment created by a computer image generator and displayed using an immersive HMD or helmet, with occluded or see-through optics that provide a wide field-of-view. These techniques are used typically for computer gaming or simulation and training applications. VR has been effective in treating post-traumatic stress syndrome and other phobia disorders.

Image compression, coding and watermarking for transmission, storage and security: Image compression is used for reducing the file size to be transmitted/stored and image coding and watermarking is used for providing security and authenticity/ownership. Image compression technique may be lossy or lossless. Lossless compression is preferred for archival purposes and often for medical imaging, technical drawings, clip art, or comics. This is because lossy compression methods, especially when used at low bit rates, introduce compression artifacts. Lossy methods are especially suitable for natural images such as photographs in applications where minor (sometimes imperceptible) loss of fidelity is acceptable to achieve a substantial reduction in bit rate. The lossy compression that produces imperceptible differences may be called visually lossless. Methods for lossless image compression are: Run length encoding (used as default method in PCX and as one of possible in BMP, TGA, TIFF image file formats), DPCM and Predictive Coding, Entropy encoding, Adaptive dictionary algorithms such as LZW (used in GIF and TIFF), Deflation (used in PNG, MNG, and TIFF), Chain codes, etc. Methods for lossy compression are: Reducing the color space to the most common colors in the image, Chroma subsampling by dropping some of the chrominance information in the image, Transform coding using Discrete Cosine Transform (DCT) or wavelet transform. Research is still going on in chaos-based and diffusion-based image encryption techniques and also in visible and invisible watermarking techniques [6].

Simulating imaging devices: Medical image simulation is useful for biological modeling, image analysis, and designing new imaging devices, but it is not widely available due to the complexity of simulators, the scarcity of object models, and the heaviness of the associated computations.

Fig. 8.2 The standard test image used in digital image processing. (Photograph of the Swedish model—Lenna Soderberg, taken in 1972)

Figure 8.2 is a very good, widely used photograph of the Swedish model—Lenna (Lena) Soderberg, taken in 1972. Originally, it has 512 × 512 pixels. It is used as a standard test image by most of the researchers and academicians for testing/comparing the image processing algorithms/techniques developed by them. Most of the image processing algorithms are coded/tested using MATLAB language/tool [4–6].

8.7 Future Research Scope in Digital Image Processing

The applications of digital image processing and analysis are continuously expanding through all areas of science and industry, including:

- Medicine: detecting cancer/tumor/wound in an MRI/CT/ultrasonic scan
- Microscopy: counting the germs in a swab.
- Remote sensing: detecting intruders, producing land cover/land use maps.
- Astronomy: calculating the size of a planet.
- Materials science: determining cracks in a metal.
- Machine vision: automatically count items in a factory conveyor belt.
- Forensics and security: detecting a person [4, 5], eye color or hair color.
- Robotics: to avoid steering into an obstacle.
- Optical character recognition: automatic license plate detection.
- Microplate reading: detecting where a chemical was manufactured.
- Metallography: determining the mineral content of a rock sample.

- Military, defence, TV broadcasting, Film industry, and commercial field.
- Multimedia communication [6], computer graphics/animation/simulation.
- Biometric and pattern recognition applications [4, 5].

Though tremendous research work has been done in all these application areas of image processing, still there is a lot of scope for research with the advancement of technology in all the fields of Engineering, Science and Technology. Research work can be done on embedded system design for these applications, VLSI chip design for specific digital electronics circuits/processors/functions used in the devices to be designed for these areas and also novel/efficient/fast image processing techniques/algorithms using signal processing, soft computing and computational geometric techniques. Advanced/Automated embedded systems, including special purpose ICs, SoCs for these applications/operations can be designed, patented, fabricated and marketed. Without manual intervention/operation, the systems like driverless cars, vision-operated devices/applications [5], and computer systems are being designed because of continuous research work in this area. Finally, it would be very interesting if one can predict how a human being or animal would be after "m" years from now and also how they would have been "n" years back (Refer to the example in the last page of this paper).

8.8 Conclusion

In this paper, some fundamental issues in images, imaging devices and image processing are discussed. The image processing operations, image analysis and its applications in various streams of science and technology, engineering, commercial field, medical field and industry are elaborated. Finally, the research scope in this field is mentioned for motivating the upcoming researchers and industrialists to develop interest in this area and to do good research and design for the growth and welfare of the society.

References

1. Yaroslavsky L (2004) Digital holography and digital image processing: principles, methods and algorithms. Kluwer Scientific Publishers, Dordrecht
2. Gonzalvez RC, Woods RE (2011) Digital image processing, 2nd edn. Addison Wesley, Reading
3. http://www.google.co.in
4. Vijaya PA, Shivakumar G (2012) An improved artificial neural network based emotion classification system for expressive facial images. Int J Comput Inf Syst 48–57
5. Sheela SV, Vijaya PA (2011) Appearance based method for eye gaze tracking. J Comput Sci 7(8):1194–1203

6. Harshita KM, Vijaya PA (2012) Secure data hiding algorithm using encrypted secret message. Int J Sci Res Publ 2(6):1–4
7. Jain AK (2006) Fundamentals of digital image processing, 2nd edn. PHI, Englewood Cliffs
8. http://en.wikipedia.org/wiki/Digital_imaging
9. Duda RO, Hart PE, Stork DG (2000) Pattern classification. Wiley, NY
10. Proakis JG (2007) Digital Signal Processing, 4th edn. Pearson Education, India

Part II
VLSI

Chapter 9
System Verilog Based SOC Verification Environment for FLASH MEMORY

J. Dinesh Reddy

Abstract This paper discusses a generic flow on how an automated SV-based test bench environment which is randomized with constraints can verify a SOC effectively for its functionality and code coverage [1]. Today, in the era of multimillion gate ASICs, reusable intellectual property (IP), and system-on-a-chip (SoC) designs, verification [1] consumes about 70 % of the design effort. Automation lets you do something else while a machine completes a task autonomously, faster and with predictable results. Automation requires standard processes with well-defined inputs and outputs. Not all processes can be automated. Because of the variety of functions, interfaces, protocols, and transformations that must be verified, it is not possible to provide a general purpose automation solution for verification, given today's technology. It is possible to automate some portion of the verification process, especially when applied to a narrow application domain. Tools automating various portions of the verification process are being introduced. Here, we have a SOC with a ARM multicore processor which talks to one of the peripherals, which is a flash memory (CODE FLASH and a DATA FLASH). The content of the paper discusses about the methodology used to verify such a SOC-based environment. Cadence OVM libraries are explored for the solution of this problem. We can take this as a state of art approach in verifying SOC environments. The goal of this paper is to emphasize the methodology [2] implemented in system verilog for SOC verification.

Keywords Assertion-based verification · Bus functional models · Constrained randomization · Coverage driven verification · SOC

J. Dinesh Reddy (✉)
Department of ECE, BMS College of Engineering, Bangalore, India
e-mail: dineshreddy.ece@bmsce.ac.in

V. Chakravarthi et al. (eds.), *Proceedings of International Conference on VLSI, Communication, Advanced Devices, Signals & Systems and Networking (VCASAN-2013)*, Lecture Notes in Electrical Engineering 258, DOI: 10.1007/978-81-322-1524-0_9,

9.1 Introduction

This paper defines the scope of verification and specifies its solution. The verification problem is quantified in a set of coverage models derived from architectural and implementation features. This serves as a starting point for the verification engineer to build the verification environment and execute the Verification plan. All the bus functional models, coverage monitors in the verification environment are described in detail.

It is required to get familiar with the macro for understanding the verification plan. It is a FLASH Memory macro which concentrates on Flash data Read. It has some set of functionalities involved during this process.

The Key functions of this block are detailed below.

1. Bootswap control Process
2. Address Generation
3. Variable Fetch Wait control
4. Read Data Control
5. Standby Control

9.1.1 Objective of the Verification

The following are the objectives of the verification activity

a. Protocol Accuracy on Bus interface
b. Assertion checks
c. Functional coverage
d. Verification of Functionality

9.1.1.1 Directed Approach Versus Constrained Random Test Bench

Directed:

1. One test case per scenario identified
2. Checks specific to test case
3. Dependency on code coverage to measure verification completeness
4. Huge number of test cases to maintain

Constrained random test bench

1. Scenario generator capable of generating random scenarios.
2. Test case writer will have the ability to constrain the scenarios
3. Checker will contain data and protocol checks for all scenarios
4. Functional coverage used to measure verification completeness
5. Lesser number of test cases to maintain

9.2 OVM Methodology [2]

Reuse methodology which codifies the best practices to develop a test bench. It provides a library of classes which enable faster and efficient implementation of test bench components and defines a structure for a test bench.

9.2.1 OVM: Test Bench Components

a. **Driver**—This is the component that connects to the DUT interface and drives stimulus. Can be equated to a Bus Functional Model.
b. **Monitor**—A Passive entity that samples the DUT interface and extracts information like transactions, scenarios, and events
c. **Sequencer**—Generates stimulus data and passes it on to the driver
d. **Sequence Item**—This is the basic transaction on which the driver operates. The driver will look at the sequence item and drive the interface accordingly

Sequence—This could be a single transaction or a group of transactions on the interface. One sequence can use a single sequence item or multiple sequence items

An active uVC (universal verification component shown in Fig. 9.1) will contain a sequencer and a driver. A sequence will create one or more sequence items. These sequence items are passed to the driver by the sequencer. The test case can call multiple sequences to create a specific scenario

The sequence item will have the basic set of constraints for the various parameters of the transaction. This can be constrained further from the sequence.

Directed tests can be written by providing values for all parameters of the transaction. Whereas in random tests, the parameters are randomized within the

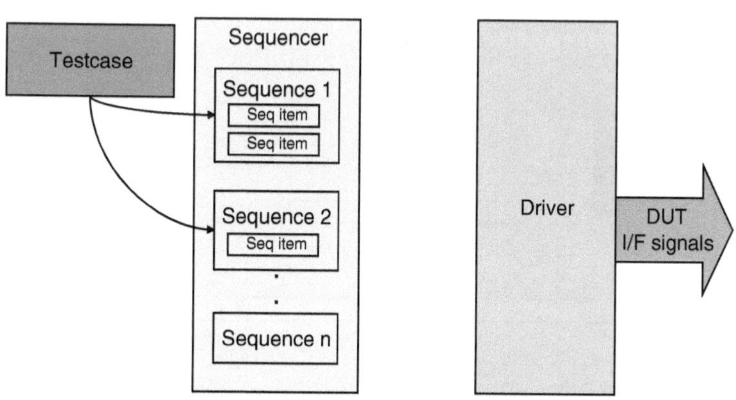

Fig. 9.1 An active uVC component

range of interest. We can Use the Functional combinational matrix, to identify all possible constraints for a specific mode.

The monitor (shown in Fig. 9.2) will collect information from the interface and decode it into transactions, events, and data items and Will pass required information to the checker and coverage modules. Checker will perform protocol and data integrity checks. Coverage module will collect functional coverage information

9.2.1.1 Functional Coverage

All features of the DUT to be verified are captured. Scenarios of interest are also identified. They are implemented as functional coverage points.

Monitor events and data structures are used to collect coverage. And when a particular scenario happens on the interface, it is marked as covered. Test plan will be the reference for coverage

9.2.2 Architecture of Flash Memory Verification Environment

Application of this environment on Flash memory (which is our Design Under Test) we have the environment given in Fig. 9.3.

If we go for directed approach, number of test cases will be more because of the one test case per scenario approach. So a test case be written for all possible

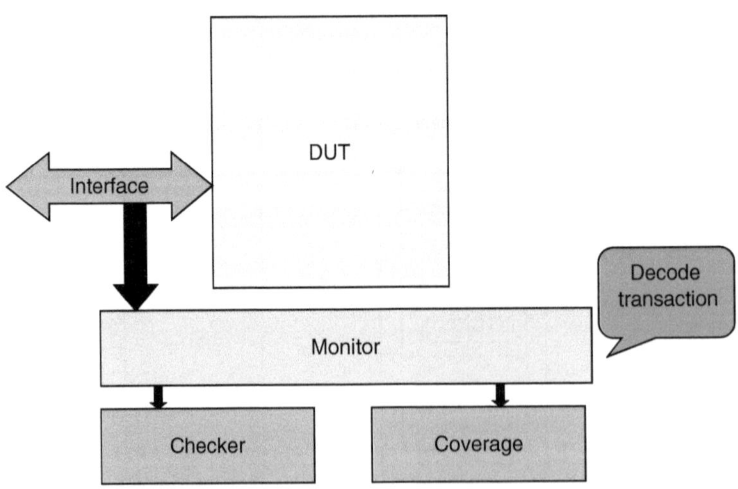

Fig. 9.2 Monitor component in uVC

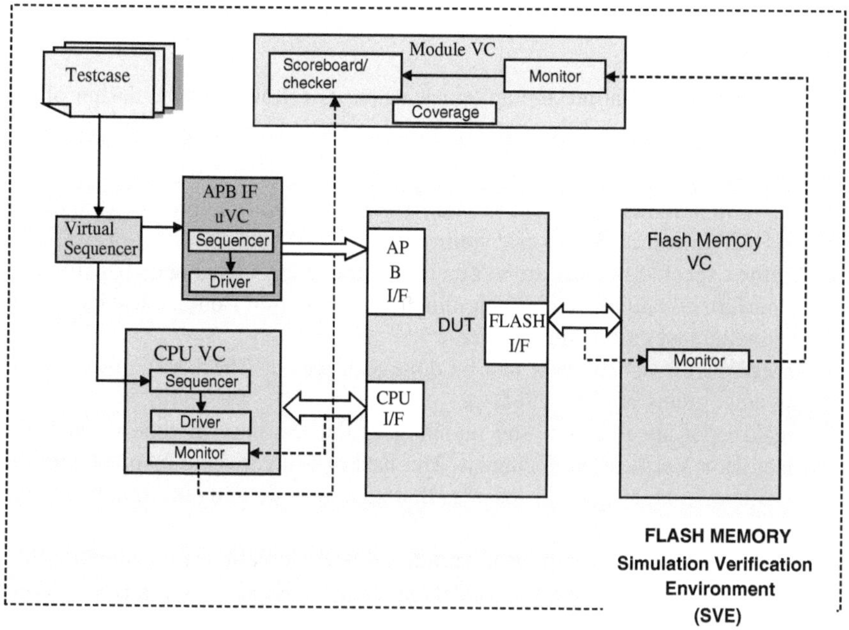

Fig. 9.3 High-level interface details of verification environment

scenarios so that it can be verified with lesser no of test cases using randomization approach. The randomizer will take care of all possible scenarios and will randomize the signal within the range of interest.

Documents such as black box, white box Functional combinational Matrix, and Verification Document can be used to design Driver, Checker, and Coverage modules.

Random test bench flow—as an example is shown below.

Data Read from the Flash in a particular DIV mode
The Parameters that can be randomized are: Address and Transaction Data Size
Coverage points are: Operation—Read and Address
Error operation in case of different Data size
The Random Test bench Would check
If operation is to Read from Flash, the output from the Flash should be read
If Address input given with the Request, it should be received at the output
If there is a different data size specified for the transaction, the read request should not be processed.

The Fig. 9.3 shown is the high-level architecture of the test bench. Top level is a Flash memory Simulation verification Environment which contain all the basic verification components as per the OVM methodology. All the blocks will be defined in detail in later section of this paper.

9.2.3 Verification Environment Design Approach

The following steps outline the approach which is followed in the design of the verification environment [3], verification list, assertions, and functional coverage items:

(a) The verification environment should have the flexibility to provide stimulus to the DUT to test all kinds of conditions.
(b) All the expected outputs are to be automatically calculated and checking will be performed automatically with minimal intervention from the test case. This makes the test case coding easier.
(c) Assertions are added. This will be done as a part of Checker items.
(d) Coverage items are identified.
(e) Break-up of specification and mapping to the test case is done through the Black Box Verification document. The lists of test cases are mapped with the specification break-up. This ensures that no item is missed out from the Design specification document.
(f) List of all possible bit settings in various registers in various functionalities like boot swap control, variable fetch wait control, Read data control, Address decoding and memory interfacing, and Standby control are emphasized under Functional Combinational Matrix document. This will help the test case writer to generate all possible stimuli for DUT.

All the test scenario as captured in Black Box verification document will be randomly generated with constraints. The coverage block will take care that all the functionality is covered, and we get 100 % functional coverage against the identified coverage items. This will be achieved by running test cases in many iteration.

9.2.4 Test Bench Description

Following are the basic test bench component as per OVM terminology [2].

1. Test case
2. Virtual sequencer
3. APB uVC (Universal verification component)
4. CPU VC (Verification component)
5. Flash VC
6. Module VC
7. Coverage

The detailed descriptions of the individual components of the test bench are described below.

TEST CASE

The test case writer can generate one or more random scenarios with or without constraints. This will be achieved by calling sequences from virtual sequence library [4]. As per the required operation, Virtual Sequence will use CPU VC sequencer and APB uVC sequencer to trigger required transaction.

Figure 9.4 illustrate how test case calls sequences from virtual sequence library.

Virtual Sequencer

Virtual Sequences coordinate invoke APB sequences as well as CPU VC sequences as per the operation required. Virtual sequences enable centralize control of APB uVC and CPU VC sequences. They invoke register configuration sequences on APB interface and input signals sequences on i/p interface. This is illustrated in Fig. 9.4

APB_uVC (APB Bus Model)

This APB universal verification component is created using OVM-based Agent architecture. This is an interface UVC which can be reusable.

The APB_UVC Components are:

APB-Sequencer,
APB-Driver.
APB-Monitor.

The APB-Sequencer is a test bench component, which controls items that are provided to the driver for execution. A sequencer can be controlled by a sequence, which provides a single transaction or a stream of transactions that forms a more structured and meaningful stimulus pattern.

The Virtual Sequencer will access the APB sequencer for the DUT register access.

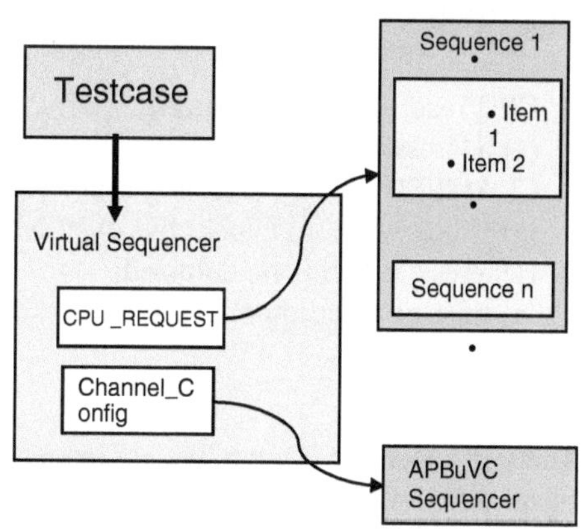

Fig. 9.4 High-level interface details of verification environment

List of Sequences:

1. APB Read
2. APB Write
3. PRESETZ

Direct address values are given for write and read sequences.

APB-Driver is an active entity that drives a data on the DUT. This driver samples and drives the DUT signals. APB-driver can operate stand-alone, but typically, it is instantiated inside an agent (component). APB-driver interacts with the DUT by driving and sampling the DUT signals. An APB-driver converts transactions into a format required by the DUT. In this case, the Driver drives the settings for the registers in the DUT.

The APB-Monitor is a passive component that monitors the bus and keeps tracks of all APB transactions with the DUT and provides the information to the module VC Scoreboard through TLM (transaction-level modeling) packet

CPU VC

The CPU VC components are:

CPU-Sequencer.
CPU-Monitor.
CPU-Driver.

The CPU-Sequencer is a test bench component, which controls items that are provided to the driver for execution. A sequencer can be controlled by a sequence, which provides a single transaction or a stream of transactions that forms a more structured and meaningful stimulus pattern. All sequences that attempt to provide items to drivers initiate requests with the APB-sequencer. The CPU-sequencer determines which sequence should be allowed to provide a data item to be processed by the driver. When a sequence is selected, its data item is constrained and randomized and then provided to the driver. Virtual Sequencer will make use of CPU-Sequencer to initiate read transactions for DUT.

List of sequences:

CPU read: fields randomized
(a) address. (b) data size.
CLAMPED values for configuring: Fields randomized
(a) Clock wait. (b) Clock div. c) FLNUM. (d) FLSIZE.
Heapresetz: Fields randomized:
(a) time of assertion of reset signal.

CPU-Monitor

The CPU-Monitor is a passive component that monitors the bus and keeps tracks of all CPU read transactions with the DUT and provides the information to the module VC Scoreboard through TLM. CPU-Monitor is a passive component and checks the protocol on bus interface. CPU-monitor also performs latency checking to make sure the delays between the signals are according to the configurations.

CPU-Driver

CPU-Driver is an active entity that drives a data on the DUT. A CPU-driver converts transactions into a format required by the DUT. This driver samples and drives the DUT signals. CPU-driver can operate stand-alone, but typically, it is instantiated inside an agent (component). A CPU-driver in a proactive agent initiates transfers on the bus according to test directives. When a CPU-driver is instructed to execute a new transfer, it drives the transfer to the DUT based on the bus protocol.

Flash VC

Monitor:
Monitor is a passive component which watches the output bus of DUT. The monitor will collect information from the interface and decode it into transactions, events, and data items. It provides required information to the checker and coverage modules. It also returns the data back to the DUT for the read request received.

Module VC

It has a monitor, scoreboard, and coverage
Monitor.
Monitor is a passive component which watches the transactions on input and output bus of DUT through TLM. The monitor will collect information from the other VCs and provides required information to the score board, checker, and coverage modules.
Score Board.
It collects the TLMs [4] from the corresponding Monitors and adds it to the CPU request queue and Flash queue which is compared in the checker.
Checker.
Module VC Checker collects information from Monitor in the form of transactions, events, and data items. It flags an error message if it detects a mismatch in the transaction.

Coverage

All functional coverage items are covered in this module [4].
The list of functional coverage items are listed in Table 9.1 as an example

Table 9.1 Functionality versus coverpoint

Functionality	Coverage_item
ECC DATA READ/WRITE.	1. CSECCLR/CDEDCLR check 2. CSECADR/CDEDADR check
Address decoding and boot swap control.	1 Configuration check from the clamped values. 2 SWAPADR output check 3 AC output check for selection of CE0/CE1
Variable fetch wait control	1. CLKDIV; CLKWAIT; check 2. RDEN and CAPEN generation check according to CLKDIV and CLKWAIT 3.FR_BX_EN_PF and FR_BX_RDY_PF
Stand by control	1.FIDLEREQ generation from CPU; FIDLEACK output from the STBC control . 2.CE0/CE1 output to high for one Clk cycle
ECC read data control	1. PSEL and PADDR setting 2. READ/WRITE operation on the 4 registers

9.3 Conclusion

For a Flash memory on SOC, it would just take your time in configuring the Cadence OVM environment for our DUT and write the BFM for the driver of this environment on all masters and slaves and proceed with a maximum of 40 test cases to verify the DUT for its at most functional coverage. Here, in the paper, it has been emphasized about an environment with a single peripheral talking to a master, this can be extended by configuring the OVM environment for multiple masters and multiple slaves with added Bus functional models. The paper talks about SOC verification but not the solution for an ASIC verification. Constrained Randomization is the best solution for verifying SOCs with some tools like synopsis VMM-libraries and Cadence OVM. Despite reporting many false errors, linting, and other static code checking technologies are still the most efficient mechanism for finding certain classes of problems. Simulators are only as good as the model they are simulating. Simulators offer many performance-enhancing options and the possibility to cosimulate with other languages or simulators. Assertion-based verification is a powerful addition to any verification methodology. This approach allows the quick identification of problems, where and when they occur.

Verification-specific SystemVerilog features offer an increase in productivity because of their specialization to the verification task and their support for coverage-driven random-based verification. Use code and functional coverage metrics to provide a quantitative assessment of your progress. Do not focus on reaching 100 percent at all cost. Do not consider the job done when you have reached your initial coverage goals. Use a source control system and an issue tracking system to manage your code and bug reports.

References

1. Bergeron J (2006) Writing testbenches using SystemVerilog. Springer, Norwell, MA
2. Website Reference for OVM http://ovmworld.s3.amazonaws.com/contributions/OVM%202.0%20Golden%20Reference%20Guide_0.pdf
3. Cadence, Inc. (2003) Hybrid RTL formal verification ensures early detection of corner-case bugs. http://cadence.com/products/magellan/magellan_wp.html
4. van der Schoot H, Bergeron J (2006) Transaction-level functional coverage in SystemVerilog. San Jose, DVCon

Chapter 10
Behavioral Modeling of LDO

Raveendra Somana, S. Bhanu Prakash and B. G. Shivaleelavathi

Abstract A 1.2 V 40 mA low-dropout regulator (LDO) for system-on-chip applications with 700 mV dropout is designed in Verilog-A. The proposed LDO provides fast line and load-transient responses with temperature-independent operation. The proposed LDO has been designed in Verilog-A using tsmc 65 nm CMOS technology and the total error of the output voltage due to line and load variations is low. The proposed LDO Design can be used to check the functional correctness of the SOC in the AMS verification flow [1].

Keywords Line regulation · Load regulation · SOC · Low-dropout regulator (LDO) · AMS verification flow

10.1 Introduction

Industry is pushing toward complete system-on-chip (SoC) design solutions that include the both analog and digital parts of the system on a single chip. This forces the industries to go for the AMS verification to make sure the functional correctness of the complete SOC. Today, there are various efficient verification methodologies which ensure 99.99 % correctness of digital designs, but the same does not hold true when it come to Analog/Mixed signal/SOC. Now, due to

R. Somana (✉) · S. B. Prakash
Department of Electronics and Communication, JSSATE, Bangalore, India
e-mail: raveendras4a5@gmail.com

S. B. Prakash
e-mail: bhanu.somisetty@gmail.com

B. G. Shivaleelavathi
JSSATE, Bangalore, India
e-mail: shivaleelavathi@yahoo.com

V. Chakravarthi et al. (eds.), *Proceedings of International Conference on VLSI, Communication, Advanced Devices, Signals & Systems and Networking (VCASAN-2013)*, Lecture Notes in Electrical Engineering 258, DOI: 10.1007/978-81-322-1524-0_10,

increase in Analog Mixed-Signal SOC's/Chips, it is necessary to use a potential Verification flow which provides the similar confidence on functional correctness as seen for Digital Design.

In a SOC, the majority portion will be carried out by the digital design and a small portion will be done in analog design. But all these SOC's use the various power supplies to operate the various parts of the system. In general, a Low-Dropout Regulator (LDO) is used to provide a constant power supply for the whole SOC. Due to the emerging need of high-performance low-voltage LDOs for low-voltage mixed-signal systems, many researchers have recently proposed advanced methods to improve the performance of LDOs. Rincon-Mora [2] proposed current-efficient voltage buffer, forward-biased power transistor, pole-zero doublets for load-regulation enhancement, and capacitance multiplication. Heisley et al. proposed using a DMOS power transistor. Chevalerias et al. proposed using an NMOS power transistor with charge-pumped gate drive. The main aims of all the proposed methods are: (1) to enable low-voltage regulation; (2) to reduce slew-rate limit at the gate drive; and (3) to improve load regulation and transient response. However, the precision of the above reveals the fact that there are limitations on the structure and frequency compensation scheme of classical LDOs, especially for the low-voltage LDO designs. Moreover, the off-chip capacitor, which is the key for stability and high LDO performance, cannot be eliminated. This off-chip capacitor is the main obstacle to fully integrating LDOs in system-on-chip designs.

As a result, low-voltage high-stability and fast-transient LDOs with, preferably, capacitor-free operation should be developed. Solving the correlated trade-offs on stability, precision, and recovery speed is the main challenge of capacitor-free LDO design [3]. In this paper, the behavioral model is targeted to meet the specifications of the improved version of the CMOS LDO in which the circuit architecture is based on a three-stage amplifier design [4], and it provides a capacitor-free feature to eliminate the need of bulky off-chip capacitor. Both fast load-transient response and high power-supply rejection ratio (PSRR) are achieved due to the fast and stable loop gain provided by the proposed LDO Structure and damping-factor-control (DFC) compensation Scheme. The power PMOS transistor in the proposed LDO operates in linear region at dropout, and hence, the required transistor size can be reduced significantly for the ease of integration and cost reduction. In addition, a novel CMOS voltage reference based on weighted difference of gate–source voltages enables full-CMOS implementation.

The behavioral model of any analog circuit concentrates on the functionality of the design; it may not need to consider the performance issues as it is not the actual transistor-level model. Hence, the functionality of the proposed LDO is majorly affected by the three parameters like line Regulation, Load regulation, and the Temperature [2]. This paper describes the Functional design of the LDO under these three parameters which imitates the actual circuit design of LDO.

10.2 Block Description

The block diagram of a generic series low-dropout regulator is as shown in Fig. 10.1. The circuit is composed of a reference and associated start-up circuit, protection circuit and associated current sense element, an error amplifier, a pass element, and a feedback network. The reference provides a stable DC bias voltage with limited current driving capabilities. This is usually a zener diode or a bandgap reference. The zener diode finds its applications in high-voltage circuits (greater than approximately 7 V) with relaxed temperature variation requirements. The band-gap, on the other hand, is better suited for low-voltage and high-accuracy applications. The protection circuitry ensures that the LDO operates in safe stable conditions. Some of its functions include over-current protection (typically a fold back current limiter), thermal shut down in case of self-heating (junction temperature increases beyond safety levels), and other similar functions. The error amplifier, the pass element, and the feedback network constitute the regulation loop. The temperature dependence of the reference and the amplifiers input offset voltage defines the overall temperature coefficient of the regulator, hence low drift references and low-input offset voltage amplifiers are preferred.

Line Regulation of LDO: It defines the ratio of output voltage deviation to a given change in the input voltage. The quantity reflects the deviation after the regulator has reached steady-state. A general line regulation equation is given in Eq. (10.1).

$$\frac{\Delta V_{out}}{\Delta V_{in}} \approx \frac{g_{mp} r_{op}}{A * \beta} + \frac{1}{\beta} \frac{\Delta V_{ref}}{\Delta V_{in}} \tag{10.1}$$

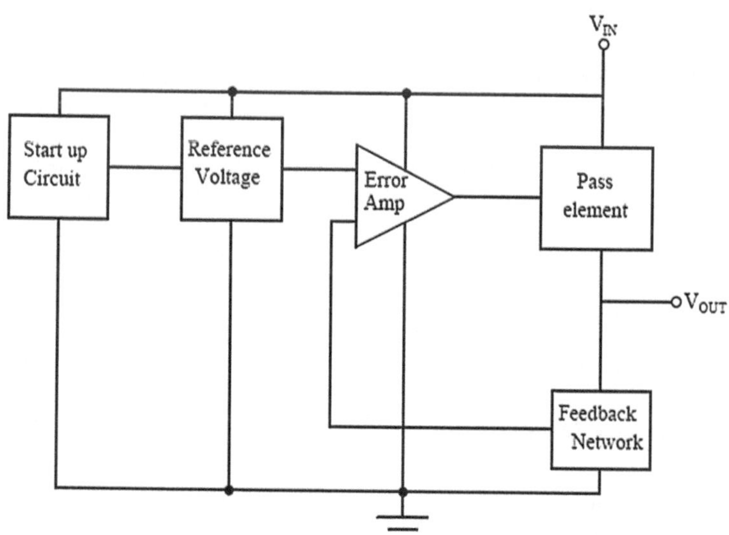

Fig. 10.1 General LDO block diagram

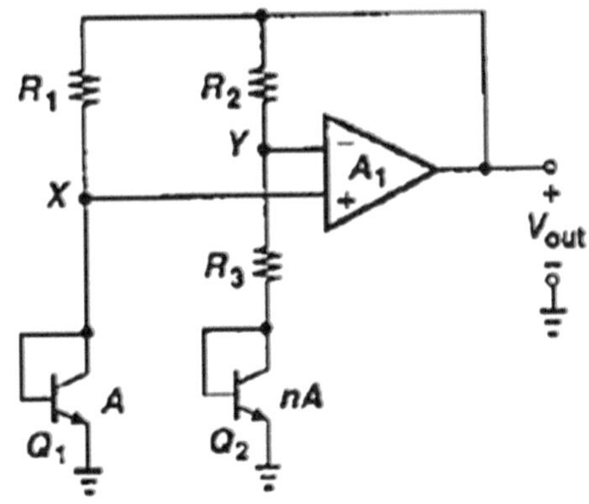

Fig. 10.2 Band-gap reference circuit

Smaller output voltage deviation for a given dc change in input voltage corresponds to a better voltage regulator. To increase the line regulation, the LDO regulator must have a sufficiently large loop gain. These quantities become clear in the LDO regulator design discussion.

Load Regulation of LDO: It is a measure of output voltage deviation during no-load and full-load current conditions. The load regulation is related to the loop gain, $A\beta$, and the pass transistor output impedance, r_{op}. This relation is given in Eq. (10.2).

$$\frac{\Delta V_{out}}{\Delta I_o} = \frac{r_{op}}{1 + A * \beta} \tag{10.2}$$

The load regulation improves as the loop gain increases and the output resistance decreases. The load regulation only applies to the LDO regulator steady-state conditions and does not include load-transient effects.

Band-Gap Reference: The basic operation of this block is to provide a reference voltage irrespective of the any temperature changes. The band-gap reference circuit diagram is as shown in the Fig. 10.2. It is a combination of negative and positive temperature coefficient [5]. The Eq. (10.3) shows band-gap reference equation.

$$V_{out} = V_{BE2} + (V_T * \ln n)\left(1 + \frac{R_2}{R_3}\right) \tag{10.3}$$

10.3 Verilog-A Description

The Verilog-A language was designed as an extension of spice to be compatible for both low and high-level abstractions levels. Similar to Verilog HDL and its ability to span the range of abstraction levels for digital descriptions, the Verilog-A

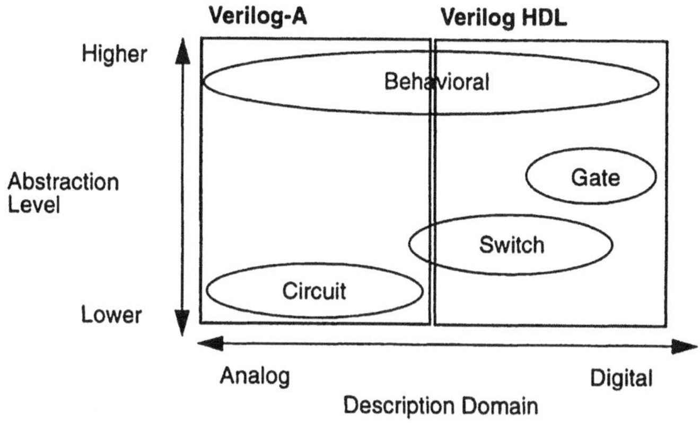

Fig. 10.3 Verilog-A abstraction level

language was designed to function just as effectively at describing high-level analog behaviors as well as circuit-level descriptions. The Abstraction levels are as shown in Fig. 10.3.

The Verilog-A allows the description of analog and/or mixed-signal systems with varying amounts of detail. The analog behavioral capability allows the designer to span the abstraction levels, allowing direct access to the underlying technology while maintaining the capability of system-level modeling and simulation. As such, the analog and mixed-signal system can be described and simulated at a high level of abstraction early in the design cycle to facilitate full-chip architectural trade-offs. The resulting Verilog-A description, as an executable specification, promotes communication and consistency throughout the design process.

The Verilog-AMS specification, currently under development by Open Verilog International, Is targeted to be a single-language solution for the specification and simulation of analog, digital, and mixed-signal systems. The objectives of the Verilog-AMS specification are to facilitate portable mixed-signal system description and simulation. In addition, a design described with the Verilog-AMS language will provide the capability to integrate system and circuit-level aspects of the design allowing the design intent to be maintained throughout the entire mixed-signal design process. Verilog-A is a subset of Verilog-AMS [6].

10.4 LDO Modeling

According to the behavior of the LDO, it can be divided into two major blocks. The top level block is named as LDO_BGR_Top in which two major blocks are there. Those two blocks are named as BGR_Top and LDO_Top. The temperature-

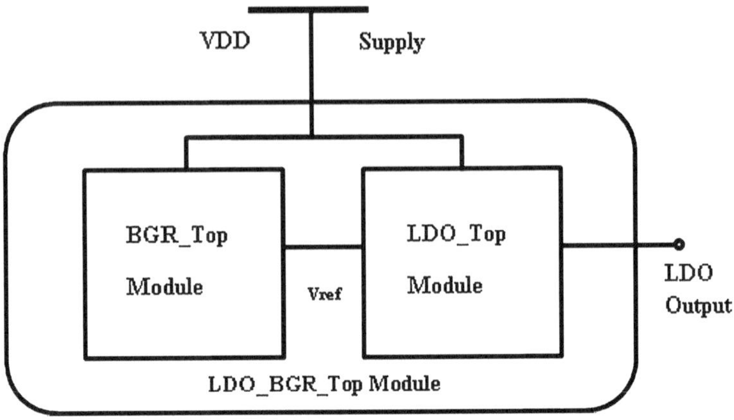

Fig. 10.4 Behavioral model block diagram of LDO

independent voltage $\left(V_{ref}\right)$ which is the output of BGR_Top is given to the LDO_Top block. The Behavioral model block diagram is as shown in Fig. 10.4.

The BGR_Top which represents the band-gap reference operation is implemented. The Table 10.1 shows a piece of Verilog-A code. As we know that the BGR is derived from the two positive- and negative-temperature coefficients, the code written in Verilog-A extracts those values. The Prepositional to the Absolute Temperature (PTAT) increases as the Temperature increases, whereas the Complementary to the Absolute Temperature (CTAT) decreases as the Temperature increases with the same magnitude.

The Major LDO block operation is to provide the constant output voltage of 1.2 V even for the change in the input voltage and the output current as these two are the major parameters which affects the output voltage. The piece of code which shows the LDO behavioral operation is as shown in Table 10.2. The behavioral model of the LDO is designed in such a way that it imitates the behavior of the actual Circuit model. That means that the line regulation and the load-regulation values should match exactly with the transistor-level design [7].

Table 10.1 BGR model

```
else if (V(vdd) >=(1.2-vtol)) begin
CTAT = (requal * $vt * ln(n));
PTAT = ($vt*ln(Io/(2*(pow($temperature,m))*(exp(-e.g./$vt)))));
V(vbg) <+ CTAT + PTAT;
end
```

Table 10.2 LDO behavioral model

```
else if (V(in) >=vreg && abs(I(out)) <=Il)
   begin
  $strobe(''LDO Output for the load less than 100u'');
  ro = 50; V(out) <+ vldo + (I(out)*ro) + val;
   end
else if (V(in) >=vreg && abs(I(out)) > Il && abs(I(out)) <= Im)
   begin
  $strobe(''LDO output for 100u > load < 1 m'');
ro = 1; V(out) <+ vldo + (I(out)*ro) + val;
   end
else if (V(in) >=vreg&&abs(I(out)) > Im && abs(I(out)) <=Ih)
   begin
    $strobe(''LDO output for the Max load <= 40 m'');
   ro = 25e-3;
   V(out) <+ vldo + (I(out)*ro) + val;
    end
```

10.5 Experimental Results

The proposed behavioral model of the LDO is designed to meet the respective line and load-regulation values as the transistor-level model values. The line regulation means the output change with respect to the input changes is calculated under 0 and 40 mA current loads. The line regulation result plot of the behavioral model LDO at maximum load of 40 mA is as shown in Fig. 10.5.

The load regulation is defined as the ratio of output change to the load current under constant input voltage and is plotted as shown in Fig. 10.6. The Fig. 10.7 gives the LDO output voltage curve with respect to the temperature range from −40 to 120 °C. The variation of the Voltage for temperature is very less approximately 0.1 mV from corner to corner.

The behavioral model results are summarized and compared with the actual transistor-level design for the line and load regulations. These compared values are noted down in the Table 10.3.

The behavioral model which has been done in Verilog-A will produce the exact values since no other parameters will effect it, where as many other parameters like area, frequency will affect the transistor-level design. The input and output values of LDO are summarized in Table 10.4.

One of the major advantages of Behavioral modeling is that the simulation time to simulate a block is very much less when compared to transistor-level design. The Designed LDO is also checked with the simulation time for both transistor-level simulation and the Behavioral simulation and the simulation times noted have been tabulated in Table 10.5. As we go for a much bigger designs, the simulation time of a circuit increases, this behavioral model is very useful in such a way to replace the transistor-level design, since it can be simulated in very less time comparatively with the circuit design.

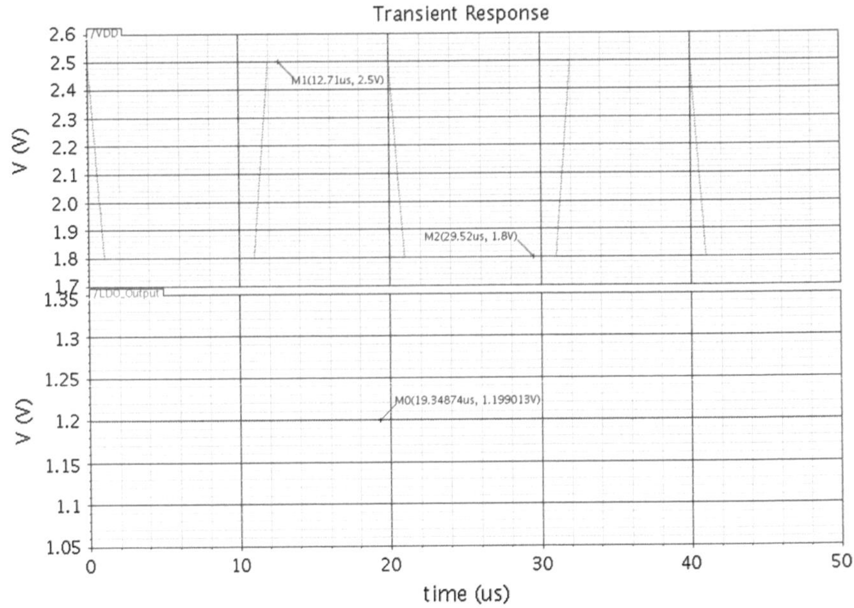

Fig. 10.5 LDO output changes with respect to the input voltage changes under load current as "40 mA" at 27 °C

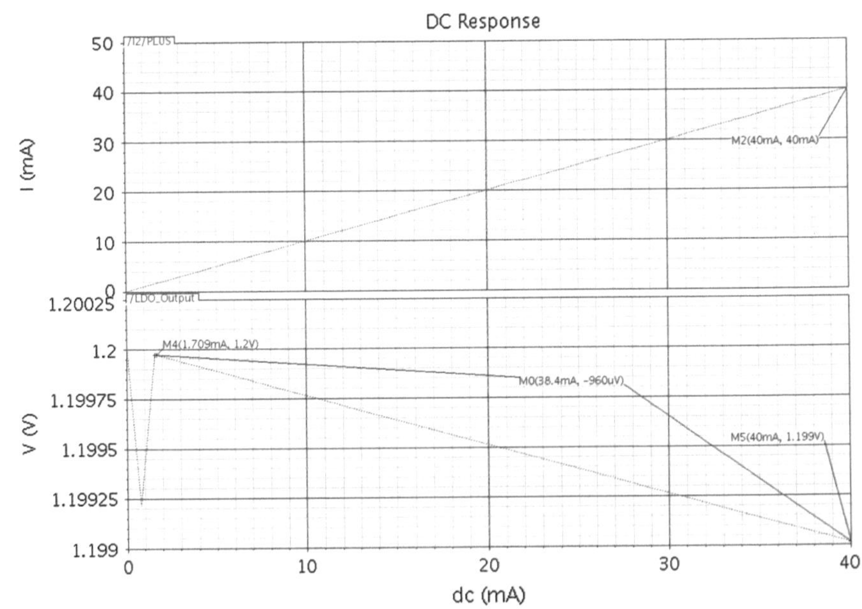

Fig. 10.6 Load-regulation *curve* for output voltage with respect to output current changes from 0 to 40 mA

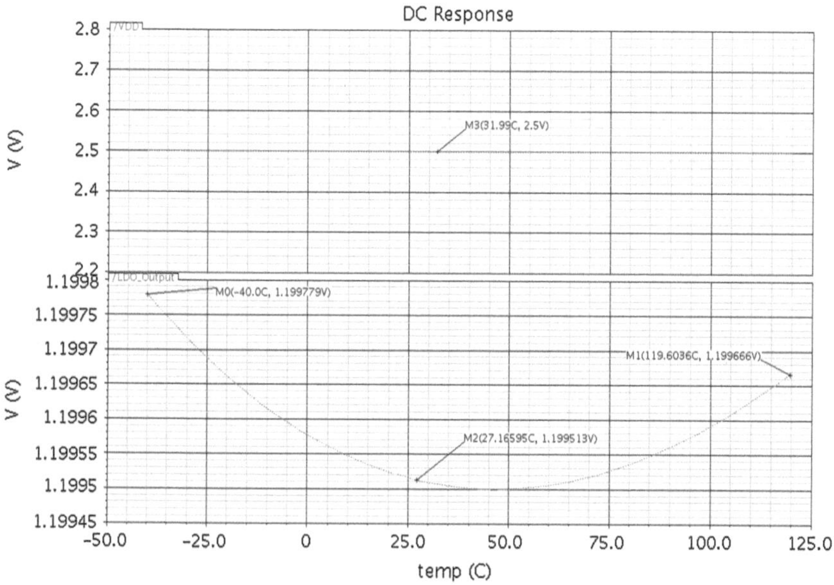

Fig. 10.7 LDO output voltage *curve* with respect to temperature at a constant input voltage

Table 10.3 Load-regulation values comparison

Output load		Simulated values			Units
		Slow (−40)	Typical (27)	Fast (120)	°C
"0 mA" load	Verilgo-A model	1.200279	1.200013	1.200168	V
	Transistor level	1.200264	1.200005	1.200153	V
"40 mA" load	Verilog-A model	1.199279	1.199013	1.199168	V
	Transistor level	1.199264	1.199005	1.199153	V

Table 10.4 Input and output voltage values

Input voltage (V_{in})	Output voltage	
	Transistor-level model (V_{out})	Verilog-A model (V_{out})
1.2	0	0.1
1.94	1.2005	1.2

Table 10.5 Simulation time

Simulation time (ms)	Transient analysis time	
	Transistor-level model (ms)	Verilog-A model (ms)
1	200	20.355
5	238	24.425
10	256	26.466

10.6 Conclusion

The Proposed behavioral model of the LDO is designed for the specification of the transistor-level design. The performance of the proposed behavioral model of LDO is proven to be much better than the transistor-level LDO model. The Output parameters of the behavioral model are exactly following the transistor-level LDO model characteristics. The comparisons of these two models are given in Tables 10.3 and 10.4.

Since the functionality of the designed behavioral model exactly matches with circuit-level model, this Behavioral model can be used in the SoC functional verification process to represent the actual LDO model to verify the whole functionality of the SoC [1]. This can be further extended to design the complete SoC including both Digital and Analog blocks, which can be programmed using the Verilog-AMS.

References

1. Analog Mixed Signal Verification Methodology (AMSVM) proposed by Divyesh Gajjar, eInfochips
2. Rincon-Mora GA, Allen PE (1998) A low-voltage, low quiescent current, low drop-out regulator. IEEE J Solid-State Circuits 33:36–44
3. Leung KN, Mok PKT (2003) A capacitor-free CMOS low-dropout regulator with damping-factor-control frequency. IEEE J Solid-State Circuits 37:1691–1701
4. Leung KN, Mok PKT, Ki WH, Sin JKO (2000) Three-stage large capacitive load amplifier with damping-factor-control frequency compensation. IEEE J Solid-State Circuits 35:221–230
5. Razavi B (2001) Design of analog mixed circuits. MAMcGraw-Hill, Boston
6. Verilog HDL (2009) Verilog-AMS language reference manual Version 2.3.1, June 2009 by Accellera
7. Milliken RJ (2007) Full on-chip CMOS low-dropout voltage regulator. IEEE Trans Circuits Syst Regul Pap 54(9):1879–1890

Chapter 11
Regulated Cascode Preamplifier-Based Front-end Readout ASIC "ANUSPARSH" for Resistive Plate Chamber Detector

Menka Sukhwani, C. K. Pithawa, V. B. Chandratre, S. Veena and T. Megha

Abstract A regulated cascode transimpendence preamplifier based low power, multichannel, fast front-end ASIC "ANUSPARSH" is designed for readout of ~3.6 million pick up strips of RPC detector of INO-ICAL experiment. This ASIC comprises eight front-end channels, each consisting of a regulated cascode preamplifier, two stages of differential amplifier providing total gain of ~7 mV/μA and a fast leading edge discriminator with LVDS output followed by a multiplexed fast analog buffer capable of driving 50 Ω cable. The regulated cascode preamplifier is used for the first time as readout of RPC detector exhibiting good impedance matching with the detector impedance of ~50 Ω over a wide frequency range. This ASIC is fabricated in 0.35 μm mixed CMOS process and tested successfully with the RPC detector. The front-end electronics requirements of INO-ICAL RPC detector and design approach for development of ANUSPARSH ASIC are presented along with test results when interfaced to the RPC detector.

Keywords India-based neutrino observatory (INO) · Iron calorimeter (ICAL) · Regulated cascode (RGC) · Resistive plate chamber (RPC)

11.1 Introduction

The magnetized Iron Calorimeter (ICAL) detector of India-based Neutrino Observatory (INO) is being setup with resistive plate chamber (RPC) detector as

M. Sukhwani · C. K. Pithawa · V. B. Chandratre (✉)
Electronics Division, Bhabha Atomic Research Centre, Trombay, Mumbai 400085, India
e-mail: vinay.chandratre@cern.ch

S. Veena · T. Megha
Electronics Corporation of India Limited, Prabhadevi, Mumbai 400028, India

V. Chakravarthi et al. (eds.), *Proceedings of International Conference on VLSI, Communication, Advanced Devices, Signals & Systems and Networking (VCASAN-2013)*, Lecture Notes in Electrical Engineering 258, DOI: 10.1007/978-81-322-1524-0_11,

active element, for precision measurement of the neutrino oscillation parameters and the order of neutrino mass levels [1]. In order to achieve these physics goals, it is required to analyze the muon track bending effects in the presence of high magnetic field in ICAL detector. The valid muon track in ICAL detector volume is reconstructed by capturing the RPC pick-up strip hit pattern with respect to a global trigger and its direction is determined by precision leading time measurement of the detector signal. The INO-ICAL RPC detector setup comprises ~3.6 million pick up strips and is being operated in avalanche mode providing signal with subnanosecond leading edge time.

The front-end readout electronics for INO RPC detector therefore requires a low-power, multichannel amplifier-discriminator solution with subnanosecond response. The ANUSPARSH ASIC architecture, as shown in Fig. 11.1, therefore comprises eight front-end channels, each having regulated cascode transimpedance preamplifier, two stages of differential amplifier followed by a fast leading edge discriminator with LVDS output. In order to analyze detector pulse profile for time walk correction, a multiplexed fast analog buffer capable of driving a 50 Ω cable is also incorporated in this ASIC. The regulated cascode transimpedance preamplifier topology is used for the first time as front-end readout of an RPC detector. This pre-amplifier topology was chosen due to its inherent low-input impedance over wide frequency range leading to good impedance matching with the RPC pick up strip impedance of ~50 Ω, besides providing large transimpedance gain and lower power consumption. The ANUSPARSH ASIC has been successfully tested and characterized with the RPC detector.

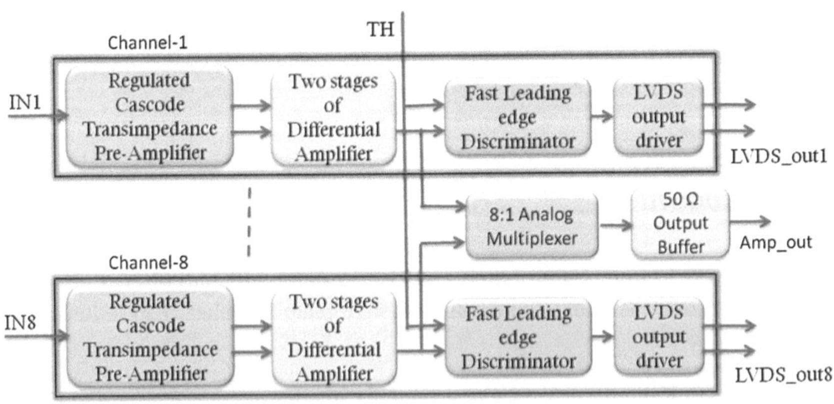

Fig. 11.1 Anusparsh ASIC architecture

11.2 Anusparsh Asic Design

11.2.1 RPC Front-end Electronics Requirements

The INO-ICAL RPC detector is a fast, planar, gaseous detector, having multiple pick up strips orthogonally placed on top and bottom sides of the detector providing complementary signals of sub-nanosecond rise time for muon track vector reconstruction. With reference to the simplified representation of INO-ICAL RPC detector shown in Fig. 11.2a, the RPC detector strip line impedance is given as [2],

$$Z = \frac{377}{\sqrt{\epsilon_r'}(x + 1.393 + 0.667\ln(x + 1.444))}\ for\, x \geq 1 \tag{1}$$

Where, $x = \frac{w}{h} \geq 1$ in this case and w is the strip width with h being the RPC pick-up panel thickness. The ϵ_r' is given by,

$$\epsilon_r' = \frac{\epsilon_r + 1}{2} + \frac{\epsilon_r - 1}{2}(1 + 10x)^{-1/2} \tag{2}$$

Here, ϵ_r is relative permittivity of the RPC pickup panel dielectric material. Taking the manufacturing materials and tolerances into account, impedance of INO-ICAL RPC detector is estimated to be $\sim 50 \pm 5\ \%\ \Omega$.

A typical charge distribution of INO-ICAL RPC detector when operated in avalanche mode with voltage amplification of 90 (equivalent trans-impedance gain of 4.5 mV/μA, as pick-up strip is terminated with 50 Ω resistor) is shown in Fig. 11.2b. This distribution has mean at $\sim$55 pC and sigma value of 15 pC. The linear input dynamic range for the frontend electronics is therefore estimated to be $\sim$100–1100 fC [(mean $\pm$ 3σ)/90] resulting in an equivalent current dynamic range of $\sim$10–110 μA with a triangular pulse approximation having base of $\sim$20 ns and t_r of $\sim$1 ns.

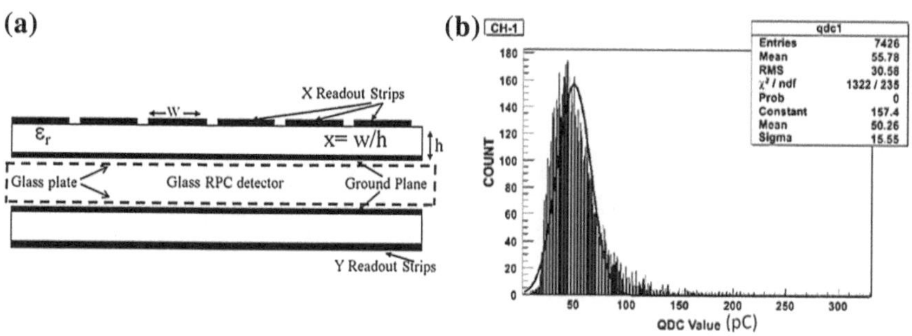

Fig. 11.2 **a** A simplified representation of INO-ICAL RPC detector. **b** Typical charge distribution of INO-ICAL RPC detector in avalanche mode

The front-end readout electronics for such detector setup is therefore required to have a fast amplifier with input impedance matched to the detector strip impedance at the frequency of operation, linear operation in both the polarities of input to cater to the complementary input signals from X–Y RPC strips and gain of ~7 mV/μA considering the input dynamic range from avalanche mode of detector operation, maximum available output swing (±800 mV in the chosen 0.35 μm CMOS technology) and maximum acceptable system noise level.

The amplifier is required to be followed by a fast leading edge discriminator with LVDS driver in order to interface a subsequent time-to-digital converter for precision leading time measurement with resolution better than 200 ps (time between two correlated events in adjacent detector layers). Further, the overall design has to be optimized for low-power consumption in order to cater to ~3.6 million pick up channels.

11.2.2 Fast Amplifier Stages

Two topologies, telescopic cascode [3] and regulated cascode (RGC) [4, 5] were analyzed with respect to their input impedance, power consumption and noise for the transimpedance pre-amplifier as shown in Fig. 11.3a, b, respectively. For both the topologies, resistive loads were used as they exhibit less parasitic capacitance

$$Z_{in} = \frac{1}{g_{m1}} \tag{3}$$

Here, g_{m1} is the transconductance of transistor M1 in Fig. 11.3a. Now, considering the small signal model of regulated cascode architecture, as shown in Fig. 11.3c, its input impedance is given by,

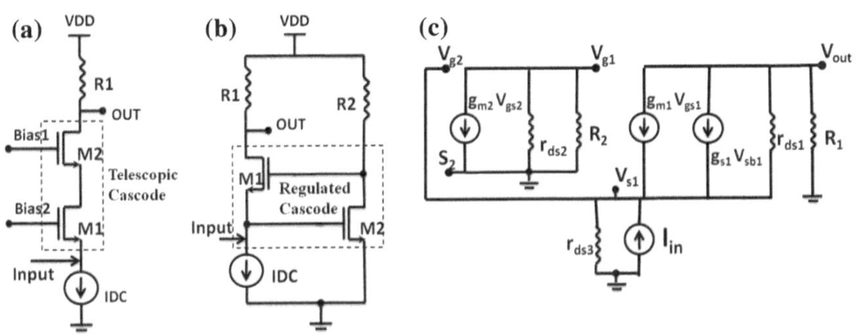

Fig. 11.3 Transimpedance pre-amplifier topologies. **a** Telescopic cascode. **b** Regulated cascode. **c** Small signal model of the regulated cascode transimpedance pre-amplifier

$$Z_{in} = \frac{1}{g_{m1} * (1 + g_{m2}/(G_2 + g_{ds2})) * G_1/(G_1 + g_{ds1})} \quad (4)$$

Here, g_{m1} (g_{m2}) is the transconductance, g_{ds1} (g_{ds2}) is the drain-to-source resistance of transistor M1 (M2), and G_1 and G_2 are load admittances in Fig. 11.3b. Neglecting g_{ds1} and g_{ds2} as compared to G_1 and G_2, Eq. (4) can be approximated as,

$$Z_{in} = \frac{1}{g_{m1}(1 + g_{m2}R_2)} \quad (5)$$

Therefore, a local feedback loop formed by transistor M2 and resistance R2 in RGC architecture reduces the input impedance by voltage gain of the loop allowing lower input impendence and better impedance matching with the RPC detector over wide frequency range. Further, the RGC architecture exhibits improved timing performance for the same bias current as compared to the telescopic architecture. In view of these merits, RGC transimpedance pre-amplifier was chosen for the ANUSPARSH ASIC.

The RGC pre-amplifier is followed by two stages of differential amplifier. An identical dummy RGC pre-amplifier stage is used to obtain a balanced DC operating point for the subsequent differential amplifier stages.

11.2.3 Fast Leading Edge Discriminator and Analog Line Driver Buffer

In order to determine the muon track direction in the INO-ICAL detector volume, a fast leading edge discriminator is implemented in each front-end channel of ANUSPARSH ASIC with *low-voltage differential signal* (*LVDS*) output driver [6] providing total propagation delay of ~1 ns, rise time of ~700 psec, sensitivity of 2 mV and adjustable common mode reference. In the INO-ICAL detector configuration, the discriminator output is required to travel ~2 m of distance for the data acquisition system. The LVDS standard for discriminator output therefore provides lower susceptibility to common mode noise maintaining the signal integrity and timing over long distance with relatively lower static power consumption.

Further, precision time tagging of the valid detector events requires corrections for the time and amplitude walks inherent in the detector signal. These corrections could be performed either by analyzing the detector signal profile or by measurement of discriminator output width over threshold. In the ANUSPARSH ASIC, the amplified detector signal is, therefore, provided through a multiplexed analog buffer [7] exhibiting subnanosecond response on a 50 Ω cable.

11.3 Test Results

The ANUSPARSH ASIC is designed, fabricated in 0.35 μm CMOS process and tested with the prototype RPC detectors of sizes 1 × 1 m and of 2 × 2 m. The ASIC has met its gain, timing, impedance matching, and power specifications with both the detector assemblies. The tested specifications of this ASIC are given in Table 11.1. The Fig. 11.4a, b show simulated results for gain, 3 dB bandwidth and

Table 11.1 Tested specifications of anusparsh ASIC

Total channel gain	~7 mV/μA
Linear input dynamic range	±100 μA
LVDS common mode voltage range	0.8 to 1.6 V
Power consumption/channel	~45 mW @ 3.3 V supply
Die area and package	3.5 mm by 3.5 mm; CLCC44

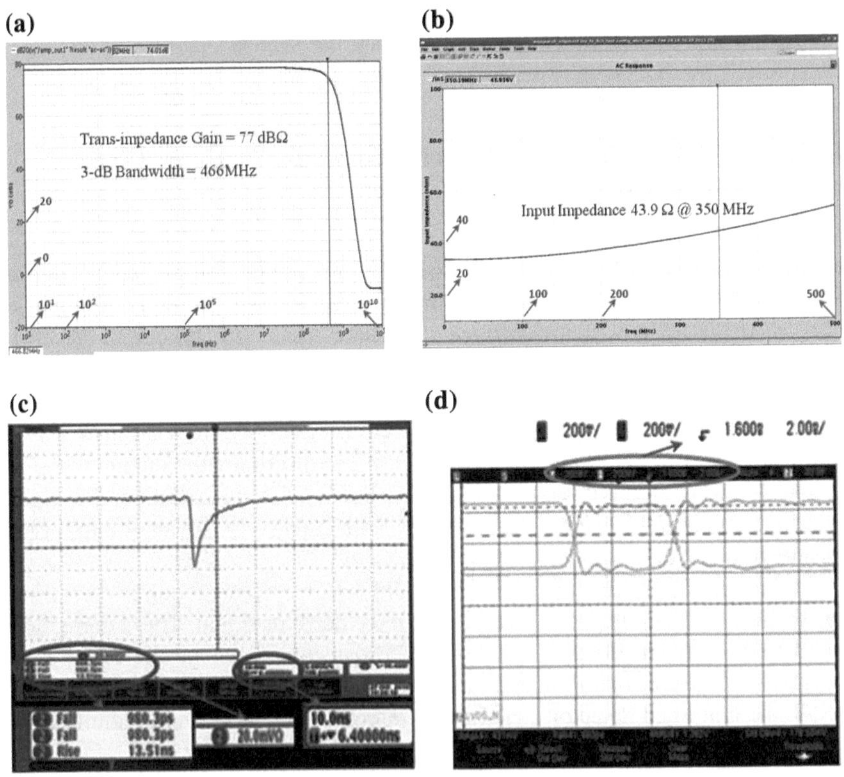

Fig. 11.4 **a** Simulation results for gain (dB) and 3 dB BW; **b** simulated input impedance (Ω) variation w.r.t. frequency (MHz); **c**, **d** test results of anusparsh ASIC

input impedance variation with respect to frequency. The Fig. 11.4c, d show test results for amplifier output with subnanosecond leading edge and discriminator LVDS output.

11.4 Conclusion

The regulated cascode transimpedance topology implemented in the ANU-SPARSH ASIC has proven to be good front-end architecture for the RPC detector with proper impedance matching at the operating frequency. The ASIC has met all the design specifications and has been successfully interfaced with the RPC detector. A two chip solution is being planned for the amplifier and discriminator to improve the switching noise isolation and amplifier stability.

Acknowledgments Authors would like to thank Shri Sekhar Basu, Director, BARC; Dr. T.S. Ananthakrishnan, Head, Electronics Division, BARC; Prof. N. K. Mondal, TIFR; Dr. B. Satyanarayana, TIFR and INO collaboration for their support and encouragement.

References

1. INO collaboration, detailed project report, 2006, 2008, and website. http://www.ino.tifr.res.in/ino//index.php
2. Paul CR (1992) Introduction to electromagnetic compatibility
3. Anghinolfi F, Jarron P, Martemiyanov AN, Usenko E, Wenninger H, Williams MCS, Zichichid A (2004) NINO: an ultra-fast and low-power front-end amplifier/discriminator ASIC designed for the multigap resistive plate chamber. J Nucl Instrum Methods Phys Res A 533:183–187
4. Park SM, Toumazou C (2000) A packaged low-noise high-speed regulated cascode transimpedance amplifier using 0.6 μm N-well CMOS technology. In: proceeding European solid-state circuit conference, Stockholm, pp 432–435
5. Mengxiong L (2007) 5 GHz optical front end in 0:35 mm CMOS. PhD thesis, The University of Nottinghum
6. Tajalli A, Leblebici Y (2007) A power-efficient LVDS driver circuit in 0.18 um CMOS technology. In: IEEE 3rd conference Ph.D. Research in microelectronics and electronics (PRIME), Bordeaux
7. Lin ZM, Lai HC (2007) A high driving capability CMOS buffer amplifier for TFT-LCD source drivers. Eng Lett 15(2):327–330

Chapter 12
Fault-Tolerant Reversible Logic for Combinational Circuits: A Survey

M. Poornima, M. S. Suma, Namita Palecha and T. Malavika

Abstract Power minimization is the most required criteria in today's world of electronics. Reversible logic provides an aid for low power. Fault-tolerant design is the one that enables a system to continue operation, possibly at a reduced level (degradation), rather than failing completely, when some part of the system fails. This helps in serving many safety critical applications. This paper provides a survey of an overview of latest advancements in research of reversible logic techniques at fault-tolerant level. It gives an overview of the methodologies used in the reversible engineering and the fault-tolerant gates used in them. An attempt is made to give a survey of the techniques used in different combinational logics and briefing them.

Keywords Reversible logic · Fault tolerance · Combinational circuits

12.1 Introduction

Conventional logic is not reversible. Irreversible logic circuits dissipate heat in the amount of [$kT \ln 2$] Joule for every bit of information that is lost, where k is the Boltzmann constant and T is the operating temperature [1]. The system will not

M. Poornima (✉) · T. Malavika
MVJ College of Engineering, Bangalore, India
e-mail: poornima80@hotmail.com

T. Malavika
e-mail: malavika24@gmail.com

M. S. Suma · N. Palecha
RV College of Engineering, Bangalore, India
e-mail: sumams@rvce.edu.in

N. Palecha
e-mail: namitapalecha@rvce.edu.in

V. Chakravarthi et al. (eds.), *Proceedings of International Conference on VLSI, Communication, Advanced Devices, Signals & Systems and Networking (VCASAN-2013)*, Lecture Notes in Electrical Engineering 258, DOI: 10.1007/978-81-322-1524-0_12,

dissipate any energy until the system is able to return to its initial stage from its final state even if anything occurs in between. This was proved by Bennett [2] in 1973. Reversible logic circuits are those where the input vector can be recovered from output vector, i.e., there is no loss of information. Hence, power dissipation of reversible circuit, under ideal physical circumstances, would be zero. Reversible computing can potentially require zero or very little energy [3].

Reversibility recovers bit loss in the circuit but is not able to detect bit error. Fault-tolerant reversible circuit is capable to prevent error at outputs [4, 5].

In fault-tolerant reversible circuit method, once the required gates have been designed and an appropriate synthesis framework (either adapted from existing methods or custom-built for the new gates) has been established, fault-tolerant implementation requires no extra expenditure in design or verification effort [6]. Reversible logic, which allows the reproduction of the circuit's inputs from observed outputs [2], finds applications in quantum computing low-power design [5, 7], nanotechnology [8, 9, 10, 11, 12] etc.

12.2 Reversible Logic

A reversible logic gate is an n-input, n-output logic device with one-to-one mapping. Hence, the vector of input states can be always reconstructed from the vector of output states. The reversible gates output vector is a permutation of the numbers, 0 to $(2^n - 1)$. The reversible gate is balanced, i.e., the output is equal to one for exactly one half of its inputs [13]. This helps to determine the outputs from the inputs but also the inputs can be uniquely recovered from the outputs. Extra inputs or outputs are added so that the number of inputs is made equal to the number of outputs whenever it is necessary. The complexity and performance of the circuit is decided, based on the following parameters [3, 4, 7, 8, 14, 19].

Garbage outputs: The number of unused (not used as primary outputs) outputs present in the reversible logic circuit. They are essential without which, reversibility cannot be achieved. An efficient design should keep the number of garbage outputs to minimum.

Number of reversible gates: Total number of reversible gates used in the circuit (gate count).

Constant inputs: The number of inputs which are maintained (or set to) constant at 0 or 1 in order to get the required function.

A reversible circuit should be designed using minimum number of reversible gates. One key requirement to achieve optimization is that the designed circuit must produce minimum number of garbage outputs. Also, they must use minimum number of constant inputs [8]. Limitation of reversible circuit is that they do not allow feedback and fan out [6, 16, 17]. If no feedback, it means that there is no loop and if no fan out, it means that an output cannot be used more than once. However, they can be achieved using additional gates [15].

It has also been shown by some researchers that fault testing can be much easier for reversible circuits than for irreversible circuits [5, 18]. This is mainly because, in a reversible circuit, gates between the fault site and the output are information-lossless, so an error at the fault site will be immediately seen at the output [18].

12.3 Basic Reversible Gates

Many reversible gates exist in literature. Commonly used reversible elements include the Toffoli gate, (TG) [3], the Fredkin gate (FRG) [14], the Feynman gate, (FG) [16], also known as controlled NOT, a special TG with only one control input, the Peres gate (PG), which combines TG and FG are depicted in Fig. 12.1. Because of their simplicity and low cost, there are design approaches and tools that incorporate them separately or in combination with each other [7, 9, 19].

The future computation circuits can make use of the design methods using existing reversible logic gates and new reversible logic gates and are very important and useful for building ultra low-power robust integrated circuits [8, 6, 20].

12.4 Fault-Tolerant Mechanism

Fault tolerance is the property that enables a system to continue operating properly in the event of any failure of some of its components. If the system itself is made of fault-tolerant mechanism, then the detection and correction of faults becomes simpler and easier. Additional, higher level fault tolerance mechanisms can be

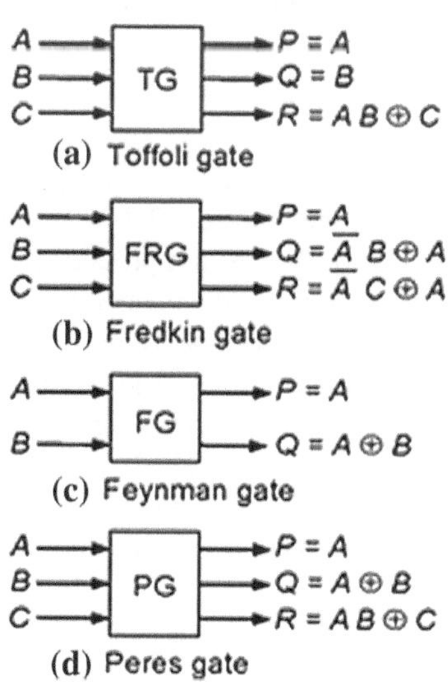

Fig. 12.1 Commonly used reversible logic gates

built on top of gate-level approach in order to increase the level of coverage beyond single-point deviations [6].

Fault-Tolerant (FT) gate is also known as Conservative Reversible Gate [13], or Parity Preserving Gate [6] logic gate. In order that the system to be fault tolerant, their gates used should be parity preserve [6]. Parity checking is one of the conventional methods to detect errors in digital logic systems. Parity preserve is the gate that the parity of input and output vectors of them is the same [6]. Making a reversible circuit robust or fault-tolerant is much more difficult than a conventional logic circuit [6]. Fault tolerance can be achieved in systems by using parity bits. Parity preserving reversible circuit design will be very important for development of fault-tolerant reversible systems in nanotechnology which is an emerging technology nowadays [12, 19]. Parity preserving reversible circuits is the future design trends to the development of fault-tolerant reversible systems [6, 7].

Reversibility can recover bit loss but is not able to detect bit error in circuit. Fault-Tolerant reversible circuit is capable to prevent error at outputs [5].

12.5 Combinational Circuits

Combinational circuits play as a part of logics in many applications. Different combinational circuits are proposed in various works by authors. Here, they are identified by the uniqueness in the work and given a brief survey on them. Some of the combinational circuits implemented by the authors are given in the below descriptions.

12.5.1 Adders/Subtractors

In a fault-tolerant adder circuit, the input parity must match the parity of the output. A fault-tolerant full adder/subtractor circuit along with design of serial binary adder/subtractor circuit is presented by Parminder Kaur et al. in their paper [9] as depicted in Fig. 12.2. Using only one single control, it can act as an adder or a subtractor. It does not produce any unnecessary garbage outputs. There are three inputs A, B, C_{in} and a control line ctrl which controls its mode of operation. When control signal ctrl is at logic 0, the circuit acts as full adder and when ctrl goes to logic 1, the circuit performs subtraction. The sum and difference line shown as S/D and its carry and borrow signal is represented by C/B. The rest of nine constant inputs are forced to logic 0, whereas there are 11 garbage signals. This can be used to realize other arithmetic circuits such as ripple carry adder, carry look-ahead adder, carry-skip logic, and multiplier/divisors [9].

(a)

(b)

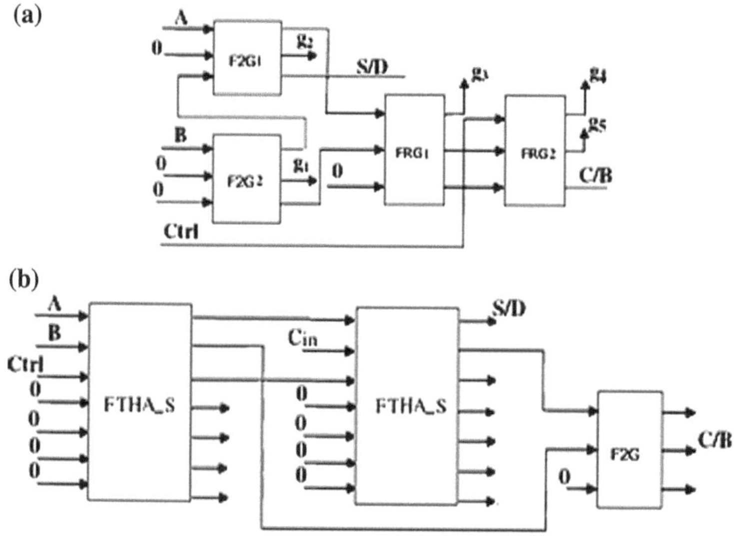

Fig. 12.2 Fault-tolerant full adder/subtractor [9]

12.5.2 Carry-Skip and Carry Look-Ahead Adders

A novel fault-tolerant carry-skip BCD adder using the reversible full adder circuit that uses the new 4*4 parity preserving reversible IG gates has been depicted in the paper by Islam and Begum [11]. The gate is one-through, which means one of the input variables is also output. The input pattern corresponding to particular output pattern can be uniquely determined. The reversible IG is parity preserving. The IG gate is universal in the sense that it can be used for implementing arbitrary Boolean functions [11]. A novel fault-tolerant reversible full-adder circuit using the proposed IG gates has also shown and is optimized in terms of gate count, number of garbage outputs and constant inputs [11]. This is depicted in Fig. 12.3.

Fault-tolerant high-speed reversible logic circuits for the BCD Carry-select adders with an efficient parity preserving reversible New Gate (NG) circuit has been presented by Bharathi and Neelima [12, 20]. Here, in the design of Carry-Select Adder, the FRG is made use as a multiplexer to select the final outputs with the enable as external carry. The reversible logic gates based implementation of carry-select BCD adder is gives the results as the number of reversible gates used is 27, and the number of garbage outputs produced is 39.

An optimized reversible BCD adder using a new reversible gate, SCL (Six Correction Logic) gate in combination with the existing reversible logic gates is proposed in the paper by Bhagyalakshmi and Venkatesha [8] for further improvement of the optimization parameters. The SCL gate is used for the correction in the BCD addition. This produces 10 garbage outputs with 6 constant inputs. A comparison of the different designs in terms of the important design

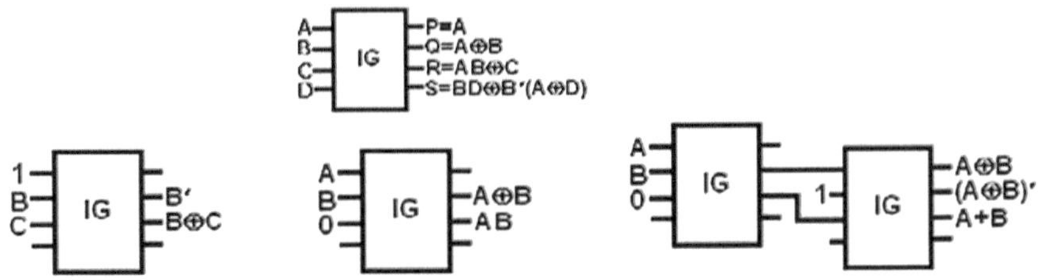

IG gate implementing XOR, AND/XOR, XOR/XNOR/OR gates

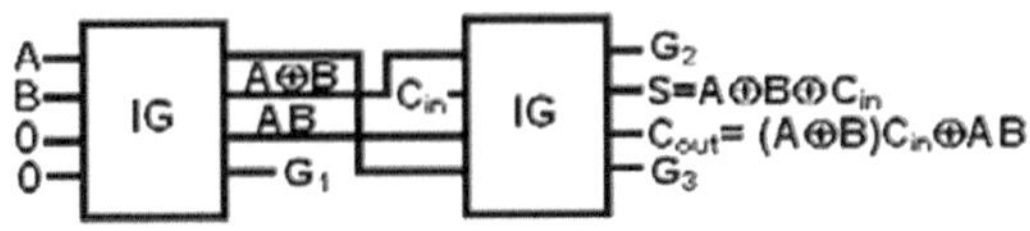

fault tolerant reversible full adder circuit.

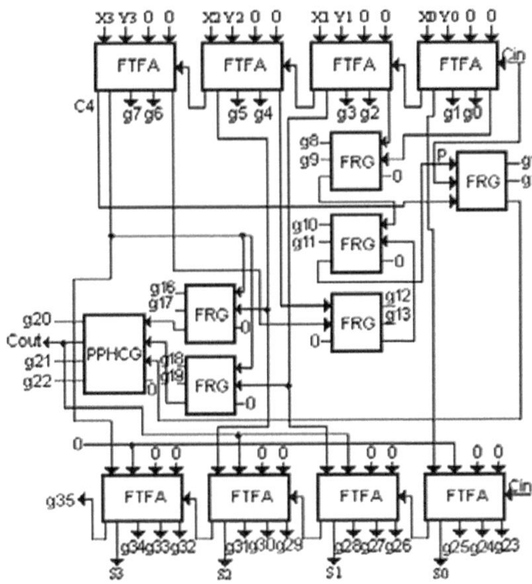

Reversible logic implementation of fault tolerant carry skip BCD adder.

Fig. 12.3 Fault-tolerant carry-skip BCD adder [11]

parameters like number of reversible gates, number of garbage outputs, and number of constant inputs in addition to the delay parameters are given [8].

12.5.3 Multiplier

A novel 4 × 4 bit reversible fault-tolerant parity preserving multiplier circuit which can multiply two 4-bit numbers is presented by Sinha and Syal [7] in their

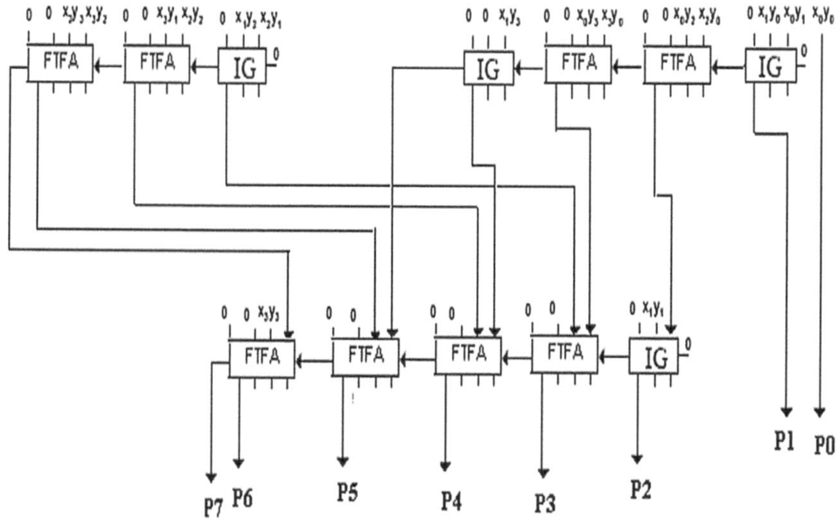

Fig. 12.4 RFTPA circuit [7]

paper. It is faster and has lower hardware complexity and better in terms of delay and power. Multiplier is implemented using FRG and IG gates. For product term generation, the FRG gate is used. The FRG gate is used to perform AND operation by forcing one constant input as logic 0, whereas it produces required product term along with two garbage outputs. Multiplier partial products are generated in parallel using 16 FRG. This will be a better circuit as it has less hardware complexity compared to other gates, and moreover, it posses parity preserving logic. Reversible Fault-Tolerant Parallel Adder (RFTPA) circuit developed by using reversible fault-tolerant full adder (FTFA) and half adder (FTHA) using the IG gates [7] is depicted in Fig. 12.4.

12.6 Conclusion and Future Work

An attempt is made in this paper to survey the various methods for developing the combinational logic circuits using the fault-tolerant reversible logic gates. Reversible circuits play a major role in future. Also the fault tolerance approach by itself and in combination with other methods can be used in various applications. A briefing is done on the conventional reversible gates and fault-tolerant methods for the combinational logic. Various papers are studied on the methods for implementing various combinational logic circuits using fault-tolerant reversible logic gates. It is hoped that this survey provides an insight for incorporating fault tolerance methods into reversible circuits without much extra design effort and with modest hardware, which can be useful for ensuring the robustness of reversible logic circuits in various application domains. More circuits can be surveyed as a future work and can also depict the importance in their application domains.

References

1. Landauer R (1961) Irreversibility and heat generation in the computing process. IBM J Res Dev 5:183–191
2. Bennett CH (1973) Logical reversibility of computation. IBMJ Res Dev 17:525–532
3. Toffoli T (1980) Reversible computing. In: automata, languages and programming, Verlag, Berlin, pp 632–644
4. Rahman R, Jamal L, Babu HMH (2011) Design of reversible fault tolerant programmable logic arrays with vector orientation. Int J Inf Commun Technol Res 1(8) (ISSN-2223-4985)
5. Mitra SK, Chowdhury AR (2012) Minimum cost fault tolerant adder circuits in reversible logic synthesis. IEEE conference 25th international conference on VLSI design (VLSID), 7–11 Jan 2012, pp 334–339
6. Parhami B (2006) Fault tolerant reversible circuits. In: proceedings of 40th Asimolar conference signals, systems, and computers, Pacific Grove, CA, pp 1726–1729
7. Sinha HP, Syal N (2012) Design of fault tolerant reversible multiplier. Int J Soft Comput Eng (IJSCE) 1(6):120–124 ISSN: 2231-2307
8. Bhagyalakshmi HR, Venkatesha MK (2010) Optimized reversible BCD adder using new reversible logic gates. J Comput 2(2):28–32 ISSN 2151-9617
9. Kaur P, Dhaliwal BS (2012) Design of fault tolerant full Adder/Subtractor using reversible gates. IEEE conference international conference on computer communication and informatics (ICCCI) Coimbatore, India, 10–12 Jan 2012, pp 1–5
10. Babazadeh S, Haghparast M (2012) Design of a nanometric fault tolerant reversible multiplier circuit. J Basic Appl Sci Res 2(2):1355–1361 ISSN 2090-4304
11. Islam MDS, Begum Z (2008) Reversible logic synthesis of fault tolerant carry skip BCD adder. J Bangladesh Acad Sci 32(2):193–200
12. Haghparast M (2008) Design of a novel fault tolerant reversible full adder for nanotechnology based systems. World Appl Sci J 3(1):114–118
13. Bruce JW, Thornton MA, Shivakumaraiah L, Kokate PS, Li X (2002) Efficient adder circuits based on a conservative reversible logic gates. In: proceedings of IEEE computer society annual symposium on VLSI, Pittsburg, PA, pp 83–88
14. E. Fredkin and T. Toffoli, "Conservative logic", Intl. Journal of Theoretical Physics, pp. 219-253, 1982
15. Ali MB (2012) Optimized design of carry skip BCD adder using new FSNG reversible logic gates. IJCSI Int J Comput Sci Issues 9(4), (3):424–431
16. Feynman R (1985) Quantum mechanical computers. Opt News 11:11–20
17. Perkowsi M, Jozwiak L, Kerntopf P, Al-Rabadi A, Coppoa A, Buller Song AX, Khan MMHA, Yanushkevich S, Shmerko VS, Chzazowska-Jeske M (2001) A general decomposition for reversible logic. In: Proceeding RM, Starkville, pp 119–138
18. Polian I, Hayes JP (2010) Advanced modeling of faults in reversible circuits. IEEE conference, design and test symposium (EWDTS), pp 376–381
19. Islam MS, Rafiqul Islam M (2005) Minimization of reversible adder circuits. Asian J Inf Technol 4(12):1146–1151
20. Bharathi M, Neelima K (2012) Scope of reversible engineering at gate-level: fault-tolerant combinational adders. Int J VLSI Des Commun Syst (VLSICS) 3(2):85–98

Chapter 13
A CMOS Standard Cell-Based Time-to-Digital Converter

K. Hari Prasad, Menka Sukhwani, Pooja Saxena, C. K. Pithawa and V. B. Chandratre

Abstract This paper presents a design of 4-channel Time-to-Digital Convertor (TDC) ASIC based on vernier ring oscillator technique. This technique implements two ring oscillators with very slight difference in time periods, which defines the resolution of TDC. The slight difference in time period is generated by using different fan-out load of the delay cell used to make respective ring oscillators. An on-chip calibration circuit provides the oscillator time period accurately for corrections, thereby reducing PVT (process, voltage, and temperature) variations. The TDC has been implemented using standard cell library of 0.35 μm commercial CMOS technology, achieving a resolution of 114 ps with a dynamic range of 1.8 μs and power consumption of 23 mW/channel.

Keywords CMOS standard cell · Digital time-period calibrator · Ring oscillator · TDC · High-energy physics experiments · Positron emission tomography

13.1 Introduction

Time-to-Digital Converter (TDC) measures the precise Time Interval (TI) between two logical events 'Start' and 'Stop.' TDC finds applications in laser range finding systems [1], instrumentation of Positron Emission Tomography imaging [2], and High-Energy Physics experiments (HEP) [3]. In the present HEP experimental scenario, having very large number of detector channels, a compact, low-cost, low-power, and precise time interval measurement circuits are needed. This TDC

K. Hari Prasad · M. Sukhwani · C. K. Pithawa · V. B. Chandratre (✉)
Electronics Division, Bhabha Atomic Research Centre, Mumbai 400085, India
e-mail: vbc@barc.gov.in

P. Saxena
Homi Bhabha National Institute, Mumbai 400094, India

V. Chakravarthi et al. (eds.), *Proceedings of International Conference on VLSI, Communication, Advanced Devices, Signals & Systems and Networking (VCASAN-2013)*, Lecture Notes in Electrical Engineering 258, DOI: 10.1007/978-81-322-1524-0_13,

design is based on Vernier ring oscillator technique [4, 5, 6], which utilize two oscillators with very slight difference in time periods. The silent features of vernier technique are small design area, less logic resources, high resolution, and theoretically infinite dynamic range. The vernier technique has design advantage over Delay Lock Loop (DLL)–based TDC in terms of area, power efficiency, and physical layout complexity. This implementation is carried out using standard CMOS cell library–based semi-custom back-end design approach. The variation in Process, Voltage, and Temperature (PVT) results a nonlinear variation in the propagation delay of the standard cells. This results in random variation of ring oscillator time periods. Therefore, an on-chip digital calibrator is used to measure the precise time periods of the ring oscillators. The key blocks of the TDC are ring oscillators, counters, time-period measurement, and calibrator and Serial Peripheral Interface (SPI). The timing critical blocks, ring oscillators, and counters have been laid-out using custom layout design tools. The automatic P and R (place and route) design tool is used to implement time-period calibrator and SPI interfaces.

13.2 Design Principle

The block diagram and timing diagram of vernier technique is illustrated in Fig. 13.1a, b. The Time Interval (TI) 'T' is measured with the help of slow and fast ring oscillators. The slow and fast oscillators start oscillating with frequencies f_1 ($T_1 = 1/f_1$) and f_2 ($T_2 = 1/f_2$) ($f_2 > f_1$) on the arrival of 'start' and 'stop' signals, respectively. As T_2 is slightly less than T_1, therefore, from the arrival of stop signal, both oscillators cover the same number of cycles until their first coincidence. On each cycle, the fast oscillator approaches the slow one by a step $\Delta T = T_1 - T_2$, which defines the LSB of TI measurement. The coarse counter counts the number 'n' of slow oscillator time-period 'T_1' and provides a long measurement range up to $(2^b - 1)_* T_1$, where 'b' is the number of bits in coarse counter. The fine counter counts number 'm' of fast oscillator time periods till coincidence. Therefore, the number of bits 'k' in fine counter should qualify '$2^k \geq m$.' The conversion equation for the evaluation of time interval 'T' is given by the Eq. (1), where '$(n - m)_* T_1$' gives the coarse and '$(m - 1)\, _*\Delta T$' gives the fine time measurement. The maximum dynamic range of this technique is given by the Eq. (2). The coincidence or leading of fast oscillator from slow oscillator is detected by the phase detector, which toggles the End Of Conversion (EOC) signal that latches the coarse and fine counter data.

$$T = (n - m)T_1 + (m - 1)\Delta T \tag{1}$$

$$DR_{max} = (2^b - 2^k - 1) * T_1 \tag{2}$$

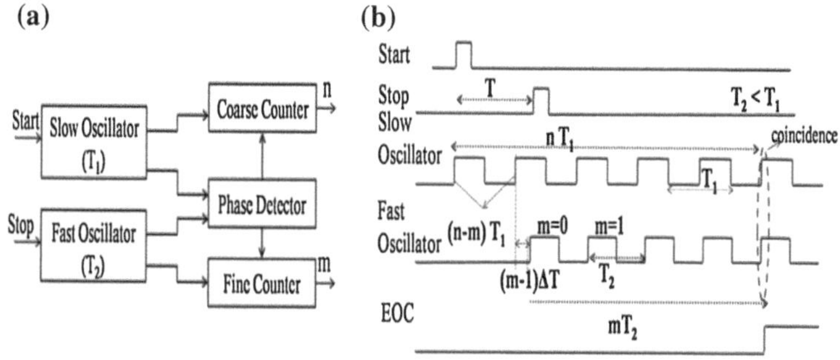

Fig. 13.1 **a** Block diagram **b** Timing diagram of vernier technique

13.3 TDC ASIC Block Diagram

In the TDC ASIC block diagram shown in Fig. 13.2, four identical TDC channels with individual start (4-*start*) and stop (4-*stop*) have been designed. These channels are interfaced through priority encoder logic with SPI slave for data readout. The priority encoder logic and SPI slave are designed using verilog HDL. At every system clock (*SYSCLK*) rising edge, the encoder in the order of one to four samples the EOC lines of the TDC channels. On active status of EOC, the TDC data are appended with its respective channel number and send to the SPI register by asserting *data_ready* signal. The 19-bit data (2-bit channel number, 8-bit fine, and 9-bit coarse) in the SPI register is serially shifted out on Master In Slave Out (*MISO*) line by SPI slave on Serial Clock (*SCK*) edges, when Slave Select (*SSEL*) is active low. The calibration channel is separated from measurement channel with separate SPI interface so that the calibration data can be read out periodically by asserting the *Cal_start* signal.

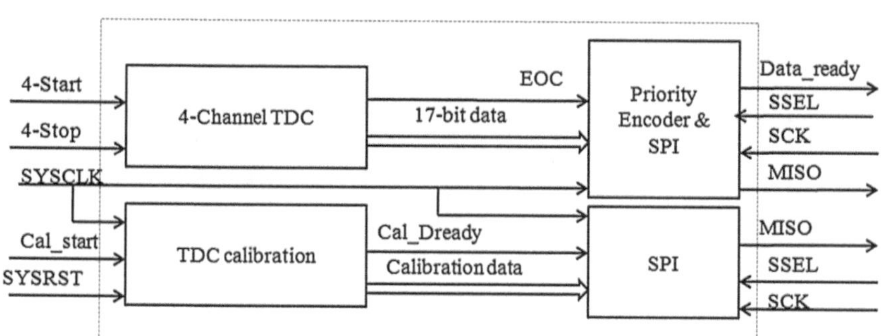

Fig. 13.2 Block diagram of 4-channel TDC

13.4 Description of Design Blocks

13.4.1 Slow and Fast Ring Oscillators and Phase Detector

The Fig. 13.3a depicts the schematic diagram of standard cell based ring oscillator. In case of slow and fast oscillators, the delay cell 'DLY' drives a fan-out of '2' (by AOI) and '1' (by NOR), respectively. This results in unequal propagation delay introduced by 'DLY' in both the ring oscillators. This produces a slight difference in frequencies of oscillators, which are fed to the phase detector as shown in Fig. 13.3b. The flip-flop (FF-1) of phase detector samples the state of slow oscillator at the rising edge of fast oscillator. The second flip-flop (FF-2) then samples the data, previously sampled by FF-1 on the next rising edge of fast oscillator. This process continues until first coincidence or leading of fast oscillator from slow oscillator, resulting in toggling of EOC signal.

13.4.2 On-Chip Digital Time Period Calibrator

The on-chip calibrator is used to calibrate the time period of ring oscillators due to variations in process, temperature, and operating voltage. The calibration is done as per Eq. (3), where 'T_0' is the calibrated time period, 't' is the calibration time window, and 'N' is the number of pulse count within a calibration window. A long calibration window of duration '80 μs' is chosen and is generated internally using 12-bit counter driven by the system clock (50 MHz). This window is long enough to calibrate the time periods with an accuracy of few picoseconds across all process corners. For instance, with calibration window of t = 80 μs duration and count of N = 11000, T_0 evaluates to 7.272 ns and with N = 11002, T_0 evaluates to be 7.271 ns. Two 14-bit calibration counters are employed to find out the value of N for both slow and fast oscillators.

$$T_0 = \frac{t}{N} \tag{3}$$

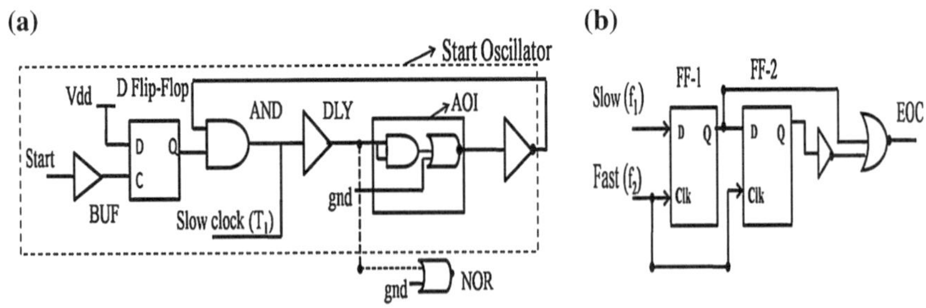

Fig. 13.3 Schematic diagram of **a** Ring oscillator **b** Phase detector

Table 13.1 Ring oscillator time periods over process corners for four channels (Ch1, Ch2, Ch3, Ch4)

Corners channels	Worst power	Typical	Worst slow	Worst one	Worst zero
Ch1: T_1 (ns)	5.392	7.382	9.804	7.465	7.074
Ch1: T_2 (ns)	5.32	7.268	9.639	7.356	6.963
ΔT (ns)	0.072	0.114	0.165	0.109	0.111
Ch2: T_1 (ns)	5.392	7.383	9.804	7.464	7.074
Ch2: T_2 (ns)	5.32	7.268	9.639	7.356	6.962
ΔT (ns)	0.072	0.115	0.165	0.108	0.112
Ch3: T_1 (ns)	5.392	7.383	9.804	7.464	7.074
Ch3: T_2 (ns)	5.32	7.268	9.639	7.358	6.962
ΔT (ns)	0.072	0.115	0.165	0.106	0.112
Ch4: T_1 (ns)	5.392	7.383	9.804	7.464	7.074
Ch4: T_2(ns)	5.32	7.268	9.639	7.358	6.962
ΔT (ns)	0.072	0.115	0.165	0.106	0.112
Calibration data: Calibrated time-period of reference oscillator					
T1 (ns)	5.394	7.384	9.808	7.464	7.076
T2 (ns)	5.318	7.265	9.634	7.353	6.959
ΔT (ns)	0.076	0.119	0.174	0.111	0.117

13.5 Simulation Results

The ASIC is implemented using 0.35 μm commercial CMOS technology and RC-extracted netlist of the design is accurately simulated using Spice models. The simulated time periods T_1 and T_2 of ring oscillators are 7.382 and 7.286 ns on typical corner. The difference of $T_1 - T_2$ gives LSB 'ΔT' of 114 ps. The maximum dynamic range of TDC with 9-bit coarse counter and 8-bit fine counter is 1.8 μs from Eq. (2). The T_1, T_2, and ΔT values of four channels and calibration

(a) (b)

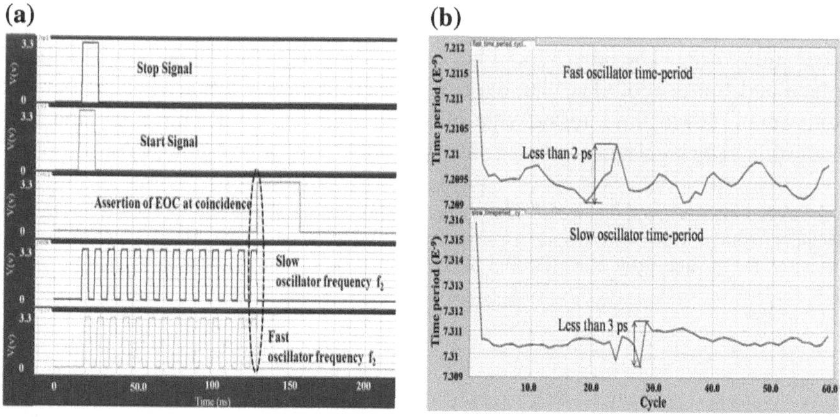

Fig. 13.4 **a** Variation of $T_1$and T_2 over cycles **b** Coincidence of slow and fast oscillators

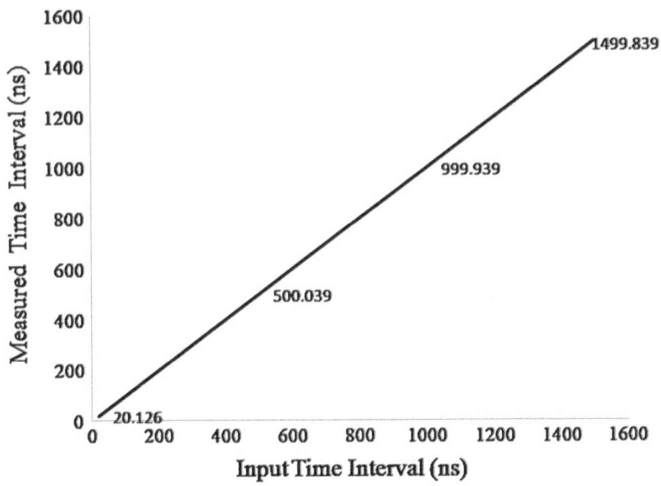

Fig. 13.5 Plot between applied time versus measured time on typical corner (simulation)

channel across five design process corners is shown in Table 13.1. The variation in LSB is less than 100 ps across process corners. Figure 13.4a shows the assertion of EOC signal at the coincidence of both oscillators. The variation in time-periods over the oscillation cycles is less than 1's of pico-second as shown in Fig. 13.4b. Figure 13.5 shows the measured time interval for input sweeps of '20 ns, 500 ns, 1 μs, and 1.5 μs' on typical corners.

13.6 Conclusion

A standard cell-based TDC has been proposed and designed using 0.35 μm CMOS technology. The timing critical blocks have been designed using manual placement and routing. The other blocks have been designed using an automatic P and R (place and route) tool. The Standard cell-based implementation of TDC benefits from small design time, less design effort, low power, small chip area, and high resolution. The significant achievement is in the reduction of power consumption.

Acknowledgments Authors would like to thank Shri Sekhar Basu, Director BARC; Dr. T.S. Ananthakrishnan, Head, Electronics Division, BARC; Dr. V. M. Datar, Head, Nuclear Physics Division, BARC and Prof. N.K. Mondal of TIFR for their support and encouragement.

References

1. Deng Z et al (2002) Design of new front end electronics for animal PET. In: nuclear science symposium and conference, vol 3, pp 1543–1546
2. Callewaert L et al (1989) Front end and signal processing electronics for detectors at high luminosity collider. IEEE Trans Nucl Sci 36:446–457
3. Kalisz J (2004) Review of time interval measurement techniques with picoseconds resolution. Metrologia 41(1):17
4. King ME, Barton RD (1971) Two vernier time-interval-digitizer. IEEE Trans Nucl Instrum Methods 359–370
5. Chen P et al (2007) PVT sensitive vernier based time-to-digital converter with extended input range and high accuracy. IEEE Trans Nucl Sci 54:294
6. Hari Prasad K, Chandratre VB, Saxena P, Pithawa CK (2011) FPGA based time-to-digital converter. DAE Symp Nucl Phys G-7

Chapter 14
Task Migration for 3 × 3 Tile-Based NoC Architecture

Maithreyi Uttarkar and H. R. Vanamala

Abstract Advancement in process technology has led to the emergence of MPSoC. At the same time, MPSoC leads to several design challenges. Thermal hot spots and temperature variations affect performance, reliability, power, and cooling cost. So, there is a need for a framework to act on elevation in temperature and temperature variations. In this paper, distributed thermal balancing migration scheme is applied for 3 × 3 tile-based NoC architecture. A balanced thermal profile is got by having workload balance on applying the migration policy only among the neighboring cores; thereby this scheme aims to have less communication overhead and less hot spot generation in the architecture.

Keywords MPSoC · NoC

14.1 Introduction

In semiconductor processing technology, the speed of integration boils down to Moore's law, leading to emergence of system on chip. Further research has lead to the development of Multiprocessor System on Chip (MPSoC).

MPSoCs are racks of processors scaled down to a single chip. They improve the performance of the system by utilizing many processor elements to execute system application task. Network on Chip (NoC) is a communication medium in MPSoC and the different cores communicate through routers. There are different types of

M. Uttarkar (✉) · H. R. Vanamala
Department of Electronics and Communication, PES Institute of Technology, 100ft ring road, BSK 3rd stage, Bengaluru 560085, India
e-mail: maiuttarkar@gmail.com

H. R. Vanamala
e-mail: vanamalahr@pes.edu

V. Chakravarthi et al. (eds.), *Proceedings of International Conference on VLSI, Communication, Advanced Devices, Signals & Systems and Networking (VCASAN-2013)*, Lecture Notes in Electrical Engineering 258, DOI: 10.1007/978-81-322-1524-0_14,

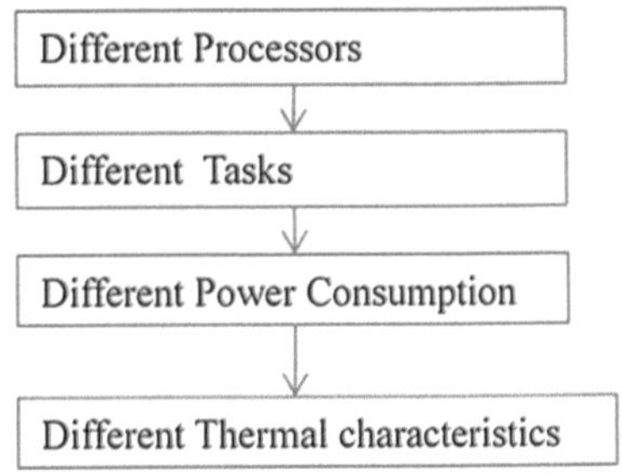

Fig. 14.1 Variation of temperature in MPSoC

NoC architectures such as Mesh type, Torus type, etc. Here, a 3 × 3 tile-based mesh type of NoC architecture is preferred, and the task migration scheme is applied for it.

MPSoC leads to several design challenges in the field of area, power and delay. High-power density causes thermal hot spots and temperature variation, thereby affecting cooling cost, leakage power, performance, reliability, and timing. Hence, MPSoCs are facing lot of power and thermal challenges. Each of the different cores in MPSoC performs different task and thereby directs MPSoC to have different thermal characteristic and hence reduces lifetime of the chip, deteriorates its performance, affects reliability and increases cooling cost. This behavior is shown in Fig. 14.1.

Therefore, there is a need for a framework to balance temperature and workload of the processor simultaneously.

In this paper, a 3 × 3 tile-based NoC architecture is targeted. A balanced thermal profile is obtained by applying distributed thermal-balancing migration (DTB-M) policy [1]. This migration policy internally depends on steady-state temperature-based migration (SSTM) scheme and temperature prediction–based migration (TPM) scheme [1]. SSTM scheme distributes tasks to core based on their various heat removal capabilities. TPM predicts the thermal impact of various workload combinations and adjusts the task allocation in a neighborhood. This helps all cores to get good mix of hot and cool tasks [1]. These proactive methods are used to get balanced thermal profile by reducing thermal hotspots in MPSoC.

14.2 Related Work

Due to drawback of centralized thermal management scheme, different DTM techniques have been carried out to achieve balanced thermal profile in MPSoC. It has been shown that proactive thermal management attained better thermal profile and performance when compared to reactive thermal management techniques such as thread migration, PID control [2]. Hardware and software techniques were combined to form hybrid DTM technique at system level in high-performance microprocessor [3]. Reactive strategies were combined with adaptive random

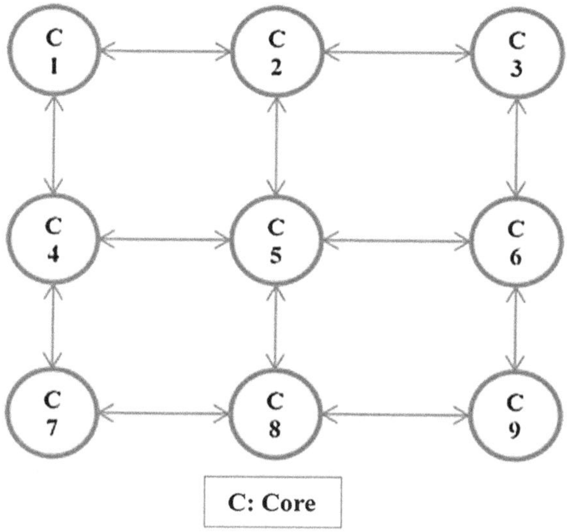

Fig. 14.2 Tile-based 3 × 3 NoC architecture

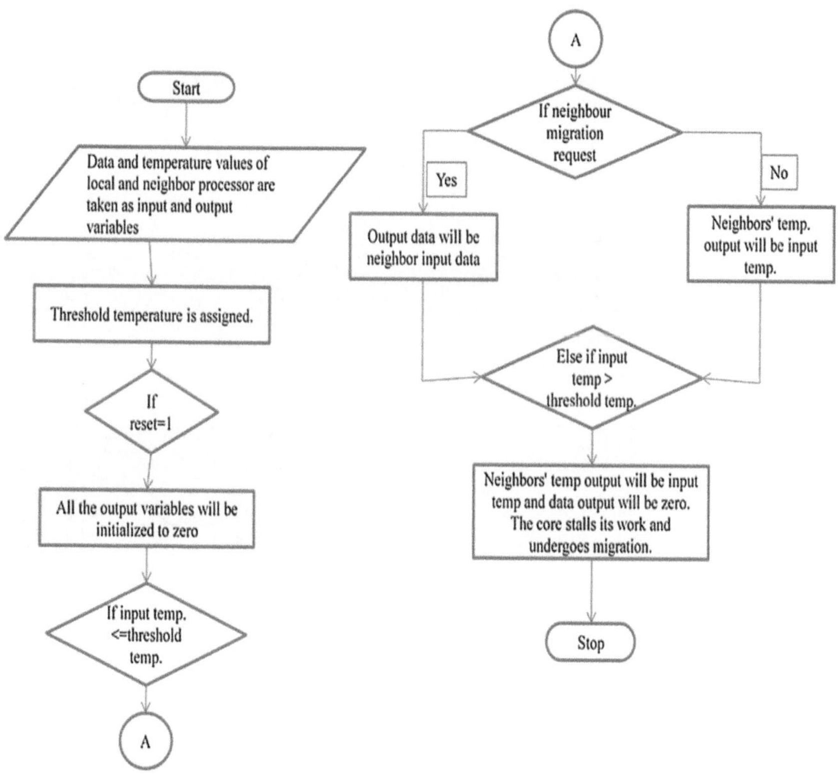

Fig. 14.3 Process for task migration

technique to enhance temperature hot spot and temperature variation reduction [4]. Stopgo, DVFS, distributed policies, global control and impact of workload balance in MPSoC have been discussed in the literature [5, 6].

14.3 Methodology

Task migration scheme is applied for 3 × 3 NoC architecture (Fig. 14.2). This distributed thermal-balancing scheme follows master-slave execution protocol and master-slave communication [1]. These architectures internally work on SSTM and TPM scheme. At this point of time, the threshold temperature is assumed. Various methods, such as ARMA [2], Neural network predictor model [1], are used to predict the threshold temperature. The DTB-M scheme is applied to single core assuming that the neighbor is present. This process represented in Fig. 14.3 can be extended to all nine cores.

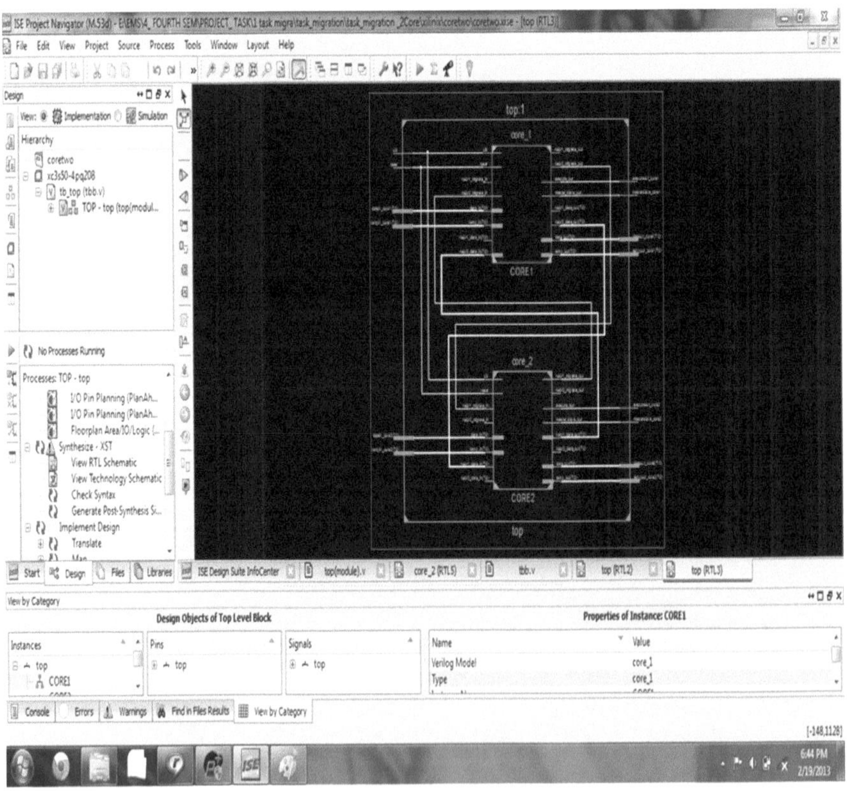

Fig. 14.4 RTL schematic of two core

14.4 Results and Discussion

The task migration scheme for a 3 × 3 tile-based NoC architecture is implemented using Xilinx ISE design suite 12.1. The core will be executing its own task if the input temperature is within the threshold temperature. If the neighbor core's input temperature exceeds the threshold temperature, then the core stalls its own task and executes the neighbor's task. This leads to less migration overhead as the process takes place among neighboring cores. Hence, we will have less hotspot by the combined action of SSTM and TPM scheme [1]. Figures 14.4 and 14.5 show RTL schematic of two cores and the output waveform representing task migration in single core, respectively. Table 14.1 summarizes the task migration between two cores.

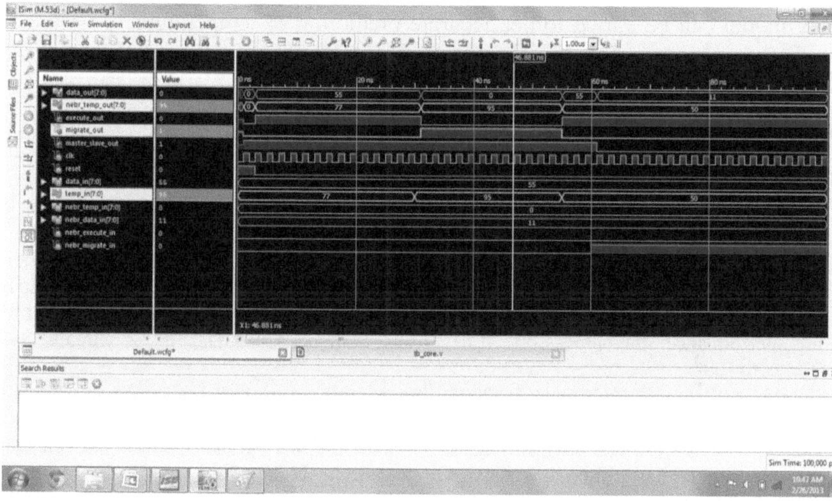

Fig. 14.5 Output waveform representing task migration in single core

Table 14.1 Task migration between two cores

	Temperature [°C]		Outcome
	Core 1	Core 2	
Case 1	80	50	No task migration between core 1 and core 2
Case 2	70	100	Task migration from core 2 to core 1
Case 3	91	40	Task migration from core 1 to core 2

Note Threshold temperature is considered as 90 °C

14.5 Conclusion and Future Scope

In this paper, we have applied distributed thermal management framework for 3×3 tile-based NoC architecture in an MPSoC platform. The framework is applied for single and two cores with the threshold temperature being assumed. The same is repeated for all the cores, the threshold temperature can be taken from temperature predictor models and the neural network temperature predictor model is the latest addition. In MPSoC, every core has an agent to supervise the core temperature. This communicates and compromises with neighboring agents to migrate and dispense task equally throughout the system. Hence, the task migration scheme applied for the NoC structure intends to reduce elevation in temperature and temperature gradients and thereby aims to achieve balanced thermal profile throughout the NoC architecture.

Further, this task migration scheme can be applied for different NoC configurations. The threshold temperature can be obtained by neural network temperature predictor model. The number of neurons and the input feature set in neural network model can be varied to get accurate prediction of temperature.

References

1. Ge Y, Qiu Q, Wu Q (2012) A multi-agent framework for thermal aware task migration in many-core systems. IEEE Trans Very Large Scale Integr (VLSI) Syst 20(10):1758–1771
2. Coskun A, Rosing T, Gross K (2009) Utilizing predictors for efficient thermal management in multiprocessor SoCs. IEEE Trans Comput Aided Des Integr Circuits Syst 28(10):1503–1516
3. Kumar A, Shang L, Peh L-S, Jha NK (2008) System-level dynamic thermal management for high-performance microprocessors. IEEE Trans Comput Aided Des Integr Circuits Syst 27(1):96–108
4. Coskun A, Rosing T, Whisnant K (2007) Temperature aware task scheduling in MPSoCs. In: proceeding design automation and Test in Europe (DATE), pp 1659–1664
5. Donald J, Martonosi M (2006) Techniques for multicore thermal management: classification and new exploration in Proc. Int. Symp. Comput. Arch. (ISCA), 2006, pp. 78–88
6. Almeida GM, Varyani S, Busseuil R, Sassatelli G, Benoit P, Torres L (2010) Evaluating the impact of task migration in multi-processor systems-on-chip. In: proceeding of 23rd symposium on integrated circuits and system design, ACM, pp 73–78

Chapter 15
Design and Implementation of an Efficient Multiplier Using Vedic Mathematics and Charge Recovery Logic

Belgudri Ritesh Appasaheb and V. S. Kanchana Bhaaskaran

Abstract Binary multiplier is one of the most time and power consuming architectures in an ALU. The performance efficiency of complex computations is determined by the multiplier algorithm used. Design of an efficient multiplier thus becomes important. An attempt has been made to implement an efficient multiplier using ancient computational techniques using charge recovery logic. This circuit is compared against the existing vedic multiplier circuits designed using conventional CMOS logic, to validate our claim. A 4×4 vedic multiplier using 2 N-2P type of charge recovery logic structure is implemented. The design and verification have been done using industry standard SPICE tools. The simulation results depict reduction in the average power consumption by 77.66 %.

Keywords Energy recovery · Charge recovery · Multiplier · Vedic multiplier

15.1 Introduction

Multipliers play an important role in today's digital signal processing and various other applications. With the advances in technology, many researchers focus and research on design of more efficient multipliers, which may offer any one or more of the following advantages, namely high speed, low-power consumption, regularity of layout and hence lesser area or combination of them in one multiplier structure. This makes them suitable for relevant high-speed, low-power, or

B. R. Appasaheb (✉) · V. S. Kanchana Bhaaskaran
School of Electronics and Communication, VIT University, Chennai, Tamil Nadu, India
e-mail: ritesh.belgudri@gmail.com

V. S. Kanchana Bhaaskaran
e-mail: vskanchana@gmail.com

V. Chakravarthi et al. (eds.), *Proceedings of International Conference on VLSI, Communication, Advanced Devices, Signals & Systems and Networking (VCASAN-2013)*, Lecture Notes in Electrical Engineering 258, DOI: 10.1007/978-81-322-1524-0_15,

compact circuit architectures. There are two major classes of multipliers, the tree and array architectures.

A brief discussion on the area, speed, and power of various architectures is as follows. The basic array multipliers, such as the Baugh-Wooley consume low-power and demonstrate relatively good performance [1]. However, their use is limited to process operands with less than 16-bit. For operand lengths of 16-bit and above, the modified Booth algorithm reduces the partial product's numbers by half. Therefore, the speed of the multiplier is reduced. Its power dissipation then becomes comparable to the Baugh-Wooley multiplier due to the circuitry overhead incurred by the Booth algorithm [1]. However, researchers aim for more enhanced circuit techniques that can cause the multiplier structures to have low-power operability. The fastest multipliers in the literature adopt the Wallace tree structure with modified Booth encoding [2]. In general, a Wallace tree leads to larger power dissipation and increased chip area due to the necessity of larger amount of interconnect wires. Hence, it is not generally recommended for low-power applications operating at higher frequency and lower technology nodes, wherein the interconnect parameters play a larger role. In such situations, the vedic multiplication proves to be an effective method that is adopted from ancient Indian Vedas. Vedic mathematics is mainly based on 16 Sutras (formulae) dealing with various branches of mathematics such as arithmetic, algebra and geometry, to name a few [3]. Multiplier using the Sutra Urdhva Triyagbyham reduces power to a greater extent compared to the Baugh-Wooley structure [4–7].

Secondly, energy recovery is proving to be a promising approach for low-power circuit design, whose advantage results from its inherent nature of deriving a constant current from the power clock and the FETs working with minimum voltage between the source and drain at any instance of time. These adiabatic circuits are classified into fully adiabatic and quasi-adiabatic circuits, based on whether full or partial energy recovery is obtained [8]. The former has no non-adiabatic loss, while the later suffers from both adiabatic and nonadiabatic energy loss components. The literatures have brought out several types and genres of such circuits, namely Efficient Charge Recovery Logic (ECRL) or 2N-2P, 2N- 2N2P, PFAL, Pass Transistor Adiabatic Logic (PAL), Clocked Adiabatic Logic (CAL), Improved Pass-gate adiabatic logic (IPGL), Adiabatic Differential Switch Logic (ADSL) [8], and Preresolved and Sense Adiabatic circuits [9]. They are designed for special functions with custom designs. Among all, 2N-2P is most power-efficient logic structure [10] for nominal frequency values. Thus, we use 2N-2P structure of adiabatic logic in our design for low-power operation.

In this paper, we implement 4 × 4 vedic multiplier using 2N-2P structure of adiabatic logic, and further, it is compared with vedic multiplier designed using CMOS logic. Section 15.2 describes the implementation of 4 × 4 vedic multiplier using CMOS logic. Section 15.3 describes the proposed vedic adiabatic multiplier, followed by Sect. 15.4 depicting simulation results, analysis and comparisons.

15.2 Vedic Multiplier

The multiplier architecture is based on the vertical and Crosswise algorithm (Urdhva Tiryakbhyam) of ancient Indian vedic mathematics. Urdhva Tiryakbhyam is a general multiplication formula applicable to all cases of multiplication [3]. The Urdhva Tiryakbhyam algorithm for 4 × 4 multiplication can be explained using line diagram as shown in Fig. 15.1. The vertical and slanting lines represent the AND or partial product generation between the operands and the horizontal line indicates the summing up of the partial products. To explain in brief, step 1 produces A0.B0 which P0, the LSB of the product. Step 2 produces A0.B1 summed with A1.B0, which produces P1 with a carry to be forwarded to step 3. In this manner the successive product bits are generated and carry bits are forwarded. The process is depicted in Eq. 15.1.

$$\left.\begin{aligned} P0 &= A0B0 \\ C0P1 &= A0B1 + A1B0 \\ C1P2 &= A0B2 + A2B0 + A1B1 + C0 \\ C2P3 &= A3B0 + A0B3 + A1B2 + A2B1 + C1 \\ C3P4 &= A3B1 + A1B3 + A2B2 + C2 \\ C4P5 &= A3B2 + A2B3 + C3 \\ V7P6 &= A3B3 + C4 \end{aligned}\right\} \quad (15.1)$$

15.2.1 2 × 2 Vedic Multiplier

In the design of the 4 × 4 vedic multiplier, the 2 × 2 block is the primary block (Basic block) [4]. The truth table shown in Table 15.1 indicates the product bits in

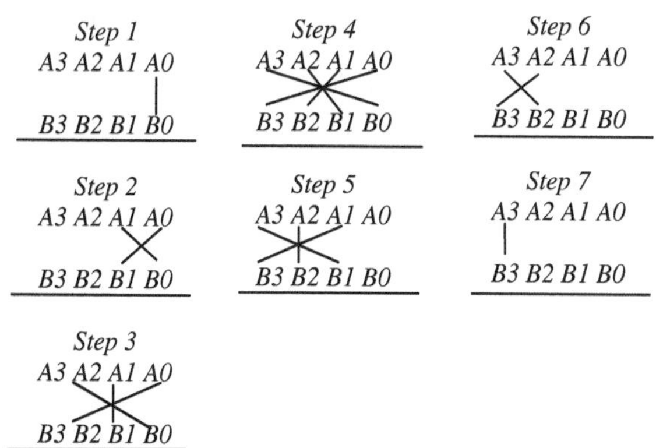

Fig. 15.1 Line diagram for 4 × 4 Multiplication using Urdhva Tiryagbhyam [5]

terms of input variables. This table is realized using K-map to generate the equations as shown in Eq. (15.2).

$$\left.\begin{aligned} \mathrm{P0} &= (\mathrm{A0\ and\ B0}) \\ \mathrm{P1} &= (\mathrm{A0\ and\ B1})\ \mathrm{xor}\ (\mathrm{B0\ and\ A1}) \\ \mathrm{P2} &= (\mathrm{A1\ and\ B1})\ \mathrm{and}\ (\mathrm{A0\ and\ B0})' \\ \mathrm{P3} &= (\mathrm{A1\ and\ B1}) \end{aligned}\right\} \quad (15.2)$$

15.2.2 4 × 4 Vedic Multiplier

The design of 4 × 4 block is a basic arrangement of the 2 × 2 blocks in an efficient and realizable manner [4]. Figure 15.2 shows the schematic of the 4 × 4 block designed using the 2 × 2 blocks.

The first step in the design of 4 × 4 block is identifying the different combinations of input bit pairs that are derived in terms of each of 2 × 2 blocks. Each input bit-pair is handled by a separate 2 × 2 combinational multiplier to produce four partial product rows. These partial products rows are then added optimally to generate final product bits. These partial products rows are subsequently added using 4-bit full adder cells. The partial products represent the Urdhva vertical and cross product terms.

Table 15.1 Truth table for 2 × 2 multiplier [4]

A1	A0	B1	B0	P3	P2	P1	P0
0	0	0	0	0	0	0	0
0	0	0	1	0	0	0	0
0	0	1	0	0	0	0	0
0	0	1	1	0	0	0	0
0	1	0	0	0	0	0	0
0	1	0	1	0	0	0	1
0	1	1	0	0	0	0	0
0	1	1	1	0	0	0	1
1	0	0	0	0	0	0	0
1	0	0	1	0	0	1	0
1	0	1	0	0	1	1	0
1	0	1	1	0	1	0	0
1	1	0	0	0	0	1	0
1	1	0	1	0	1	1	1
1	1	1	0	0	1	1	0
1	1	1	1	1	0	0	1

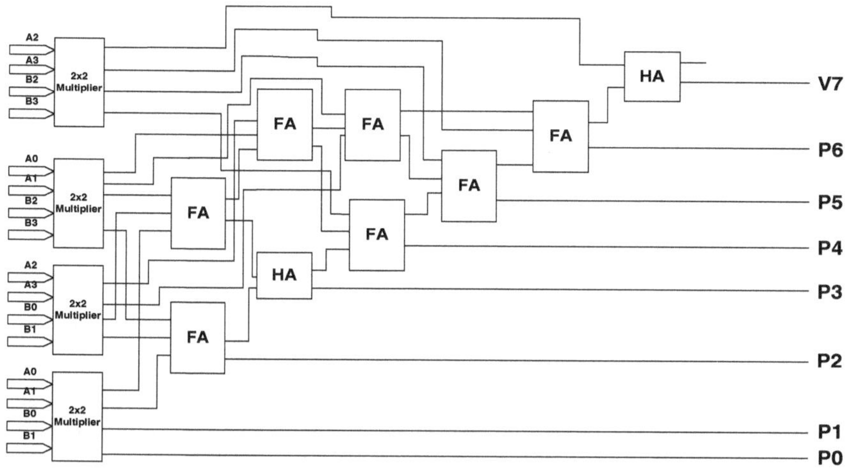

Fig. 15.2 Vedic multiplier using CMOS logic [2]

15.3 Vedic Adiabatic Multiplier

This section describes the implementation of the proposed adiabatic vedic multiplier. All the individual cells such as AND gate, half adder, and full adder are designed by the 2N-2P logic [9]. The typical AND gate realized using 2N-2P logic is shown in Fig. 15.3a. The logic has the same structure as cascade voltage swing logic (CVSL) with differential signaling [8]. The 2N-2P logic operates with four phase power-clock (PC) with *Evaluate, Hold, Recovery,* and *Wait* phases. The power-clock is so called since it powers the circuit and acts as the timing element also. The output wave form for A= B=logic 1 and/A=/B=logic 0 is shown in Fig. 15.3b. The node/ *OUT* goes *low* when both the inputs are high during *evaluate* phase of PC. This

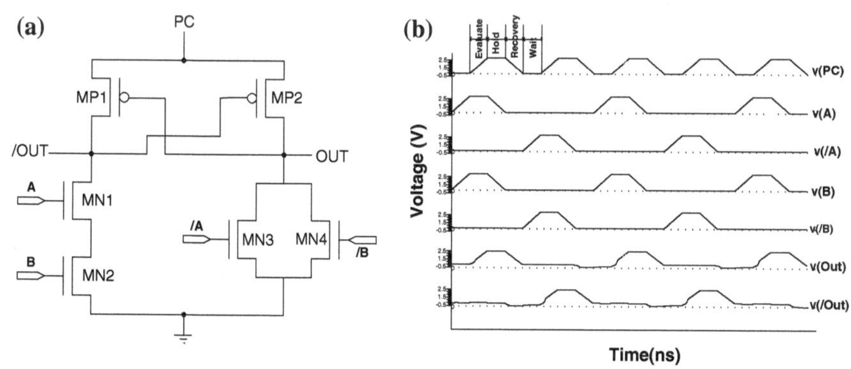

Fig. 15.3 **a** schematic. **b** power-clock, inputs of an AND Gate using 2N-2P logic of adiabatic family

switches MP2 *ON*, thus raising the *OUT* node along with the rising power-clock. During the *recovery* phase, the fully charged *OUT* node releases its charge through the conducting MP2 transistor back to PC. Thus, the *OUT* is in phase with PC as can be observed from Fig. 15.3b. During the *Hold* phase, the subsequent cell connected with the output nodes evaluates, and this process goes on with PC1, PC2, PC3, and PC4 applied to successive energy recovery stages.

Figure 15.4 shows the pipelined structure of the charge recovery vedic multiplier. Power-clocks are separated by a phase shift of 90° from each other. The input signal is applied 90° leading the first power-clock signal PC1 so that steady input can be applied during *evaluate* phase of power-clock. The output signal from the multiplier is received at the end of the 13th stage. This indicates that the time latency of the circuit is 3.25 times the power-clock duration. Clocked buffers are introduced for the product bits that arrive early to maintain the pipelining and timing. The number of buffers is determined by individual latency of respective product bits.

15.4 Simulation Results and Comparison

The two multipliers for the proposed comparison are implemented using the 250 nm TSMC Technology libraries. The average power consumed by the two methods is given in the Table 15.2. An average power of 77.66 % has been saved while using the vedic multiplier that incorporates the adiabatic logic at 25 MHz

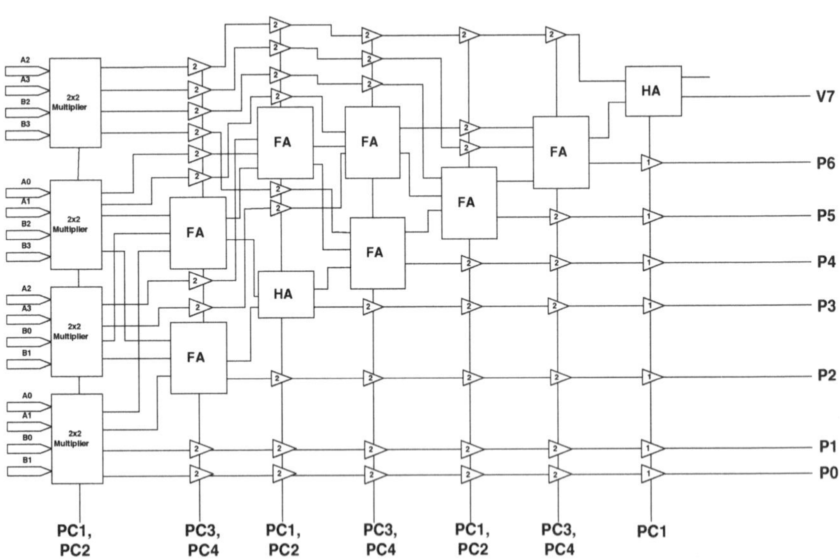

Fig. 15.4 Proposed 4 × 4 adiabatic Vedic multiplier

Table 15.2 Comparison for average power

Vedic multiplier	Average power (mW)
Using CMOS logic	2.1
Using 2N2P adiabatic logic	0.469

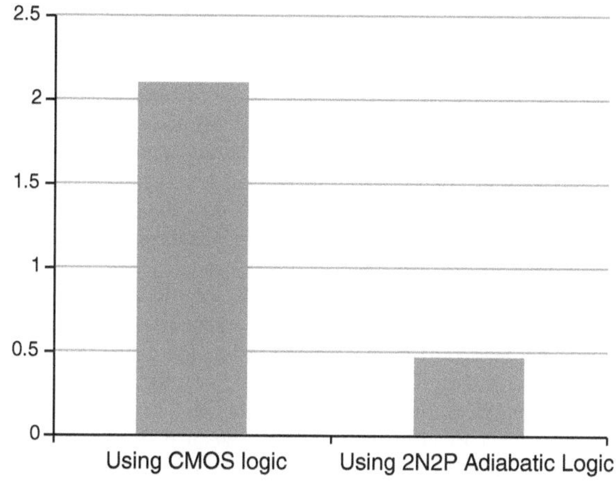

Fig. 15.5 Average power (in mW) of Vedic Multiplier using CMOS and Adiabatic logic

operating frequency. It is depicted in Fig. 15.5. This work can be extended for different genres of adiabatic family for higher operating frequency ranges and for reducing the adiabatic pipeline depth.

15.5 Conclusion

The binary 4 × 4 vedic multipliers are designed using CMOS and adiabatic circuits. The comparison results show that the vedic multiplier deployed with charge recovery logic consumes 77.66 % lesser power than its CMOS circuit counterpart. It validates that the adiabatic vedic multiplier is more power efficient than that constructed using the conventional CMOS circuit. Additional advantage is that the 4 phase-clocks serve as the synchronizing clock for the pipelining, and they also act as power supply to the circuit also. This feature can avoid the problems due to electromagnetic interference and cross talk across longer interconnect wires present in the complex design using lower technology nodes. The vedic multiplication algorithm will be implemented as an extension of this work, which would incorporate the study and analysis of the static and dynamic power components.

References

1. Rabaey JM, Chandrakasan A, Nikolic B (2002) Digital integrated circuits, a design perspective, 2nd edn. Prentice Hall, Englewood Cliffs, NJ
2. Jayaprakasan V, Vijayakumar S Kanchana Bhaaskaran VS (2011) Evaluation of the conventional versus ancient computation methodology for energy efficient arithmetic architecture, international conference on process automation, control and computing (PACC), 20–22
3. Tirthji Maharaja JSSBK (1986) Vedic mathematics, Motilal Banarsidas, Varanasi, India
4. Gurumurthy KS, Prahalad MS (2010) Fast and power efficient 16 × 16 array of array multiplier using vedic multiplication, microsystems packaging assembly and circuits technology conference (IMPACT)
5. Tiwari HD, Gankhuyag G, Kim CM, Cho YB (2008) Multiplier design based on ancient Indian Vedic Mathematics. In: Proceedings IEEE international SoC design conference, Busan, pp 65–68
6. Rudagi JM, Amble V, Munavalli V, Patil R, Sajjan V (2011) Design and implementation of efficient multiplier using vedic mathematics, 3rd international conference on advances in recent technologies in communication and computing, (ARTCom 2011)
7. Kanhe A, Das SK, Singh AK (2012) Design and Implementation of low power multiplier using vedic multiplication techniques. Int J Comput Sci Commun 3(1):131–132
8. Kanchana Bhaaskaran VS, Salivahanan S, Emmanuel DS (2006) Semi-custom design of adiabatic adder circuits. In: Proceedings of the 19th international conference on VLSI design (VLSID'06)
9. Kanchana Bhaaskaran VS, Raina JP (2012) Pre-resolve and sense adiabatic logic for 100 kHz to 500 MHz frequency classes. J Circ Syst Comput 21(5):1250045-1–1250045-20
10. Yong M, Jeong DK (1996) An efficient charge recovery logic circuit. IEEE J Solid-state Circ 31:191–196

Chapter 16
Error Detection and Correction in Embedded Memories Using Cyclic Code

S. M. Sunita, V. S. Kanchana Bhaaskaran, Deepakakumar Hegde and Pavan Dhareshwar

Abstract The test cost and yield improvement are the major factors in the era of rapidly growing memory density and capacity. The Error-Correcting Codes (ECC) is widely used to detect and correct errors in memories. The Cyclic codes are one such code which belongs to the class of ECC with algebraic structure. This paper describes the algorithm and the memory architecture required to implement error detection and correction using cyclic code. It also presents a brief comparison between the single error correction technique based on cyclic code and another single error correction technique with code based on Reed–Muller matrix. These results are also compared with a multiple error correction technique based on modified matrix code. The results validate that cyclic codes have 80 % and 90 % lesser area, 90 and 50 % more correction efficiency when compared to the code based on Reed–Muller matrix and modified matrix codes, respectively, and around 45 % less delay when compared to the two codes.

Keywords Memory testing · Error correction codes · Cyclic codes · Matrix codes · Single versus multiple error correction

---

S. M. Sunita (✉) · V. S. K. Bhaaskaran
VIT University, Chennai, India
e-mail: sunitha@pes.edu

V. S. K. Bhaaskaran
e-mail: kanchana.vs@vit.ac.in

S. M. Sunita · D. Hegde · P. Dhareshwar
PESIT, #14, 1st Floor, Kamakya Main Road, BSK 3rd Stage, 5th Block, Bangalore 560085, India
e-mail: d.k.n.hegde@gmail.com

P. Dhareshwar
e-mail: pavan.903.se@gmail.com

V. Chakravarthi et al. (eds.), *Proceedings of International Conference on VLSI, Communication, Advanced Devices, Signals & Systems and Networking (VCASAN-2013)*, Lecture Notes in Electrical Engineering 258, DOI: 10.1007/978-81-322-1524-0_16,

16.1 Introduction

High-density, low-voltage levels, small feature size, and small noise margins make the memory chips increasingly susceptible to faults or soft errors. Soft errors are the errors introduced due to high-energy particles from the external radiation hitting the silicon bulk thus flipping the value stored in a cell and causing bit inversion [1].

Soft errors are not caused by design or manufacturing defects and do not permanently damage the hardware. They only corrupt the value stored in the cell. Due to the relentless shrinkage in the device dimensions, the particles that were once considered negligible are now proving to be serious enough to cause soft errors [2]. Soft errors can be either SEU (single-event upset) or MBU (multiple-bit upset) [3, 4].

Built-in self-diagnosis techniques are used to test embedded memories. Some of the common self-diagnosis techniques used to test embedded memories is discussed in [5]. The other commonly used technique to detect and correct soft errors is the use of Error-Correcting Codes (ECC). Various error correction codes have been proposed, some of which correct multiple errors and a majority of others correct single errors. Hamming Codes are highly efficient in correcting SEUs in memory due to their smaller area and performance overhead [6]. There are many extensions of the basic SEC-DED Hamming Code such as Single-error-correcting, Double-error-detecting and Double-adjacent-error-correcting (SEC-DED-DAEC) codes [7], Single-error-correcting, Single-byte-error-correcting, Double-byte-error-detecting (SBC-DBD) codes, Double-error-correcting, Triple-error-detecting (DEC-TED) codes [8], and Double-error-detecting Single-byte-error-detecting SEC-DED-SBD codes [9, 10], Single error correction code based on the Reed–Muller matrix as proposed in [11] can correct single error in the data bits and in the check bits.

The Reed–Solomon (RS) code and Bose–Chaudhuri–Hocquenghem (BCH) Codes are capable of detecting and correcting multiple bytes of errors [8]. Hsiao et al. [12] also proposed a new class of multiple error-correcting codes, called Orthogonal Latin Square Code. The matrix codes combine the Hamming and parity codes to correct multiple-bit upsets [13]. The modified matrix codes as proposed in [5] can correct multiple errors, however, at the cost of large overhead and increased chip area.

The rest of the paper is organized as follows. Section 16.2 describes the cyclic codes and the algorithm used for error detection and correction. Section 16.3 explains the proposed memory architecture. The implementation method is explained in Sect. 16.4. Section 16.5 presents the results and discussion. Section 16.6 concludes the paper.

16.2 Cyclic Codes

The Cyclic Codes, being one of the first codes used practically for error detection and correction, possess rich algebraic structure and can be efficiently implemented using simple shift registers. The block diagram of the encoder to generate a (15, 11) cyclic code using the generator polynomial $1 + X + X^4$ is shown in Fig. 16.1.

The Eqs. (16.1–16.4) are used to generate the check bits.

$$C1(n) = D(n) \tag{16.1}$$

$$C2(n) = D(n) \text{ XOR } C1(n-1) \tag{16.2}$$

$$C3(n) = C2(n-1) \tag{16.3}$$

$$C4(n) = C3(n-1) \tag{16.4}$$

These equations are executed iteratively from D_0 to D_{10} by keeping the pointer to position 1. Then, keeping the pointer to position 2 for the next four clock cycles, the check bits are brought out.

During the read operation, the erroneous data bits along with the check bits (R_0–R_{14}) are given to the Syndrome generator circuit shown in Fig. 16.2. The syndrome bits are calculated according to the Eqs. (16.5–16.8).

$$S1(n) = R(n) \text{ XOR } S4(n-1) \tag{16.5}$$

$$S2(n) = S1(n-1) \text{ XOR } S4(n-1) \tag{16.6}$$

$$S3(n) = S2(n-1) \tag{16.7}$$

$$S4(n) = S3(n-1) \tag{16.8}$$

The equations are executed iteratively from R_0 to R_{14} by keeping Gate2 ON. Once the syndrome bits are ready after 14 clock cycles, they are brought out of the generator by making Gate1 ON and Gate2 OFF for the next 4 clock cycles. The syndrome bits identify the error position as per Table 16.1.

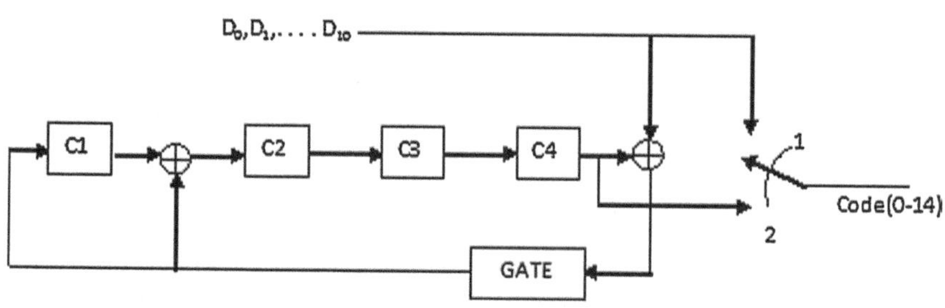

Fig. 16.1 Encoder for (15, 11) cyclic code

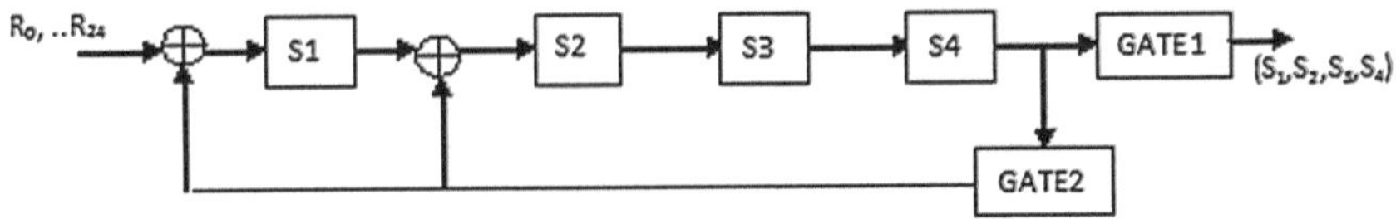

Fig. 16.2 Syndrome generator for (15, 11) cyclic code

Table 16.1 Error detection using syndrome bits

Sl. No.	Syndrome	Error position
1	0000	No error
2	0001	R_{14}
3	0010	R_{13}
4	0011	R_{10}
5	0100	R_{12}
6	0101	R_6
7	0110	R_9
8	0111	R_4
9	1000	R_{11}
10	1001	R_0
11	1010	R_5
12	1011	R_7
13	1100	R_8
14	1101	R_1
15	1110	R_3
16	1111	R_2

The algorithm for error detection and correction using cyclic code can be summarized as given below.

1. During the memory write operation, feed the data word (D_0 to D_{10}) to the encoder to generate the check bits (C_1 to C_4).
2. During the memory read operation, read the corrupted data word (R_0 to R_{10}) and its corresponding check bits (R_{11} to R_{14}).
3. Generate the syndrome bits by feeding these bits to a syndrome generator.
4. Use the syndrome bits to identify the error position as per Table 16.1.
5. Correct the error and output the corrected data word.

16.3 Memory Architecture

The block schematic employed for implementing the algorithm proposed in [5] can also be employed for implementing the cyclic codes algorithm. Figure 16.3 shows the block diagram of memory architecture for error detection and correction. During the *Memory Write* operation, the encoder generates the check bits

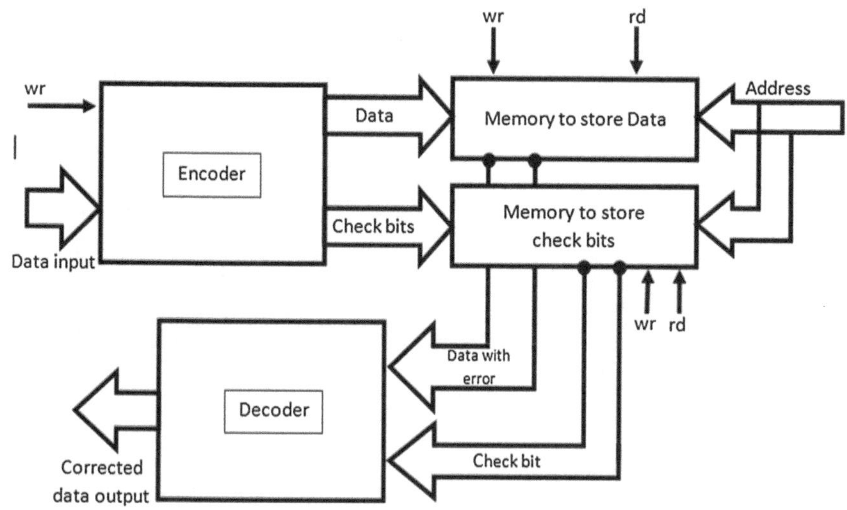

Fig. 16.3 Memory architecture for error detection and correction system

from the data bits. These are stored in the check bit memory, while the data are stored in the data memory.

During the *Memory Read* operation, the check bits are retrieved along with the data bits and fed to the syndrome generator present in the decoder. The generated syndrome bits are used to determine the error position. The error, if any, is corrected, and the corrected data are sent out of the decoder.

16.4 Implementation

The cyclic code algorithm described in Sect. 16.3 is coded in VHDL. Two other error-correcting codes, namely the single error-correcting code based on the Reed–Muller matrix as proposed in [11] and the multiple error correction code based on modified Matrix Codes as proposed in [5] are also coded in VHDL. The design is simulated using Xilinx ISim Simulator for both 16-bit and 32-bit data for each of the codes. The correct functionality was tested by giving various test bench inputs. The architectures were synthesized on Spartan 6 FPGA. Xilinx Timing analyzer tool was used to estimate the delay.

16.5 Results

Figure 16.4 shows the simulator outputs for (15, 11) cyclic code during a read operation. Here, "*data*" is the 11-bit data (D_0 to D_{10}) whose value is 7FF H in the waveforms shown. "*code*" is the 15-bit word with 4 check bits and the same 11

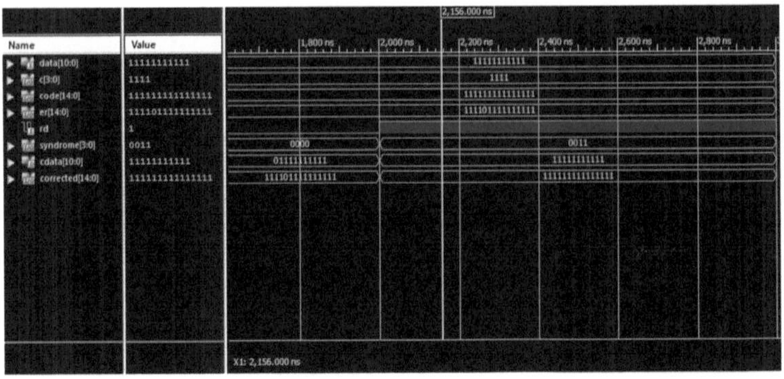

Fig. 16.4 Simulated outputs for memory read and correct using cyclic code

data bits. "*er*" is the same 15-bit word (R_0–R_{14}) with one error in bit position 10. On the positive edge of the *read* signal "*rd*," the 15-bit data word (R_0–R_{14}) is read and syndrome bits are calculated. Using these syndrome bits, the error is corrected. Finally, the corrected data "*cdata*" realizes the value 7FF H.

A performance metric, namely Correction efficiency, is defined to compare the performance of the three codes.

$$\text{Correction efficiency} = [\text{Number of errors corrected}/(\text{Area} \times \text{Delay})] \times 100\,\%$$

Here, area is defined in terms of the number of LUT's required. The delay refers to the time taken for the corrected word data to appear at the output of the decoder, once the read signal goes high. Table 16.2 shows the comparison between the three types of codes for both 16-bit data and 32-bit data. The simulation results are shown in the charts in Fig. 16.5.

The results indicate that the cyclic codes have 80 and 90 % lesser area when compared to the code based on Reed–Muller matrix and modified Matrix Codes, respectively. The delay is found to be approximately 45 % lesser than the other two codes. Hence, it has the highest correction efficiency which is about 90 and 50 % more when compared to the other two codes. Furthermore, it can be observed that the correction efficiency decreases as the number of bits increases.

Table 16.2 Comparison of results

Parameter	Cyclic code		Code based on Reed-Muller matrix		Modified matrix code	
	(15,11)	(31,26)	16-bit	32-bit	16-bit	32-bit
No. of errors corrected	1	1	1	1	9	11
No. of redundant bits	4	5	6	7	18	28
Area (No. of 6-input LUT's)	8	17	50	94	86	192
Delay in ns	11.53	14.8	15.26	27	23.4	27
Correction efficiency (%)	1.08	0.39	0.13	0.04	0.44	0.21

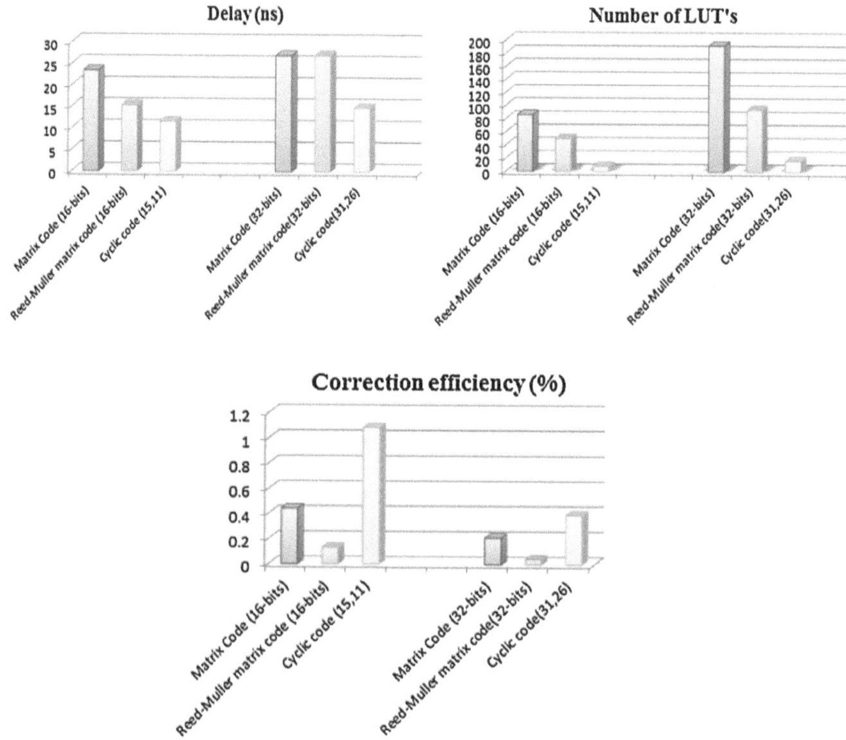

Fig. 16.5 Comparison charts for the three different codes

16.6 Conclusion

This paper presented the cyclic code technique for single error detection and correction in memories. This cyclic code technique is proved highly efficient than the other single error correction code based on Reed–Muller matrix. It is also found to be more efficient than the multiple error correction technique using the modified matrix code. Since the probability of occurrence of single errors is more, this technique can be used along with the interleaving technique to correct all errors in memory.

References

1. Shivkumar P, Kristler M, Keckler SW, Burger D, Alvisi L (2002) Modeling the effect of technology trends on the soft error rate of combinational logic. In: Proceedings of the international conference on dependable systems and networks, pp 389–398
2. Hazucha P, Svenson C (2000) Impact of CMOS technology scaling on the atmospheric neutron soft error rate. IEEE Trans Nucl Sci 47(6):2586–2594

3. Satoh S, Tosaka Y, Wender SA (2000) Geometric effect of multiple-bit soft errors induced by cosmic-ray neutrons on DRAMs. In: Proceedings of IEEE international electronic device meeting, pp 310–312, Jun 2000
4. Makhira A et al (2000) Analysis of single-ion multiple-bit upset in high-density DRAMS. IEEE Trans Nucl Sci 47(6):2400–2404
5. Sunita MS, Kanchana Bhaaskaran VS (2013) Matrix code based multiple error correction technique for n-bit memory data. Int JVLSI Des Commun Syst (VLSICS) 4(1):29–37
6. Hsiao MY (1970) A class of optimal minimum odd-weight-column SEC-DED codes. IBM J Res Dev 14:395–401
7. Dutta A, Touba NA (2007) Multiple bit upset tolerant memory using a selective cycle avoidance based SEC-DED-DAEC code. In: Proceedings of IEEE VLSI test symposium (VTS)
8. Houghton AD (1997) The engineer's error coding handbook. Chapman and Hall, London, UK
9. Reddy SM (1978) A class of linear codes for error control in byte-per-package organized memory systems. IEEE Trans Comput C-27:455–458
10. Chen CL (1983) Error correcting codes with byte error detection capability. IEEE Trans Comput C-32:615–621
11. Khalid F (2011) Design error detection and correction system based on Reed_Muller matrix for memory protection. Int J Comput Appl 34(8):42–48
12. Hsiao MY, Bossen DC, Chien RT (1970) Orthogonal Latin square codes. IBM J Res Dev 14:390–394
13. Argryides C, Pradhan DK, Kocak T (2011) Matrix codes for reliable and cost efficient memory chips. IEEE Trans VLSI Syst 19:420–428

Chapter 17
Determining Standard Cell Drive Strength Based on On-Chip Load Assessment

D. N. Krishna Kumar, Ramya S. Rajan and Veena S. Chakravarthi

Abstract Primarily, research in the design of high-performance standard cell libraries has been focused on drive strength selection of various logic gates. The paper proposes a novel technique of assessing load using on-chip path delay measurement and hence determines the drive strength of a standard cell in the paths iteratively. This approach is a three-phase approach viz. Synthesizing with standard default configuration of multidrive strengths, Drive strength assessment using load assessment by on-chip path delay measurement and reconfiguring the cells for proper drive strength by the synthesis process. The approach needs a multidrive standard cells library. This work focuses on determining the Drive strength of a standard cell by determining the load based on on-chip delay measurement though the other two steps explained in brief to give the complete perspective.

Keywords Logic synthesis · On-chip load · Load assessment · Path delay · Drive strength

17.1 Introduction

One of the most important steps in ASIC design is the synthesis phase. Synthesis can be applied at different levels, corresponding to the model under consideration. Library binding is the back-end of logic synthesis and constructs an

D. N. Krishna Kumar · R. S. Rajan (✉) · V. S. Chakravarthi
Department of ECE, BNMIT, Bangalore, India
e-mail: ramyasunderrajan@gmail.com

D. N. Krishna Kumar
e-mail: daskumar_in@yahoo.com

V. S. Chakravarthi
e-mail: scveena@bnmit.org

V. Chakravarthi et al. (eds.), *Proceedings of International Conference on VLSI, Communication, Advanced Devices, Signals & Systems and Networking (VCASAN-2013)*, Lecture Notes in Electrical Engineering 258, DOI: 10.1007/978-81-322-1524-0_17,

interconnection of cell instances from a given library starting from multilevel logic network. Emphasis is placed on the algorithmic approach to library binding. Cell-library binding is the task of transforming a logic network, described by a set of logic equations, into an interconnection of components that are cell instances of a given library. Library binding allows us to retarget logic designs to different design goals. For this reason, it is also often called technology mapping [1].

The first part of this work is synthesizing with standard default configuration of multidrive strengths. This is done in two phases. Initially, by selecting a particular type of cells like cell of a particular drive strength and fine tunes it by picking required cells for a design goal. The library contains the set of logic primitives that are available in the desired design style. Hence, the binding process must exploit the features of such a library, in the search for the best possible implementation. Typical optimization objectives are either the minimization of the critical delay or the minimization of power consumption (or area) under delay constraints [2].

In cell-library design and cell layout, the drive strengths for each cell function are selected to provide a predetermined scaling of total transistor active area within a cell. For example, the x2 cell has twice transistor active area of x1 cell. This granularity of cell drive strengths leads to over-design. Let us consider another example for a delay corresponding to drive strength of x1.2 is required, and the choices are x1 and x2, x2 will be chosen in order to meet the requirement. This increases the power consumption unnecessarily. Both switching power and leakage power are undesirably increased by such quantization. It is a known fact that the load, a standard cell sees on chip is reflected by the path delay in a circuit [3]. As technology scales to DSM, process variations will present significant impact on path delay. This trend makes the deviation between simulated path delay and actual path delay in a manufactured chip more significant. When on-chip path delays fall below hundreds of pico seconds, path delay measurement performed by off-chip equipments will be dominated by parasitic capacitance, resistance of probe and transmission line impedance fluctuations. Therefore, precise on-chip delay measurement methodologies are of high demand in semiconductor industry for rapid silicon debug [4]. Here, we apply this on-chip technique to assess path delay of a circuit and hence determine the drive strength of the cells in a path.

This paper focuses on part 2 of the flow diagram shown in Fig. 17.1, i.e., drive strength assessment using load measurements by on-chip path delay measurement

17.1.1 On-Chip Path Delay Assessment

A new method to estimate the on-chip path delay using the *Vernier Concept* is explained in this section. This is similar to a vernier scale used for scientific measurements. The vernier scale consists of a coarsely marked "fixed" scale which is used to measure between two of the scale's smallest graduations. The user then reads the finer vernier scale which measures between the smallest graduations on the fixed scale—providing much greater accuracy [5]. Similarly, in this paper,

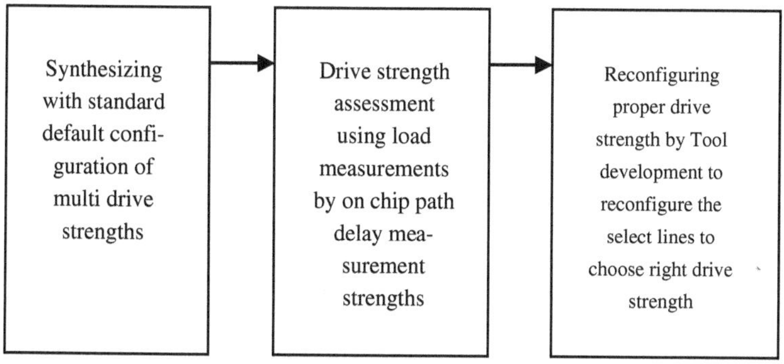

Fig. 17.1 Flow diagram of the logic synthesis as a concept

we have adopted this method to measure the on-chip path delay measurement. Here, we consider a pair of tapped delay lines with a flip-flop at each corresponding pair of taps. A stop signal propagates through one of the delay chain, while the start signal propagates through the other, clocking the flip-flop at each stage. The difference between the start and stop propagation delays determines the timing between adjacent stages. The difference between the start and stop propagation delays determines the timing between adjacent stages [6]. The vernier measurement technique is based on a path delay difference between two chains. The chain is realized with delay cells connected in cascade. For example, let us consider a path as shown in the Fig. 17.2.

This concept is applied to measure the register to register path by using coarse and fine measurements. The coarse counter measures the coarse timing and fine clock starts counting after the coarse counter has stopped. For operating system clock frequency of 1 GHz, the coarse counter is clocked with pulses of 256 ps duration and its final value corresponds to coarse time interval measurement. The fine counter is clocked with pulses of 2 ps duration.

The path delay for the considered case for simulation is around 700 ps and that measured using the vernier concept is around 598 ps. This wide difference is observed because the concepts are validated on the FPGA platform with the available DLL on FPGA. It is possible to accurately measure the actual path delays of the order of 300–400 ps using the high-frequency and high-accuracy PLL cores in VLSI technology.

Fig. 17.2 Timing diagram indicating path considered for measuring delay [courtesy 6]

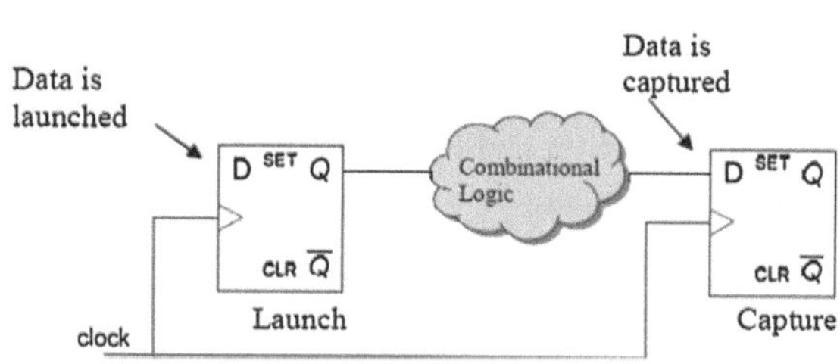

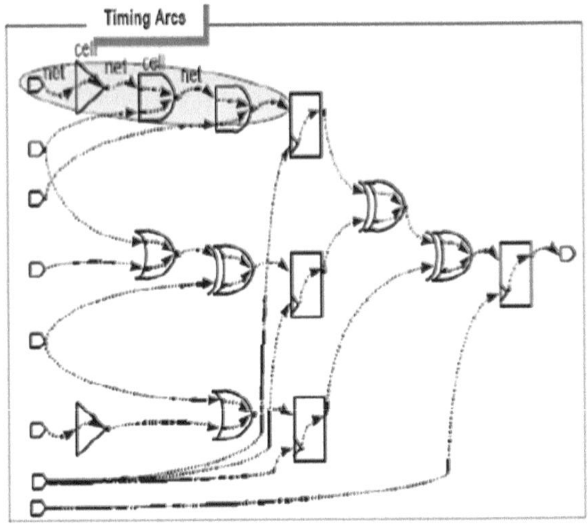

Fig. 17.3 Circuit with multiple paths [courtesy 6]

This technique holds good for any type of path like Input to Register, Register to Output. For a path with multiple gates as shown in the Fig. 17.3, we estimate the path delay by breaking each path into timing arcs. Each timing arc contributes to net or cell delay. The total path delay is the addition of all the net and cell delay. This path delay is calculated using the logical effort method. An electrical logical effort model is made by replacing each logic element with a simple electrical model and retaining the wiring topology of the original circuit.

A circuit simulation of the logical effort model produces voltages proportional to desired transistor widths [7].

In Fig. 17.3, the multiple paths are divided into separate timing arcs, from which we estimate the path delay. The path delay of a transition through series of CMOS gates is a function of cell delays; the load the cell sees and interconnects.

17.1.2 Load Assessment

The path delay can be seen as a function of load for a set of cells performing the same logical function but having different drive strengths. The computed path delay is used to assess the load of the circuit. The computed on-chip path delay is again simulated in cadence specter tool for variable load iteratively. The standard cells used for simulation are taken from Standard cell library that has variable drive strength with tappings. The current is estimated for variable load which can be used to estimate the drive strength of the circuit.

17.1.3 Drive Strength Assessment

By increasing the drive strength, a larger load can be driven. The drive strength of the cell is estimated by fixing the transistor sizes. The transistor sizes have to be accurate else the circuit will not be efficient in terms of area, power, and delay [8]. In the example considered, the required drive strengths are selected the choosing proper select lines of the MUX according to the load of the circuit. The standard cells are characterized with proper aspect ratios to drive the appropriate load. The MUX associated with the cell enable the choice of the drive strengths. The proper selection of the drive strengths through MUX select lines is carried out by a synthesis tool with the information of the load connected to it. The algorithm is beyond the scope of this paper.

The drive strength is fixed according to the load the circuit can drive. For variable load, the drive strength is fixed accordingly.

17.1.4 Standard Cell-Library Development

The Conventional Drive Strength of Standard Cells have 1X, 2X,...,32X without tappings. The standard cell library proposed has Standard cells with variable drive strength with tappings. These tappings provide variable drive strengths. The cells are designed by assessing their on-chip path delay, load, and drive strengths. These standard cell-library cells are used based on required drive strength. The select lines can be selected with the help of a multiplexer provided which will help to select depending on the type of the load and its drive strength. The Fig. 17.4 shows a 2 input NOR gate with variable drive strength with tappings.

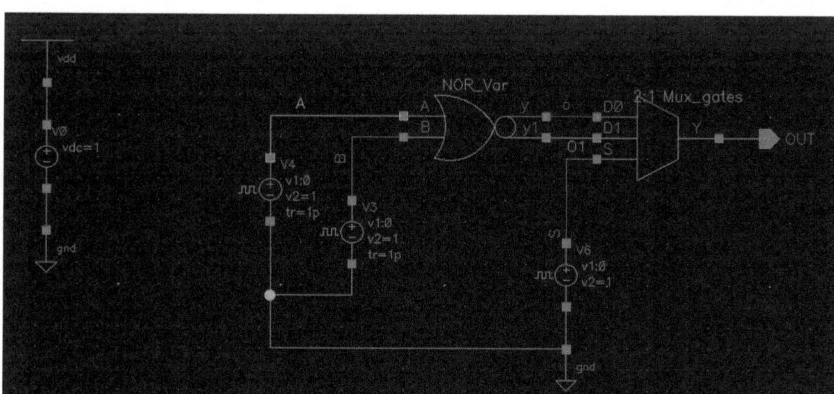

Fig. 17.4 2 Input NOR gate with variable drive strength selectable with the help of a 2:1 MUX

17.2 Synthesis Algorithm to Reconfigure Proper Drive Strength

The required drive strength can be configured by the logic synthesis algorithm which takes in the load information in addition to the constraint and HDL source file. The select lines are used to choose right drive strength. This explanation is out of scope of this current paper.

17.3 Results and Conclusion

Here, we have demonstrated the technique to determine the drive strength of the standard cell by assessing the load connected to a particular cell by measuring the on-chip path delay. The complexity of the work is in integrating the two concepts into synthesis and optimization algorithm as path delay is a circuit parameter and drive strength is the standard cell parameter. Also, accuracy of the path delay measurement depends on the precision and accuracy of on-chip PLL design as vernier concept is based on the counter method. Future plan of action will be to integrate these two and study the feasibility of designing a single cell of adaptive drive strength based on the load connected to it as against existing multi cells standard cell-library per type. This work redefines the logic synthesis flow which will result into optimum power logic circuits without any over shoots and undershoots.

Acknowledgments Authors wish to acknowledge Management of BNM Institute of Technology and VTU for the support and encouragement given to us in this work.

References

1. De Micheli G (1994) Synthesis and Optimization of Digital Circuits, 1st edn. Jan 1. ISBN-10: 0070163332, ISBN-13: 978-0070163331
2. Zhang X, Wang MX Standard cell library having cell drive strengths selected according to delay. US Patent No. 7254802, Stanford, CA
3. Kerman JP et al Performing drive strength adjust optimization in a circuit design. US Patent No. 5724250, Unisys Corporation, Blue Bell
4. Pan DZ, Halpin B, Ren H Timing driven placement. http://www.cerc.utexas.edu/utda/publications/book_tdp.pdf
5. Jovanović GS, Stojčev MK (2009) Vernier's Delay Line Time–to–Digital Converter. A: Appl Math Inform Mech, Scientific Publications of the state University of NOVI PAZAR SER, 1(1):11–20
6. Wikipedia, Vernier scale measurement. http://en.wikipedia.org/wiki/Vernier_scale
7. Dudeck P et al (2000) A high–resolution CMOS time–to–digital converter utilizing a vernier delay line. IEEE J Solid–State Circuits 35(2):240–246
8. Yeap GK (2002) Practical Low Power Digital VLSI Design, KAP

Chapter 18
Design of 12-Bit Cyclic Vernier Ring Time-to-Digital Converter

P. Prabhavathi, N. B. Mahesha and Subodhkumar Panda

Abstract Time-to-digital converter is used to digitize the delay difference between two signals. In this paper, an implementation of 12-bit, 10 ps Vernier Ring Time-to-digital Converter (VRTDC) is presented. It uses Vernier delay cells and arbiters, placed in ring format and reuse them for the measurement of the input time interval. A prelogic unit is developed to decide leading and lagging signals, through which it is possible to measure both positive and negative phase difference. It is possible to achieve large detectable range, fine time measurement, small die size, and low-power consumption with the proposed VRTDC. The design is modeled using Verilog and synthesized using RTL Compiler targeting the design to 180 nm standard cell library.

Keywords Vernier ring · VRTDC · Time-to-digital converter · TDC · Frequency synthesis

18.1 Introduction

Time-to-digital converter is an indispensible system block of many sensitive instruments which measure phase difference between two signals. It can also be used to recognize events and provide a digital representation of the time at which they occur. For example, a TDC is used to indicate the time of arrival for each incoming pulse. Some applications measure the time interval and convert it into

P. Prabhavathi (✉) · S. Panda
Department of ECE, B N M Institute of Technology, Bangalore, India
e-mail: prabha@bnmit.org

S. Panda
e-mail: subodhpanda@yahoo.com

N. B. Mahesha
VLSI and ES, B N M Institute of Technology, Bangalore, India
e-mail: nbmahesh512@gmail.com

V. Chakravarthi et al. (eds.), *Proceedings of International Conference on VLSI, Communication, Advanced Devices, Signals & Systems and Networking (VCASAN-2013)*, Lecture Notes in Electrical Engineering 258, DOI: 10.1007/978-81-322-1524-0_18,

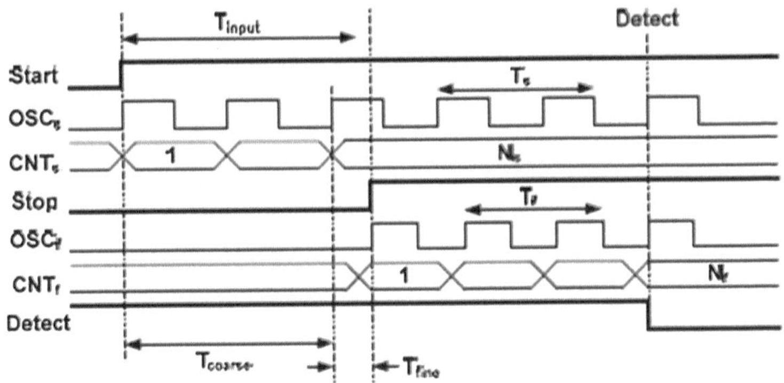

Fig. 18.1 Measurement of time difference using TDC

digital output. TDCs find applications in many physical experiments, like time-of-flight and life time measurements in atomic and high-energy physics.

TDCs are also used in measurement of time interval between two signals. Measurement is initiated at the rising edge of start pulse or falling edge of stop pulse as indicated in Fig. 18.1. The start and stop pulses may indicate the beginning and termination of events.

An inverter chain-based TDC was first implemented for a Bluetooth radio application [1]. Since then, a variety of TDC architectures have been proposed with improved resolution and detectable range. Although there are many ways to digitize the input time interval, the digital inverter delay line is still an appealing structure due to its digital-intensive design approach.

The proposed TDC is implemented as a digital system block, as such an implementation offers benefits like decrease in die area, high-frequency operation, low-power consumption and less sensitive to process, voltage and temperature variations. Figure 18.2a shows the conventional inverter delay line based TDC. Its time resolution is the propagation delay of each individual inverter, while its detectable range is proportional to the number of delay stages used.

A Vernier delay line is well known for its fine time resolution [2]. Figure 18.2b illustrates a simplified Vernier inverter delay line TDC. It employs two inverter/buffer chains with different delays of T_s and T_f, respectively. Time resolution of the Vernier TDC now becomes the delay difference of two delay lines, namely $T_s - T_f$. Where T_S and T_f are delay of slow and fast inverter cells, respectively.

Table 18.1 gives the comparison of different TDC architectures with respect to resolution and detectable range proposed in literature. A multipath gated ring oscillator (GRO) based TDC achieves a detectable range of 11 bits [3]. A two-level interpolation TDC with this parallel structure achieved a time resolution of 12.2 ps in 0.35 μm CMOS technology [4]. The proposed VRTDC implementation is an attempt to achieve fine resolution with reasonable detectable range.

It was also observed that by placing delay cells in a ring format, delay chains can be reused for measuring large time intervals. Coarse and fine counters are used

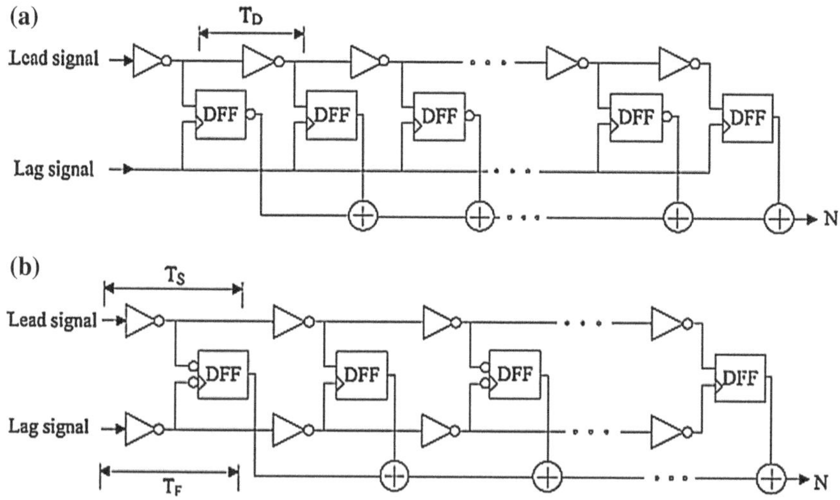

Fig. 18.2 Vernier delay line using **a** Single and **b** Two inverter chains

Table 18.1 Comparison of different TDC architectures

Reference	[1]	[2]	[3]	[4]	This paper
Time resolution (ps)	24	20	6	12.2	10
Measurement range in bits	8	5	11	14	12

to monitor the number of laps; the signals propagate along the ring. Arbiters are used to record the location where the lag signal catches up with the lead signal.

This paper is organized as follows: Sect. 18.2 gives implementation of critical building blocks of the VRTDC. Vernier ring Time-to-digital converter implementation is presented in Sect. 18.3. Results and conclusion are presented in Sect. 18.4 and references are presented in the next section.

18.2 Vernier Ring Time-to-Digital Converter System

The architecture of 12-bit VRTDC system composed of the VRTDC core, pre-logic unit, and thermometer-to-binary encoder, 6-bit fine counter and 6-bit coarse counter which is shown in Fig. 18.3.

The outputs of 30 arbiters are combined to form 30-bit thermometer code "TH" and are translated into a 5-bit binary code by a thermometer-to-binary encoder. The total amount of delay is composed of four elements: the sign bit, the coarse counter value, the fine counter value, and the thermometer code.

$$N = \pm 30\,(N_f - N_C) + TH + 30\,N_C t_s/R$$

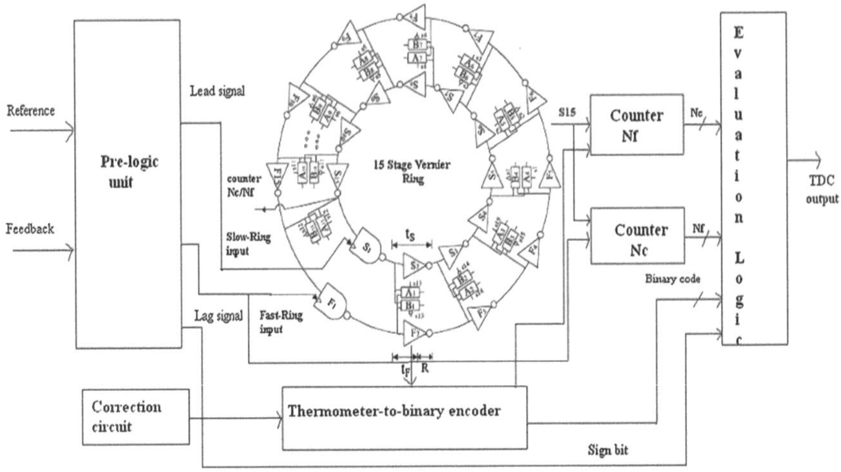

Fig. 18.3 VRTDC architecture

where N is the TDC output, N_C is the coarse counter output, N_f is the fine counter output, TH is thermometer-to-binary encoder output, t_s is the sampling time and R is the resolution.

18.3 VRTDC Implementation

The detailed description of the VRTDC architecture which consists of pre-logic unit, two types of arbiters with edge detectors, the thermometer-to-binary encoder, and correction circuit is presented in this section.

18.3.1 Prelogic Unit

The reference and feedback signals are applied to the prelogic unit; the lead signal is steered to the slow ring, while the lag signal goes to the fast ring (Fig. 18.4).

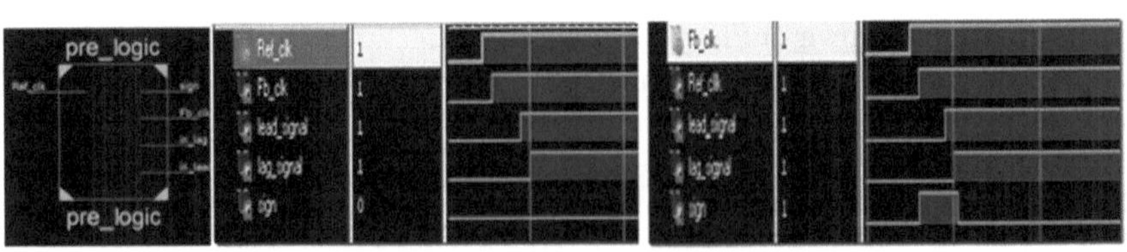

Fig. 18.4 Prelogic unit indicating positive and negative phase differences

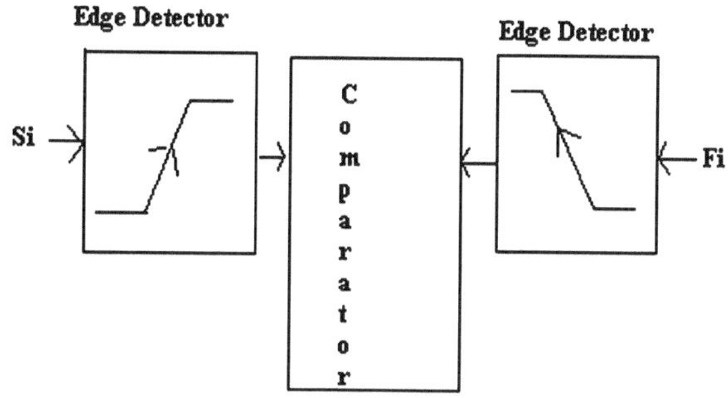

Fig. 18.5 Arbiter A

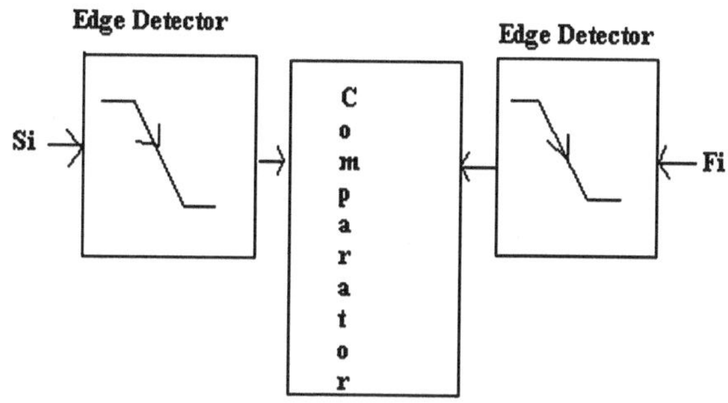

Fig. 18.6 Arbiter B

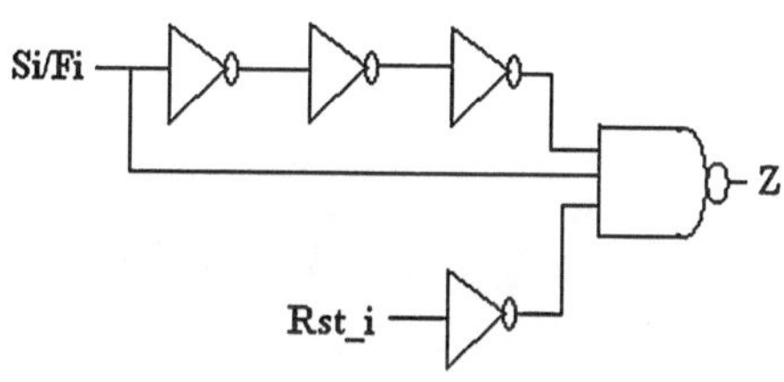

Fig. 18.7 Rising edge detector

It consists of delay path, multiplexers, arbiter, and reset path. Arbiter decides the sign bit, and multiplexers guide the lead signal to slow ring and lag signal to fast ring.

18.3.2 Arbiters and Edge Detectors

Arbiter A and Arbiter B are triggered by rising and falling edges, respectively, uses respective edge detectors. Earlier implementation of TDC had transistor-based arbiters. This work proposes fully digital implementation of arbiters as in Figs. 18.5, 18.6.

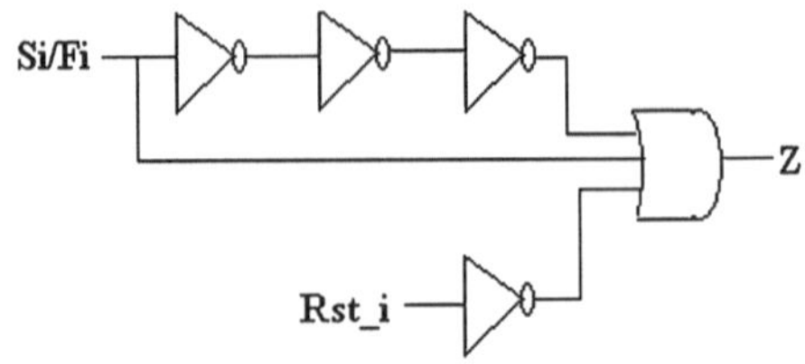

Fig. 18.8 Falling edge detector

When lag signal catches up with the lead signal, the arbiters identify the event by setting its output high. Two sets of arbiters are used to identify such events, alternatively in odd and even laps (Figs. 18.7, 18.8, 18.9).

18.3.3 Thermometer-to-Binary Encoder

When lag signal catches the lead signal, the arbiter outputs "1". The output of all 30-arbiters is combined to form Thermometer code "TH". This 30-bit Thermometer code is converted to 5-bit binary code using Thermometer-to-binary encoder, as shown in Fig. 18.10.

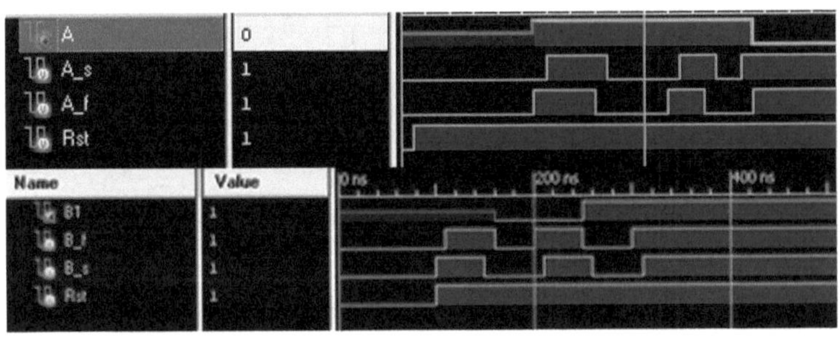

Fig. 18.9 Arbiter A and Arbiter B simulation results

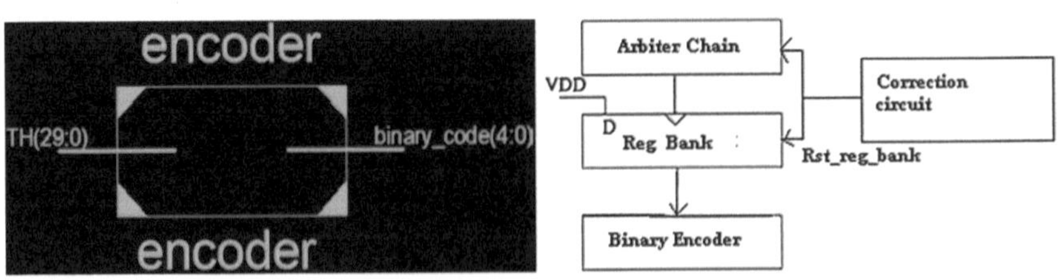

Fig. 18.10 Thermometer-to-binary encoder

18.3.4 Correction Circuit

In the process of chasing lead signal by the lag signal, an unexpected "01" transition may occur as shown in Fig. 18.11. Edge C is supposed to be compared with edge a. Unfortunately, the edge is going to be compared with the next falling edge at b. Arbiter B1 and the following few arbiters will be set to "1". Moreover, arbiters A15 and B14 have been set to "0" before lag signal propagates in the fast ring. A "001" transition will be erroneously detected at the least significant bit of thermometer code. The VRTDC would have mistakenly judged that the lag signal had caught up with the lead signal without proper error detection (Figs. 18.12 and 18.13).

The correction circuit will screen the "001" detection signal and keep edge c of the lag signal chasing edge a till the next catch-up happens. Small phase error detection circuit is used to measure the minute phase difference ($\ll$0.2 ps).

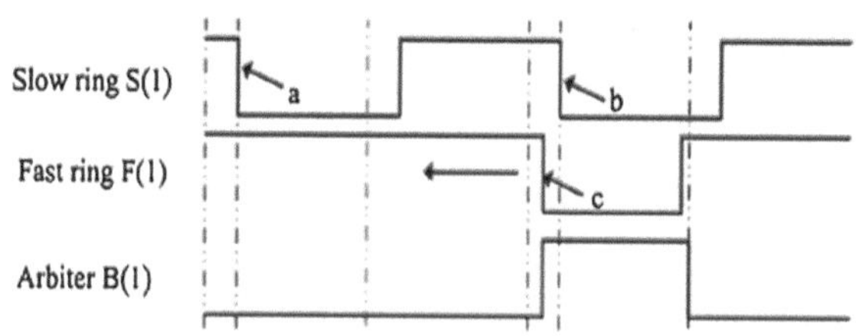

Fig. 18.11 Correction circuit [5]

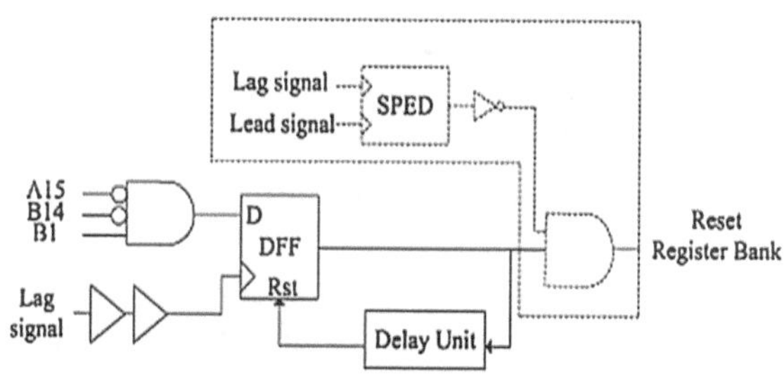

Fig. 18.12 Correction circuit [5]

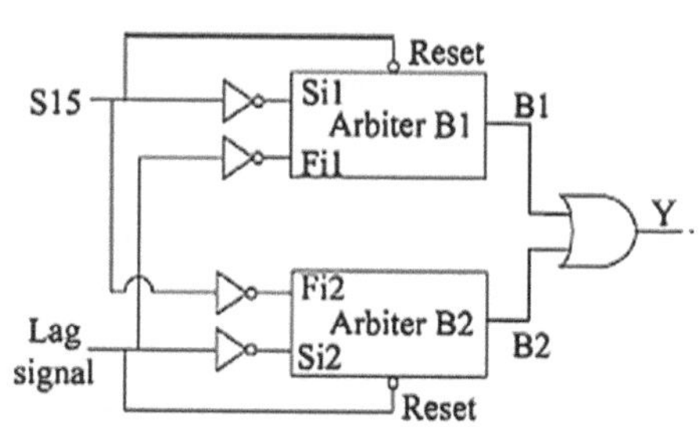

Fig. 18.13 SPED [5] (small phase error detection)

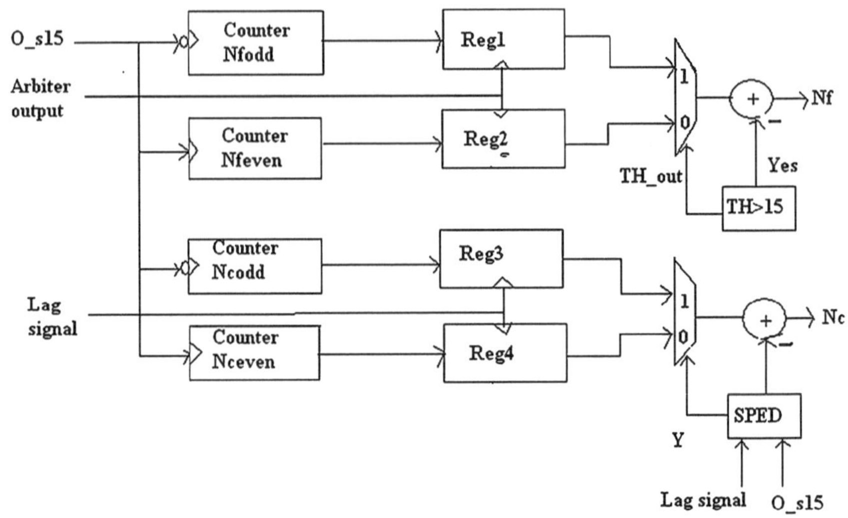

Fig. 18.14 Counter

18.3.5 Coarse and Fine Counters

Before lag signal enters the ring, VRTDC will be in the coarse measurement mode, measuring the number of periods the lead signal is ahead of lag signal (N_C). It switches to fine measurement mode when lag signal enters, where the fraction of period is measured (Fig. 18.14).

18.4 Results and Conclusion

Proposed VRTDC was synthesized separately using RTL compiler and Xilinx ISE. The design was targeted to 180 nm standard cell library in RTL Compiler. The synthesis report is as given in Table 18.2.

Table 18.2 Synthesis report

Power report

Instance	Cells	Leakage power (nW)	Dynamic power (nW)	Total power (nW)
tdc_top	204	4991.879	22945.153	27937.032

Area report

	Cells	Cell area	Net area	Wireload
tdc_top	204	1721	0	<none> (D)

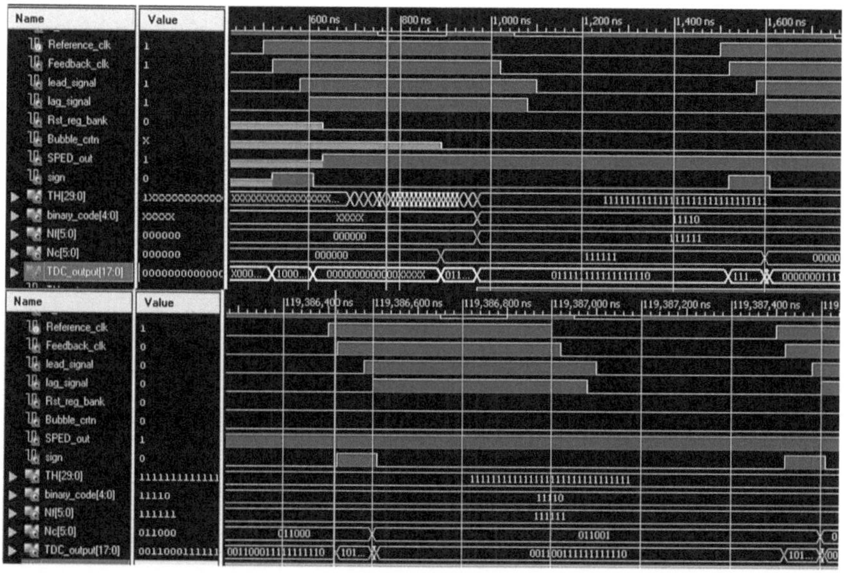

Fig. 18.15 Simulation results of VRTDC at different times

The simulation results have indicated 10 ps resolution for 12-bit measurement. Further, it is planned to use this VRTDC in All digital phase locked loop (ADPLL) for phase correction (Fig. 18.15).

References

1. Staszewski RB, Hung CM, Maggio K, Wallberg J, Leipold D, Balsara PT (2004) All-digital phase-domain TX frequency synthesizer for Bluetooth radios in 0.13 um CMOS. In: IEEE ISSCC digital technical papers, vol. 1. pp 272–527, February 2004
2. Dudek P, SzcZepanski S, Hatfield J (2000) A high resolution CMOS Time-to-digital converter utilizing a vernier delay line. IEEE J Solid-State Circ 35(2)240–247
3. Straayer MZ, Perrot MH (2008) An efficient high-resolution 11-bit noise-shaping multipath gated ring oscillator TDC. In: Symposium, VLSI circuits digital technical papers, pp 82–83, June 2008
4. Ramakrishnan V, Balsara PT (2006) A wide-range, high-resolution, compact CMOS Time-to-digital converter. In: Proceedings of 19th international conference VLSI design (VLSID'06), p 6, January 2006
5. Yu J, Dai FF, Jaeger RC (2010) A 12-bit vernier ring time-to-digital converter n 0.13 um CMOS technology. IEEE J Solid State Circ 45(4):830–842

Chapter 19
Adder-Based Address Generation for Embedded MBIST

Yasha Jyothi M. Shirur, Veena S. Chakravarthi and R. Varchaswini

Abstract Today's System on Chips (SoC) are undoubtedly memory dominant, and it is predicted that the amount of space they occupy on the die will continue to increase, reaching up to 70 % by 2017 [1]. Built in self-test (BIST) has been the traditional technique for testing embedded memories over the years. Traditional BIST circuitry includes counter-based address generator which can be replaced by Adder-based address generator. The Adder-based address generator includes simple adder circuit to generate address and data for embedded MBIST. In this paper, adder-based address generator logic in BIST controller is proposed. This new idea for generating address and data has resulted in reduced area occupied by 40–68 % and the power dissipation by 83–86 % when compared with the traditional implementations.

Keywords BIST · MBIST · Embedded memories · Memory dominant SoCs

19.1 Introduction

The increasing percentage of area used for embedded memories contributes significantly to the overall performance of the system and its failure. The yield of a SoC being largely dependent on memory yield, and it is important to implement techniques to improve it. A way to improve yield is by reducing defect occurrences

---

Y. J. M. Shirur (✉) · V. S. Chakravarthi · R. Varchaswini
Department of ECE, BNM Institute of Technology, Bangalore, India
e-mail: yashamallik@bnmit.org

V. S. Chakravarthi
e-mail: scveena@bnmit.org

R. Varchaswini
e-mail: varchaswinir@gmail.com

V. Chakravarthi et al. (eds.), *Proceedings of International Conference on VLSI, Communication, Advanced Devices, Signals & Systems and Networking (VCASAN-2013)*, Lecture Notes in Electrical Engineering 258, DOI: 10.1007/978-81-322-1524-0_19,

in both Memories and Logic. BIST has been most effective testing methods for embedded memories over the years [1]. One of the critical components of memory BIST (MBIST) is the address generator (AG). In order to detect delay/bridging faults, the address generator has to generate different address sequences to allow for correct address changeovers. Its complexity is a major design issue since it requires large area, power and limits the BIST speed. It has been shown that the relative area occupied and power consumed by the address generator for the traditional schemes in [2–4] varies between 26 and 33 % of the memory BIST scheme.

19.2 Traditional Approaches

Traditionally, address generator implementations are based on counter modules. Two major counter-based address generation techniques are linear or address compliment techniques. The linear (LI) counting method increments or decrements the address sequence during up count and down count, respectively. The linear counting method is used for detecting single-cell and coupling faults. The address complement (AC) counting method specifies an address sequence such that present address is the one's complement of the previous address [5]. Both the techniques stress the address generator logic of the BIST thus dissipating more power as address bit toggling is not controlled. Address generator implementations that can combine linear and address complement counting methods have been proposed in [6]. In [7] the implementation proposed in [6] was integrated into a more generic scheme to generate more counting methods [8].

19.3 Novel Adder-Based Address Generator

In this work, address is generated by using inexpensive adder and register combination of logic. The logic structure is shown in Fig. 19.1. The register is initialized to 0 and for every tick of *up/down* signal, the register content is incremented. New address thus generated is latched at the toggle of enable signal.

Above scheme can be easily changed to generate the address by initializing the register to all 1 s, thus realizing the linear step down address.

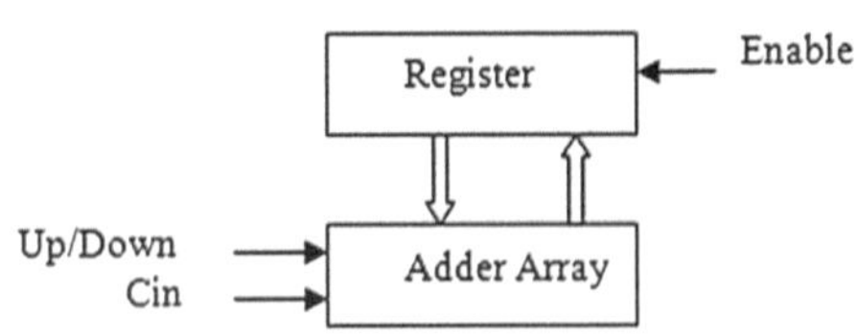

Fig. 19.1 Adder-based address generator structure

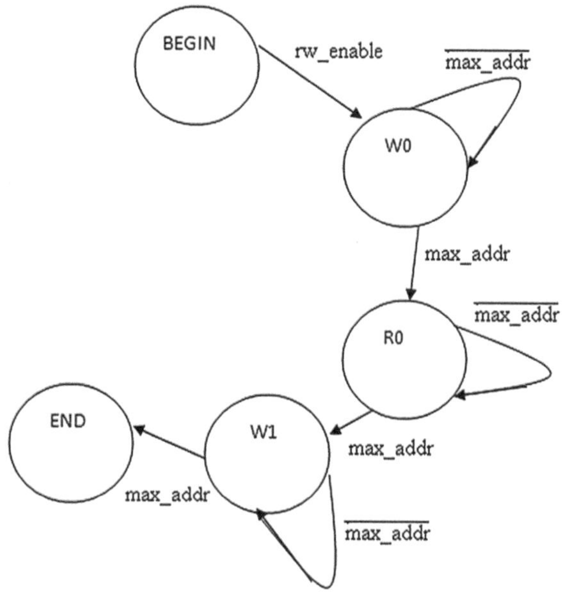

Fig. 19.2 State diagram of FSM-based address and data generator [9]

19.4 Implementation

The traditional MBIST is as shown in Fig. 19.3. In traditional MBIST, counter and decoder is used to generate the address and data. In adder-based generator, simple adder circuit is used with a latch to store the generated address which is also used as data to be stored in that particular address. Thus, address generator block in traditional MBIST can be replaced by a novel adder-based address generator. The new method has been implemented for different memory capacity to study the area and power requirement. The synthesis report obtained from cadence RTL complier tool is used to arrive at the conclusion. In meanwhile, the finite state machine (FSM)–based address generator has been also considered for comparison. In FSM-based address generator, address to memory and data generations is based on MARCH C-algorithm. This architecture has different states to write and read data from memory. The pattern generated by this algorithm is writing "0" to all locations, reading the same "0" from all locations, and writing "1" to all locations. The data written and read into the memory is compared in the comparator to detect the faults. The comparator output indicates low fail signal if data generated and ramout are equal else it raises high fail signal. The state diagram of FSM-based architecture is shown in Fig. 19.2.

In conclusion, it is observed that FSM-based address generator when compared with traditional does not leads to high reduction in area and power.

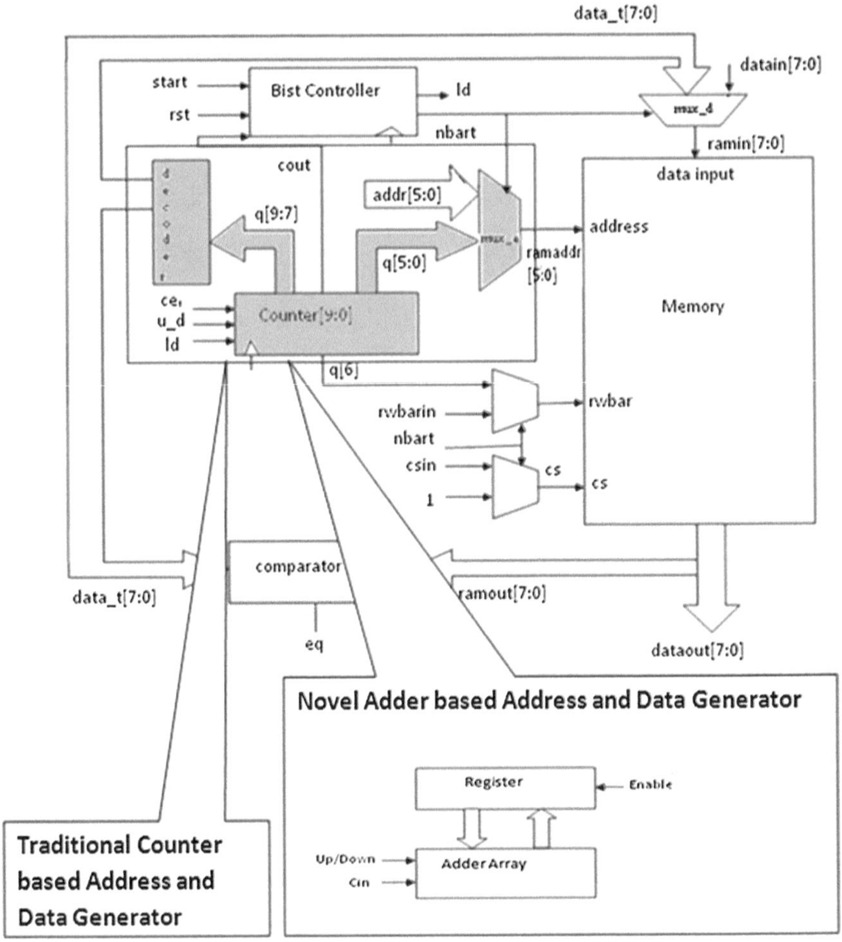

Fig. 19.3 Traditional Memory BIST replaced by Novel Adder-based address generator structure

19.5 Result and Discussions

The novel adder-based address generator is coded in Verilog and simulated in Cadence environment using ncsim simulator. The concept verifications were carried out for different memory capacities Viz., 16, 32, 64, 128, and 256. The consolidated synthesis report obtained for adder-based address generator and counter-based address generator from cadence RTL Complier for area and power is depicted in Tables 19.1 and 19.2, respectively. It is observed that the new idea used to generate address for MBIST reduces the area occupied by 40–68 % and the power dissipation by 83–86 % when compared with the traditional one. The FSM-based address generator in comparison with traditional based address generator is also depicted in Table 19.3. In FSM-based address generator, area and power reduction is of the order 0.5–5 % and 20–30 % when compared with the adder based address generator of same memory capacity (Table 19.4).

Table 19.1 Consolidated synthesis report obtained for area comparison of traditional and adder-based address generator

Cell area comparison

Memory capacity	Counter-based address generator for MBIST	Adder-based address generator for MBIST	Percentage of area reduced
Memory_16	3,316	1,054	68.2147
Memory_32	6,122	2,417	60.5194
Memory_64	11,648	5,452	53.1936
Memory_128	22,675	12,143	46.4476
Memory_256	44,208	26,267	40.5831

Table 19.2 Consolidated synthesis report obtained for power dissipation of traditional and adder-based address generator

Power dissipation comparison in nano watt

Memory capacity	Counter-based address generator for MBIST	Adder-based address generator for MBIST	Percentage of power dissipation reduced
Memory_16	68,918.714	8,966.426	86.9898
Memory_32	126,558.32	16,652.62	86.8419
Memory_64	254,255.517	36,067.012	85.8146
Memory_128	465,104.378	75,029.816	83.8681
Memory_256	975,177.937	15,5350.492	84.0695

Table 19.3 Consolidated synthesis report obtained for area of traditional and FSM-based address generator

Cell area comparison

Memory capacity	Counter-based address generator for MBIST	FSM-based address generator for MBIST	Percentage of area reduced
Memory_16	3,316	3,142	5.247285887
Memory_32	6,122	5,951	2.793204835
Memory_64	11,648	11,453	1.674107143
Memory_128	22,675	22,476	0.877618523
Memory_256	44,208	43,991	0.490861383

Table 19.4 Consolidated synthesis report obtained for power dissipation of traditional and FSM-based address generator

Power dissipation comparison in nano watts

Memory capacity	Counter-based address generator for MBIST	FSM-based address generator for MBIST	Percentage of power dissipation reduced
Memory_16	68,918.714	53,801.348	21.9350
Memory_32	126,558.32	98,213.726	22.3964
Memory_64	254,255.517	181,773.716	28.5074
Memory_128	465,104.378	356,192.198	23.4167
Memory_256	975,177.937	745,619.814	23.5401

19.6 Conclusion

The Novel adder-based address and data generator is a novel technique which is used to generate address and data required in MBIST Controller to test the embedded memories. This method reduces the area overhead by 40–68 % and the power dissipation by 83–86 % when compared with the traditional one. The reduction of area and power will be large for embedded memories of greater size. The results obtained are ensures that the new method can be used for address and data generator in embedded memory.

Acknowledgments Authors wish to thank the management of BNM Institute of Technology for their support and encouragement rendered to them in this work.

References

1. Zorian Y (2002) Embedded memory test and repair: infrastructure IP for SOC yield. In: Proceedings of Int'l test conference (ITC), Baltmore, 340–349, October 2002
2. Du X, Mukherjee N, Cheng WT, Reddy SM (2005) Full-speed field-programmable memory BIST architecture. In: Proceedings of international test conference, 1173–1182
3. Park Y, Park J, Han T, Kang S (2009) An effective programmable memory BIST for embedded memory. IEICE Trans Inf Syst E92-D(12):2508–2511
4. Kukner H (2010) Generic and orthogonal march element based memory BIST engine, Master Thesis, CE-MS-2010-01, Delft University of Technology, Delft
5. van de Goor AJ (1998) Testing semiconductor memories: theory and practice. Com Tex Publishing, the Netherlands (Ad.vd.Goor@kpnplanet.nl)
6. van de Goor AJ, Kukner H, Hamdioui S (2011) Optimizing memory BIST address generator implementations. In: Proceedings of the 2011 conference on design and technology of integrated systems in nano scale era, Athens, Greece
7. Hamdioui S, van de Goor AJ, Reyes JD, Rodgers M (2006) Memory test experiment: industrial results and data. IEEE Proc Comput Digit Tech 153(1):1–8
8. Voyiatzis I, Efstathiou C, Hamdioui S, Sgouropoulou C (2012) ALU based address generation for RAMs. International conference on design and technology of integrated systems in nano scale era
9. Adams RD (2002) High performance memory testing design principles, fault modeling and self-test. In: Frontiers in electronic testing, Vol. 22A XIII, 247 p

Chapter 20
Circuit Design Methodologies for Test Power Reduction in Nano-Scaled Technologies

Veena S. Chakravarthi and Swaroop Ghosh

Abstract Test power has emerged as an important design concern in nano-scaled technologies. The BIST circuitry for periodic self-test consumes significant power in hand-held electronic devices to increase battery lifetime. Reduced test power of a module allows parallel testing of multiple embedded cores in an IC. Peak and average power reduction during test contribute to enhanced reliability and improved yield. In this paper, we present circuit design methodologies to reduce test power in nano-scaled technologies. In addition to this advantage of reduced supply, testing concept is mentioned for initial testing which will give dual benefit of power and test time reduction.

Keywords Leakage power · Scan-Latch reordering · Shannon's expansion based synthesis (SBS) · Dynamic supply gating (DSG) · First-level supply gating (FLS)

20.1 Introduction

Power dissipation is a major concern in nano-meter technologies. Smaller feature size (due to scaled dimensions) reduces the switching capacitance per transistor but it also allows integration of more components in the same footprint. Therefore, aggregate switching power increases from one technology generation to next. At the same time, leakage power starts dominating due to new leakage mechanisms, e.g., subthreshold leakage, gate leakage, junction leakage, and GIDL [1]. Figure 20.1

V. S. Chakravarthi (✉)
Department of ECE, BNM Institute of Technology, Bangalore, India
e-mail: scveena@bnmit.org

S. Ghosh
Computer Science and Engineering, University of South Florida, ENB-118, Tampa, USA
e-mail: sghosh@cse.usf.edu

V. Chakravarthi et al. (eds.), *Proceedings of International Conference on VLSI, Communication, Advanced Devices, Signals & Systems and Networking (VCASAN-2013)*, Lecture Notes in Electrical Engineering 258, DOI: 10.1007/978-81-322-1524-0_20,

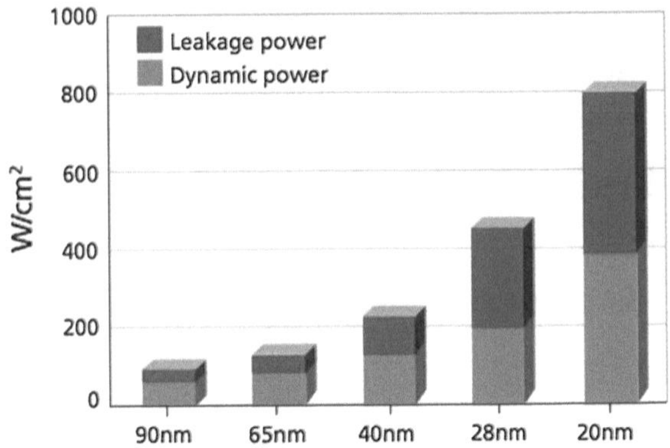

Fig. 20.1 Leakage power is becoming a significant component of total power consumption in nano-scaled technologies [1]

shows the breakdown of dynamic and leakage power consumption in scaled technologies. These sources of power consumption plays an important role in reducing the battery lifetime of handheld devices like smart phones, tablets, net books, etc.

Another important component that limits the battery life is the power consumed during periodic self-test by the built-in self-test circuitry (BIST) that is executed to ensure the functional correctness of the system during boot-up. Test power can be significantly higher than the functional power, since the functional patterns are usually strongly correlated compared to test patterns that are statistically independent of each other. It has been shown [2] that the test power could be twice as high as the power consumed during the functional mode. Lower test power is not only important to increase the battery life time but also to improve test cost, since reduced test power of a module allows parallel testing of multiple embedded cores in an IC [3]. Peak and average power reduction during test contributes to enhance reliability of test and hence, to improve yield. One example of yield loss has been shown in [4] where the increased leakage during test mode can cause damage to the chip.

Power dissipation during test can be categorized into two parts: (a) power in the combinational block and (b) power in the scan chain. Several research has been conducted in past to explore efficient techniques to reduce test power in scan-based circuits. An automatic test pattern generation technique to reduce power dissipation during scan testing was presented in [5]. Scan-latch reordering [6] or input vector reordering [7] techniques have been proposed for reduction in test power. In [3], a solution for average and peak power dissipation by transforming conventional scan architecture into desired number of selectable separate scan paths. Each scan path is in turn filled with stimulus and emptied of response.

Another solution has been proposed in [8] to address the peak power problem during external testing by selectively disabling the scan chain. These techniques target reducing number of switching in the scan chain but cannot completely prevent redundant power loss in the combinational logic.

Inserting gating logic into the stimulus path of the scan cells to prevent propagation of scan-ripple effect to logic gates offers a simple and effective solution to significantly reduce test power in the combinational logic. A NOR or NAND gating method have been proposed in [9]. Gating logic (NOR or NAND) are controlled by the test enable signal and the stimulus paths remain fixed at either logic '0' or logic '1' during the entire scan-shift operation. Multiplexers at the output of the scan cells, which hold the previous state of the scan register during shifting, have been presented in [9] to prevent activity in the combinational logic. Another method for reducing combinational power using gating is to use a scan hold circuit as a sequential element. This method helps in delay fault testing by allowing application of an arbitrary two-pattern test. The main difficulty with gating logic to prevent redundant toggling is that they add significant delay in the signal propagation path from the scan flip-flops to logic gates. Moreover, they incur large overhead in terms of area and switching power during functional mode.

In this paper, we discuss two strong methods of reducing the test power in nano-scaled technologies. In particular, we describe Shannon's expansion-based synthesis (SBS) [10], first-level supply gating (FLS) [11] and second reduced supply testing considering the quadratic reduction on power consumption. SBS essentially decomposes a circuit into a set of disjoint logic blocks with only one active block for a given stimulus. This particular structure of circuit synthesis results in lower test power. FLS is achieved by inserting a supply gating transistor in the first level of logic connected to the scan cell outputs, which essentially "gates" the VDD or GND line. FLS is as effective as the other blocking methods in terms of reducing peak power and total energy dissipation during scan testing but with significantly less delay penalty. Another method to take quadratic power reduction advantage

is to run test at reduced Supply voltage. Generally, BIST is run at speed, and there is a new trend to run scan at speed. So care must be taken to identify an optimal supply voltage at which the test can be run. Running scan at speed has additional complexity of time closure of the path at the clock source to Test start points. Running at reduced supply gives the advantage of reduced power so that one can even speed up the testing thus reducing test time. Shannon's expansion and supply gating for test power reduction. The paper is organized as follows. Section 20.2 introduces Shannon's expansion and supply gating for test power reduction. First-level hold is discussed in Sect. 20.3. Testing at low-power supply is discussed in Sect. 20.4. Finally, the conclusions are drawn in Sect. 20.5.

20.2 Low-Power Testing Using Shannon's Based Synthesis

20.2.1 Shannon Expansion

Shannon expansion has been used in logic synthesis for logic simplification and optimization [12]. It partitions any Boolean expression into disjoint subexpressions as shown in Eqs. 20.1 and 20.2.

$$\begin{aligned} f(x_1,\ldots,x_i,\ldots x_n) &= x_i \cdot f(x_1,\ldots,x_i=1,\ldots,x_n) + \bar{x}_i \cdot f(x_1,\ldots,x_i=0,\ldots,x_n) \\ &= x_i \cdot CF_1 + \bar{x}_i \cdot CF_2 \end{aligned} \tag{20.1}$$

where $CF_1 = f(x_1,\ldots,x_i=1,\ldots,x_n)$, $CF_2 = f(x_1,\ldots,x_i=0,\ldots,x_n)$ where, x_i is called the control variable, and CF_1 and CF_2 are called cofactors. From the above expression, it is clear that depending on the state of the control variable (x_i), the computed output of only one of the cofactors (CF_1 or CF_2) is required at any given instant. The output of CF_1 and CF_2 are combined using a multiplexer (MUX), which is controlled by xi. If the Boolean expression f contains subexpressions independent of control variable xi, then we may also have a shared Co-factor (sCF). Shared cofactor performs active computation irrespective of the state of the control variable. The output of the MUX (which directs the output of the active cofactor) must be OR-ed (for a sum-of-product logic representation) with the output of the sCF to derive the final output. The overall circuit after Shannon expansion is shown in Fig. 20.2.

20.2.2 Dynamic Supply Gating Scheme Using Shannon-Based Synthesis

The expression in previous subsection implies that only one cofactor performs active computation, while the other cofactor does redundant computations and leaks at any given time instant. This provides an opportunity for gating the supply of the idle cofactors to reduce power due to redundant computations and leakage current. Shannon's theorem can be utilized to identify the active/idle sections of a circuit for dynamic supply gating (DSG). The proposed DSG scheme using

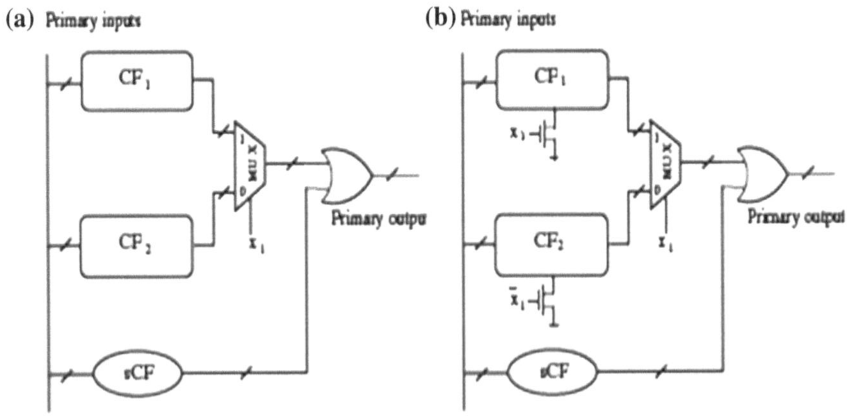

Fig. 20.2 Final circuit after Shannon's expansion (one level): **a** Basic idea, **b** After supply gating

Shannon's expansion is illustrated in Fig. 20.2b for one level of expansion. The supply gating transistors of sCF_1 and CF_2 are controlled by x_i and $\bar{x}_i$, respectively, where x_i is the control variable. This procedure can be extended hierarchically for multiple levels of expansion.

20.2.3 Selection of Control Variable

The methodology to select the control variable is detailed in [13]. A modified control variable selection method as shown below is used to balance the cofactor sizes and improve the test power reduction.

$$\begin{aligned} M_i =& \frac{a+b}{|a-b|} \quad \forall a! = b \\ =& a+b \quad \text{for } a = b \end{aligned} \tag{20.2}$$

where a(b) is the number of literals associated with $x_i(\bar{x}_i)$, and M_i is control variable selection metric.

An example of Shannon-based circuit synthesis and DSG for two-level expansion is illustrated in Fig. 20.3. After one-level expansion with respect to control variable x_i, the cofactors are CF1, CF2, and sCF (as shown by dotted rectangle). Second-level expansion of CF1 with respect to variable xj results in cofactors CF11, CF21, and sCF11. The cofactors CF11 (CF21) is gated with variables x_i and x_j (x_i and $\bar{\mathbf{x}}_\mathbf{j}$) where as sCF11 is gated only by x_i. The corresponding MUX and OR logic are also gated with xi to save active power. Similarly, CF2 is expanded with respect to variable $\mathrm{x_k}$ resulting in cofactors CF12, CF22, and sCF12. Again, cofactors CF12 (CF22) is gated with variables $\bar{\mathrm{x}}_i$ and x_k ($\bar{x}_i$and $\bar{\mathrm{x}}_k$), where as CF12 is gated by $\bar{\mathbf{x}}_i$. The corresponding MUX and OR logic are gated with $\bar{x}_1$. Finally, sCF is expanded with respect to x_l producing cofactors CF13, CF23 and sCF13. Note that CF13 and CF23 are gated by x_l, whereas sCF13 remains un-gated.

20.2.4 Sources of Test Power

There are mainly two components of test power in standard scan-based design namely, (a) power consumed in sequential elements, i.e., the scan flip-flops and, (b) power consumed in combinational circuit. The test power can be further decomposed into switching power and leakage power. In scan-based testing, around 78 % of total test energy is dissipated in the combinational block alone [9]. Hence, it is important to address the issue of power dissipation in the combinational block for low-power test application. The SBS technique can be very useful in reducing the both components (switching as well as standby) of test power in

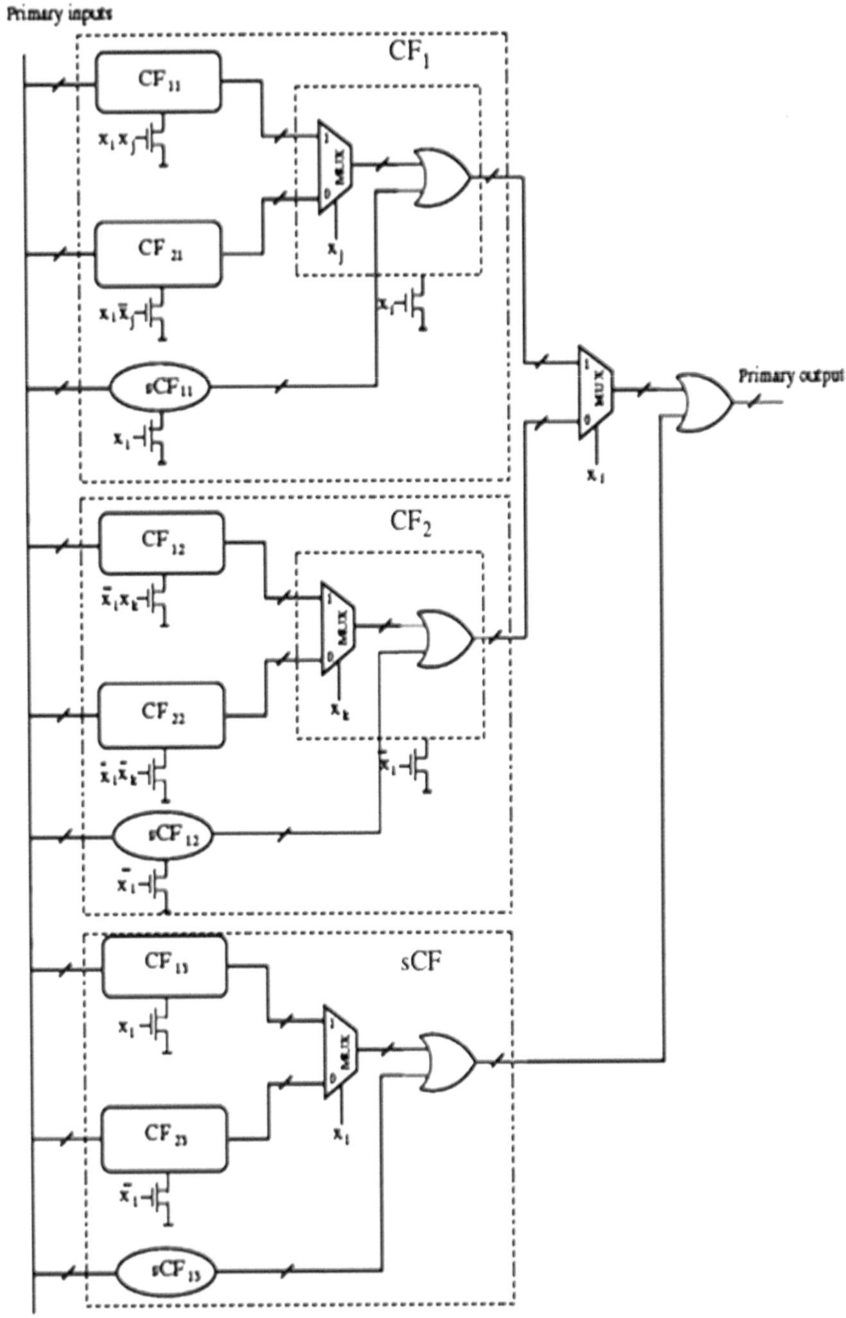

Fig. 20.3 Block diagram of a circuit after two-level Shannon's expansion

combinational block. This is again due to the structure of the circuits that inherently limits the switching in only one cofactor during each cycle of scan-shifting (and gates the other idle cofactors). SBS has significant advantage over other techniques for reducing power dissipation. For example, ATPG-based method needs redesigning of test vectors [5], while scan-latch or input vector reordering techniques can reduce the switching only at the outputs of scan registers, not in the combinational block. On the other hand, techniques proposed by [9] incorporates additional hardware that increases the area, delay, and power (in normal mode) during test synthesis. Additionally, note that test power improvement by Shannon decomposition based synthesis is significantly different from test power reduction technique proposed in [3] by scan partitioning. The advantages of SBS are that it does not require any change in the scan register and test application procedure. It can reduce both switching and leakage power. At-speed testing can be performed without any problem.

Unlike the technique in [3], SBS scheme cannot reduce power in the scan flip-flops and clock lines. However, the scan partitioning technique in [3] can be easily integrated with SBS to further improve the test power.

20.2.5 Results and Discussion

To observe the test power of ORG and SBS circuits, the circuits are modeled in Hspice with BPTM 70 nm technology. A set of random patterns are generated and applied in a manner such that it imitates the shifting of patterns in scan chain. The simulation is performed in Hspice and the power results are depicted in Table 20.1. It can be noted that as much as 74.1 % of test power can be saved with the proposed SBS technique. For most of the benchmarks, except x2, we observe significant test power saving. The diminished saving in x2 can again be attributed to the shared logic and MUXes, which switch all the time and reduces the savings obtained from gating the cofactors. The average improvement in test power over all the benchmarks is about 50.5 %.

Table 20.1 Improvement in test power (power in μw)

Circuit	P(ORG)	P(SBS)	% Improved
cht	51.5	27.5	46.60194175
CM150a	16.1	4.73	70.62111801
MUX	20.07	5.68	71.69905331
SCT	26.42	14.58	44.81453444
DECODE	12.22	9.33	23.6497545
ALU	121.7	65.38	46.27773213
COUNTER	121.1	31.25	74.19488026
PCIE	44.1	11.8	73.24263039
X2	13.01	12.56	3.458877786
Avg.			50.50672473

20.3 Test Power Reduction Using First-Level Supply Gating

Dynamic power dissipation in the combinational circuit can be reduced by lowering the switching activity of the circuit. In this section, we have described a first-level supply gating (FLS) [11] to reduce power dissipation in the combinational circuit during the scan-shifting.

Supply Gating Transistor for Reduced Switching Activity in Scan Mode Global supply gating can potentially prevent propagation of switching activity in the combinational block; however, it would result in considerable area and delay overhead due to larger gating transistor. To overcome this overhead, a novel FLS gating technique has been proposed in [11], where only the first-level logic gates connected to the scan flip-flops are gated using supply gating transistors (Fig. 20.4a). Insertion of the supply gating transistor in the first-level logic screens the rest of the combinational logic from the scan input transitions (except a 1 → 0 transition if NMOS supply gating is employed and 0 → 1 if PMOS supply gating is used). This is illustrated in Fig. 20.4b where the first transition at the input IN from "1" to "0" charges OUT1 to VDD. This transition propagates throughout the inverter chain. However, any further transition in the input (i.e., from "0" to "1") does not propagate, as the OUT1 cannot be discharged (Fig. 20.4b). This significantly reduces the redundant activity of the circuit during scan-shifting. The primary issue associated with FLS scheme is that the outputs of the first-level gates

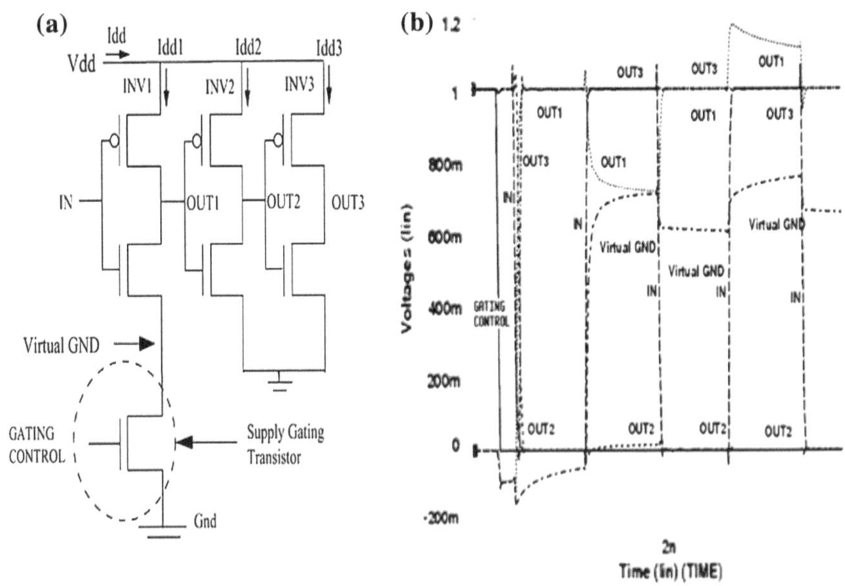

Fig. 20.4 **a** Use of FLS gating transistor in combinational part. **b** Transient response in 70 nm

are floating if they are at logic "0" (connected to the virtual ground). The voltage of a floated output is determined by the leakage between the pull-up PMOS and pull-down NMOS network of the gate. Further, the virtual ground is also susceptible to cross talk noise or transient effect due to soft error. If the voltage of the output of a first-level gate is not exactly at VDD or GND, this could result in short circuit current in the successive stages that are being driven by the first-level gate. This particularly becomes more of an issue in deep submicron technologies due to increased leakage and noise. In order to avoid this, the outputs of the first-level gates need to be forced at VDD or zero in the supply gating mode. If the GND is gated as in Fig. 20.5, then the outputs of the first-level gates can be forced to VDD by a pull-up PMOS driven by the Gating Control (GC) signal. If the VDD is gated, then the outputs of the first-level gates can be forced to ground using NMOS pull-down transistors driven by the GC signal. The general schemes of the proposed supply gating are shown in Fig. 20.5. In order to evaluate and compare these two schemes (Fig. 20.5a and b), they are applied to NAND and NOR gates. The pull-up (pull-down) transistor is kept at minimum size to optimize its impact on circuit delay and power during normal mode of operation.

20.3.1 Results

Table 20.2 compares the area, delay, and power overhead among several gating techniques, including FLS. We can observe from Table 20.2 that compared to latch and MUX-based gating, FLS is superior with respect to all three design parameters (area, delay, and power).

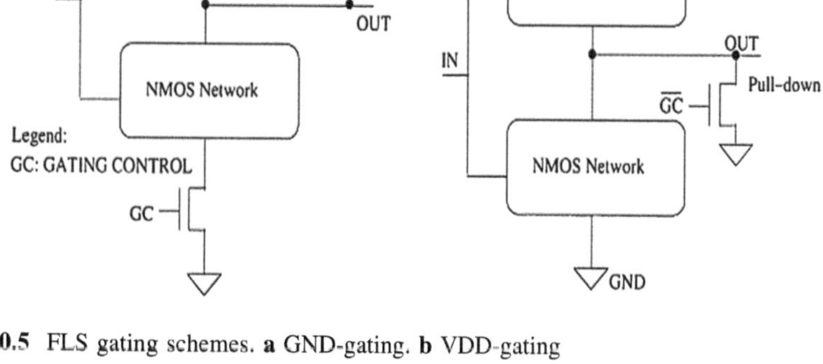

Fig. 20.5 FLS gating schemes. **a** GND-gating. **b** VDD-gating

Table 20.2 Comparison of area, delay, and power among alternative gating techniques applied to a single inverter

	Latch	MUX	NOR	FLS
Area (μm^2)	0.49	0.31	0.27	0.19
Delay (pS)	59.91	90.38	42.56	22.47
Power (μW)	22.23	19.37	12.4	7.93

20.4 Testing at Reduced Power Supply

Testing at low voltage in addition to reducing the test power consumption, makes interconnection bridging faults, and gate oxide shorts observable [14]. However, reducing power supply to near threshold voltage will increase delay by a large factor. Hence, it is essential to determine the optimal supply voltage at which the test can be run without affecting the delay performance. This optimum voltage id the voltage at which the test on CUT fails on a golden good sample. The failure here is because of the reduced voltage not able to charge the load to the fullest extent and hence results into critical path. The circuits is said to be structurally constrained at this point. This procedure is suitable for the initial wafer sort where results have shown reducing the test time to the extent of 50 % in addition to the obvious reduction in power savings.

20.5 Conclusions

We discussed two novel circuit methodologies to reduce the test power consumption in nano-scaled technologies. We described Shannon's expansion and supply gating and demonstrated that simple modification in the synthesis process can result in circuits that consume lower power (both switching and leakage) while being intrinsically more test able than designs produced by standard logic synthesis tools. Existing DFT techniques to improve test power can be easily combined with SBS to further lower the test power consumption. We discussed First-Level Supply gating methodology to reduce the scan power. FLS can be combined with input vector control to further reduce the test power. In addition, testing at reduced supply voltage is proposed to get dual advantage of reduction of power consumption and test time during initial testing.

References

1. Roy K, Mukhopadhyay S, Mahmoodi-Meimand H (2003) Leakage current mechanisms and leakage reduction techniques in deep-sub micrometer CMOS circuits. Proc IEEE 91.2:305–327
2. Zorian Y (1993) A distributed BIST control scheme for complex VLSI devices, VTS, pp 4–9
3. Whetsel L (2000) Adapting scan architectures for low power operation, ITC, pp 863–872

4. Meterelliyoz M, Mahmoodi H, and Roy K (2005) A leakage control system for thermal stability during burn-in test. In: Proceedings of ITC
5. Wang S et al (1998) ATPG for heat dissipation minimization during test application, TComp, pp 256–262
6. Dabholkar V et al (1998) Techniques for minimizing power dissipation in scan and combinational circuits during test application, IEEE TCAD, pp 1325–1333
7. Girard P et al (1998) Reducing power consumption during test application by test vector ordering, IEEE ISCS, pp 296–299
8. Sankaralingam R et al (2001) Reducing power dissipation during test using scan chain disable, VTS, pp 319–324
9. Gerstendrfer S et al (1999) Minimized power consumption for scan-based BIST, ITC, pp 77–84
10. Ghosh S, Bhunia S, Roy K (2005) Shannon expansion based supply-gated logic for improved power and testability. Test symposium, 2005. In: Proceedings of 14th Asian, IEEE
11. Bhunia S et al (2005) Low-power scan design using first-level supply gating. IEEE Trans Very Large Scale Integr (VLSI) Syst 13.3:384–395
12. Lavagnoet L et al (1995) Timed Shannon circuits: a power-efficient design style and synthesis tool, DAC, pp 254–260
13. Bhunia S et al (2005) A novel synthesis approach for active leakage power reduction using dynamic supply gating, DAC
14. Hao H, McCluskey EJ (1993) Very-low-voltage testing for weak CMOS logic ICs. In: Proceedings of international test conference, pp 275–284, October 1993

Part III
Communication

Chapter 21
Study of Continuous Nonlinear FM Pulse Compression Technique

K. Vijaya and K. N. Madhusudhan

Abstract A rectangular pulse-compression waveform having low-time side-lobes and zero mismatch loss is simulated. The waveform that is compressed with matched filter is continuous nonlinear frequency modulated (LFM), rectangular pulses. This is capable of achieving range side-lobe levels of better than−70 dB, suitable for use with satellite-borne precipitation radar.

Keywords Pulse compression · Matched filter · NFM · Range side lobes

21.1 Introduction

Pulse Compression techniques simultaneously provide maximum radar range and high resolution. Major disadvantage of this technique is time side-lobes associated with the compressed pulse. During pulse compression, undesirable time range side lobes are generated in addition to the main lobe representing the pulse. Side lobes hide weaker but important target returns. Also, any clutter in side lobes may leak into range cell of interest time side lobes associated with compressed pulse are usually minimized by applying spectral weighting in the radar receiver. It has been found that compressed pulse of nonlinear FM (NFM) has much lower side lobes than linear FM (LFM) and nonlinear FM codes usually do not require weighting. An attempt is made here to design a continuous nonlinear FM waveform and compress it with matched filter for achieving side lobe levels of −70 dB with good

K. Vijaya · K. N. Madhusudhan (✉)
Department of ECE, BMS College of Engineering, Bangalore, India
e-mail: madhusudhankn.ece@bmsce.ac.in

K. Vijaya
e-mail: vijaya.ece@bmsce.ac.in

V. Chakravarthi et al. (eds.), *Proceedings of International Conference on VLSI, Communication, Advanced Devices, Signals & Systems and Networking (VCASAN-2013)*, Lecture Notes in Electrical Engineering 258, DOI: 10.1007/978-81-322-1524-0_21,

Doppler tolerance as this may find many applications in modern Radar systems such as satellite borne precipitation Radar and other tracking systems.

The first major issue is design of continuous nonlinear FM waveform. A continuous NFM waveform is first designed and simulated in MATLAB. Next is Pulse compression. Pulse compression is achieved using matched filter. Pulse compression techniques used in radars normally produce compressed pulse with precursors and followers of reduced amplitude called range time side lobes. These side lobes are objectionable because a time side lobe of a strong echo may hide a weaker, though more important target return. Hence, reducing time side lobes is essential. Here, side lobe levels better than −70 dB are to be achieved.

The present work is to study ultra low side lobe nonlinear FM pulse compression technique for radars.

21.2 Linear FM Wave Form

One of the popular Pulse compression waveform is Linear Frequency Modulated (LFM) pulse or the chirp pulse (Fig. 21.1).

An LFM chirp waveform has nearly a rectangular Power Spectral Density (PSD), its autocorrelation function (ACF) exhibits a sinc() function shape, with its undesirable side lobes Generation of LFM is simple and compressed pulse shape and Signal-to-Noise Ratio (SNR) are fairly insensitive to doppler shifts. But a shift in doppler causes an apparent change in range and vice versa. Hence, either range or doppler must be known or determined. Weighting is usually required to reduce the time side lobes of the compressed pulse to an acceptable level.

21.3 NFM Wave Form

The nonlinear waveform is most useful in a tracking system where range and Doppler are approximately known. NFM waveforms have quite good tolerance to Doppler shifts, i.e., maintaining their desirable side lobe properties when Doppler shifted. Nonlinear FM waveforms offer a radar matched filter output with inherently low range side lobes. This yields a 1–2 dB advantage in Signal-to-Noise Ratio over the output of a Linear FM waveform with equivalent side lobe filtering.

Design of s NFM: LFM waveform needs a constant but nonzero chirp rate, a NFM waveform needs a nonconstant chirp rate. Therefore, some mechanism for adjusting chirp rate as a function of time is required. Since instantaneous frequency is also a function of time, the chirp rate could be effectively adjusted as some function of instantaneous frequency. The chirp-rate generating function may be either continuous or discontinuous; here, the design is based on a continuous nonlinear FM function [1]. The waveform is simulated in MATLAB.

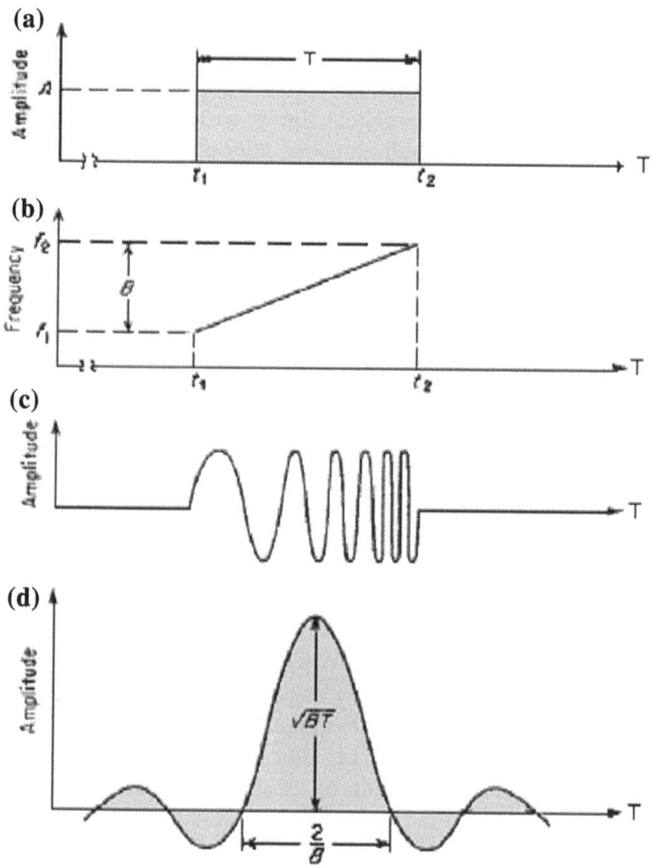

Fig. 21.1 **a** Transmitted waveform. **b** Frequency of the transmitted waveform. **c** Representation. **d** Output of the pulse-compression filter

21.4 Pulse Compression

Radar range resolution depends on the bandwidth of the received signal

$$\rho = \frac{c\tau}{2} = \frac{c}{2B} \qquad \begin{aligned} c &= \text{speed of the light}, \rho = \text{range resolution}, \\ \tau &= \text{pulse duration}, \text{B} = \text{signal bandwidth} \end{aligned}$$

The bandwidth of a time-gated sinusoid is inversely proportional to the pulse duration. Hence, short pulses are better for range resolution. Received signal strength is proportional to the pulse duration. Hence, long pulses are better for signal reception. High-power transmitters require high-voltage power supplies (kV) and present problems such as Reliability and Safety issues. In addition, they are larger, heavier, and costlier. Pulse compression is the compromise which enables Transmit a long pulse that has a bandwidth corresponding to a short pulse.

Energy content of long-duration, low-power pulse will be comparable to that of the short-duration, high-power pulse. The transmitted pulse must be modulated or coded to have sufficient bandwidth, B and can be processed to provide the desired range resolution, ρ. Pulse compression allows us to use a reduced transmitter power and still achieve the desired range resolution. The costs of applying pulse compression include added transmitter and receiver complexity and associated time side lobes. The advantages generally outweigh the disadvantages so pulse compression is used widely.

MATCHED Filter: The most unique characteristic of the matched filter is that it produces Maximum achievable instantaneous SNR at its output when a signal plus additive noise are present at the input. Noise need not be Gaussian. Peak instantaneous power divided by average noise power at the output of matched filter is equal to twice the input signal energy divided by the input noise power regardless of the waveform used by the RADAR. This is why matched Filter is referred as optimum filter in SNR sense. Given transmitter pulse shape $g(t)$ of duration T, matched filter is given by $h(t) = k\, g^*(T - t)$, 1 k. is scaling factor. Duration and shape of impulse response of the optimal filter is determined by pulse shape $g(t)$. $h(t)$ is scaled, time-reversed, and shifted version of $g(t)$. A filter that is matched to a signal **g(t)** of duration **T**, has an impulse response that is a time-reversed and delayed version of the input **g(t)**.

When a signal is input to a Matched Filter (matched to the input signal) then the output of the filter is the autocorrelation function of the signal. The autocorrelation function is the Fourier Transform of the signal's Power.

Spectral Density (PSD). A Matched Filter provides optimum (maximum) Signal-to-Noise Ratio (SNR) at the peak of its autocorrelation function and is consequently optimum for detecting the signal in noise.

NFM Signal can be generated in MATLAB as an array. The signal values can be stored in a HEX file so that it can be processed by an HDL code. An HDL code can be written for matched filter implementation.

21.5 Results and Conclusion

The study of NFM pulse compression technique reveals that the continuous NFM waveform designed produces time side lobe levels of better than −70 dB. Non-Linear FM (NLFM) chirp modulation can advantageously shape the PSD such that the autocorrelation function exhibits substantially reduced side lobes in comparison with LFM. Consequently, no additional filtering is required and maximum SNR performance. However, precision NFM chirps are more difficult to design, produce, and process. The disadvantages of the nonlinear/FM waveform are greater system complexity, limited development of nonlinear/FM generation devices, and the necessity for a separate FM modulation design for each amplitude spectrum to achieve the required side lobe level (Fig. 21.2).

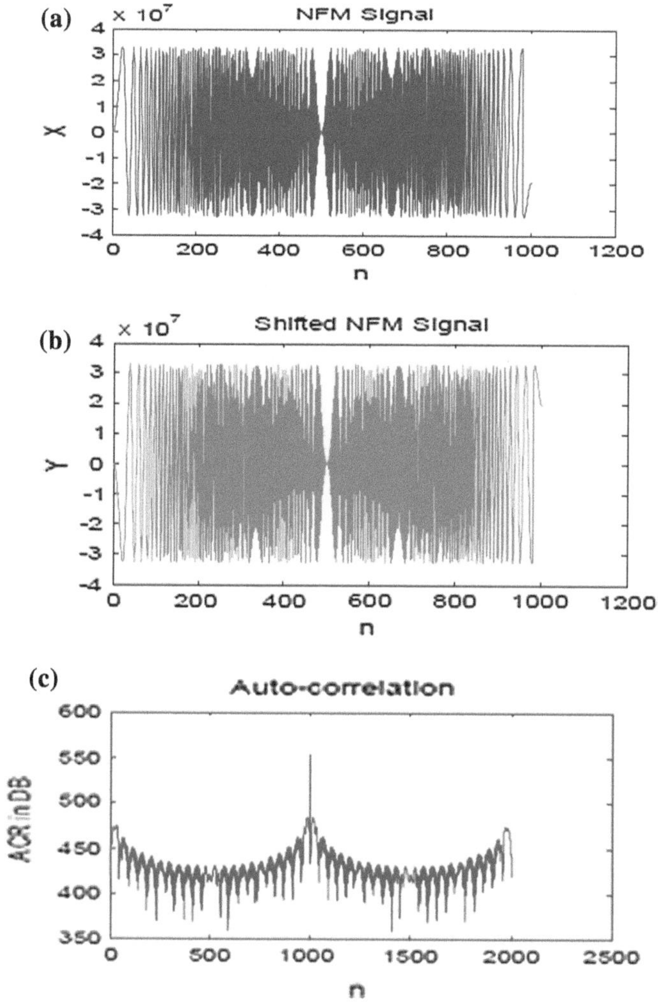

Fig. 21.2 Pulse compression; Simulation using MATLAB. **a** Nonlinear FM signal. **b** Complement of NFM signal. **c** MATLAB simulation of ACF of NFM

Reference

1. De Witte E, Griffiths HD (2004) Improved ultra-low range sidelobe pulse compression waveform design. Electron Lett 40(22):1448–1450

Chapter 22
Performance Evaluation of New Multilevel Spreading Codes for Synchronous DS-CDMA Communication

K. Usha and K. Jaya Sankar

Abstract Synchronous Code Division Multiple Access (CDMA) systems possess the advantages of efficient use of the spectral band width, resistance to co-channel interference and adaptability to variable traffic patterns. These advantages result from the usage of orthogonal spreading codes. The popularly used binary (2-level) spreading codes are Walsh, Gold, and Kasami codes. This paper proposes new multilevel (ML) orthogonal spreading codes constructed using ternary and quaternary Gray and Inverse Gray codes for multi-user Direct Sequence Code Division Multiple Access systems. The methodology explained in this paper allows to construct r-level 2n-length user codes. Multilevel spreading codes discussed in this paper are nonzero mean, varying power codes. An attempt is made to analyze these multilevel user codes through auto- and cross-correlation properties and bit error rate. Bit error rate performance of the proposed codes over Gaussian channel and their comparison with those of Walsh and Gold codes is presented in this paper. The proposed codes are found to be more suitable than the existing binary spreading codes.

Keywords Generalized gray codes · Inverse gray codes · Multilevel spreading codes · Auto- and cross-correlation metrics · DS-CDMA communication · AWGN channel

K. Usha (✉)
Department of ECE, MVSR Engineering College, Hyderabad, India
e-mail: u_kamle@yahoo.com

K. Jaya Sankar
Department of ECE, Vasavi College of Engineering, Hyderabad, India
e-mail: kottareddyjs@gmail.com

V. Chakravarthi et al. (eds.), *Proceedings of International Conference on VLSI, Communication, Advanced Devices, Signals & Systems and Networking (VCASAN-2013)*, Lecture Notes in Electrical Engineering 258, DOI: 10.1007/978-81-322-1524-0_22,

22.1 Introduction

Code Division Multiple Access (CDMA) is an important class of multiple access techniques that allows sharing of spectrum resources simultaneously by large number of system users. The Code Division Multiple Access (CDMA) technique is widely used in the existing wireless communication systems such as W-CDMA and CDMA2000. In multi-user CDMA systems, spreading codes that are used to distinguish users and spread the signal, play an important role. Binary valued Walsh, Gold, and Kasami codes are widely used as spreading codes in wireless CDMA systems. Spreading codes are broadly categorized into orthogonal and nonorthogonal codes. Walsh codes are perfectly orthogonal binary codes and are ideal for synchronous CDMA communications [1].

Different types of spreading sequences exhibit different properties, such as code length, code set size, and auto- and cross-correlation values. Most of the available code sets yield degraded performance in multi-user conditions, especially when the system is heavily loaded and the channel is dispersive. Binary code sets have the chip levels (+1, −1) and generate constant envelope or fixed power modulating signal, which enables efficient use of the available RF power in CDMA [2]. While this is true in the case of single user scenario, in a multi-user case, this is not true when multiple binary signals are added together. Amplifiers usually operate as a linear device under small signal conditions and become more nonlinear and distorted with increase in input drive level. Increasing the input drive signal in turn increases the efficiency of the RF amplifier and thus increases the total transmitted power. Current research on RF amplifier design is aimed at increasing the linear range of RF amplifiers with lower distortion levels and higher efficiency. With such technological advances, implementation of varying power codes is becoming feasible for wireless and radio communications. The limitations of binary code sets discussed above lead to the design of new multilevel (ML) (varying power) user codes.

There are few papers which focus their discussion on the design of multilevel user codes in the recent past. In [3] a method for the construction of multilevel (ML), Hadamard matrices are proposed. Usage of multilevel integer valued orthogonal codes for CDMA communications is initially proposed Poluri [4], wherein an exhaustive search scheme to construct N-length multilevel user codes is discussed. In this method, the multilevel representations of integers from 1 to 2^{N-1} are checked for orthogonality to obtain a code set. This method is proved to be tedious as the length of the code increases. Multilevel integer valued orthogonal code sets constructed by Poluri [4] are shown in Table 22.1.

There are few works addressing the design methods for multiple level user codes and their application to MC-CDMA systems [5, 6].

Gray code, a unit distance code of n-bit, is defined as a list of all 2^n bit strings such that successive code words differ in only one bit position [7]. An algorithm to construct generalized Gray codes is discussed in [8]. An “n” bit Inverse Gray Code is defined exactly opposite to Gray code, and it is a list of all 2^n bit strings of

Table 22.1 Multilevel integer-valued orthogonal codes proposed by Poluri [4]

3-level, 6-length codes (RP 6)						4-level, 8-length codes (RP 8)							
−1	−1	−1	1	0	0	−3	−3	−3	−1	−1	−3	1	−1
0	0	1	1	−1	−1	−3	−1	1	3	−1	3	3	1
0	0	1	1	1	1	−3	1	3	−3	−1	1	−1	−3
1	−1	0	0	−1	1	−3	3	−1	1	−1	−1	−3	3
1	−1	0	0	1	−1	−1	−3	3	−1	3	−1	−1	3
1	1	−1	1	0	0	−1	−1	−1	3	3	1	−3	−3
						−1	1	−3	−3	3	3	1	1
						−1	3	1	1	3	−3	3	−1

length "n" each, such that successive code words differ in (n − 1) bit positions [9, 10]. By concatenating binary Gray and Inverse Gray codes, binary user codes are constructed. The procedure for constructing the codes and the performance of these codes have been reported earlier [11]. A similar technique for the construction of multilevel user codes using ternary and quaternary Gray and Inverse Gray codes is proposed in the present work.

Rest of the paper is organized in the following manner: Section 22.2 discusses the proposed technique for the construction of 3-level and 4-level user code sets from n-git ternary and quaternary Gray and Inverse Gray codes, respectively. In Sect. 22.3, the auto- and cross-correlation properties of the proposed user codes are analyzed. Performance analysis of the proposed codes over AWGN channel is presented in Sect. 22.4. And finally, Sect. 22.5 gives the conclusions.

22.2 MultiLevel Spreading Codes

In multi-user Direct Sequence CDMA systems, user codes or spreading sequences are used to distinguish users and spread signals. Spreading codes are also called as user codes. Binary user codes (2-level) are designed by mapping radix-2 elements {0, 1} to chip amplitude levels {−1, 1}. Similar procedure is adopted to construct multilevel (ML) (>2) orthogonal codes. In ML user codes, discrete amplitude levels, similar to pulse amplitude modulation (PAM) levels, are used as chip signals for spread spectrum codes. Chip amplitudes are chosen such that they have zero mean. In order to minimize the average transmitted energy and to obtain M signal amplitude levels symmetric about zero and equally spaced, the following formula is used.

$$A_m = (2m - 1 - M),\ m = 1, 2, \ldots, M.$$

For example, for a 3-level coding, radix-3 (ternary) coding elements {0, 1, 2} are mapped to chip amplitudes levels {−1, 0, 1} and for a 4-level coding, chip amplitudes levels {−3, −1, 1, 3} are obtained by mapping the coding elements of quaternary (radix-4) number system {0, 1, 2, 3}. The procedure to construct r-level, 2n-length orthogonal spreading code sets is summarized as follows:

Step 1: Generate n-git (generalized digit) ternary or quaternary Gray code using the algorithm discussed in [8] with any permutation.
Step 2: Using the same permutation generate n-git Inverse Gray code [10].
Step 3: Append Inverse Gray Code to the Gray Code to result in 2n-length Gray Inverse Gray (GIG) code. This GIG code comprises of "r^n" code words (r = 3 for ternary and r = 4 for quaternary) of 2n-length each.
Step 4: Each row of this 2n-length GIG code is mapped to the "r" chip amplitude levels to construct multilevel code set.
Step 5: Select any one codeword as the first basis function of the orthogonal code set. A next code word is added to this orthogonal code set by checking the orthogonality with this first basis function.
Step 6: This search process is continued with remaining $r^n - 1$ code words to obtain m-level, 2n-length orthogonal spreading code set.

This procedure is repeated by selecting a different codeword as the first basis function to obtain another user code set. Number of unique code sets can be obtained using this procedure. Table 22.2 displays the construction of 3-level, 6-length GIG code set using the permutation {3, 2, 1}. Similarly, 4-level 8-length user code sets can be obtained using Quaternary GIG codes. Table 22.3 displays one such 3-level, 6-length user code set and 4-level 8-length user code sets along with their decimal representations.

Weights of the coding elements in radix-3 for n-length code are $\{3^{n-1}, 3^{n-2}, \ldots, 3^1, 3^0\}$ and similarly in radix-4, element weights are $\{4^{n-1}, 4^{n-2}, \ldots, 4^1, 4^0\}$. For example, a 3-level, 6-length ternary code {1, 2, 0, 2, 1, 0} is equivalent to $\{3^5.1 + 3^4.2 + 3^3.0 + 3^2.2 + 3^1.1 + 3^0.0\} = 426$ in decimal notation. Considering the 6 and 8—length code sets as basic sets, code sets of greater lengths (12, 16, 24, 32...) which are even multiples of lengths 6 and 8 can be constructed recursively using the following relationship (M = 6 or 8).

$$C_{2M} = \begin{bmatrix} C_M \, C_M \\ C_M \, \overline{C_M} \end{bmatrix}$$

22.3 Auto- and Cross-Correlation Properties

The performance of spreading codes is evaluated by various auto- and cross-correlation metrics. The ability of a DS-CDMA receiver to detect the desired signal relies to a great extent on the autocorrelation properties of the spreading codes and on the other hand multi-user interference rejection depends on cross-correlation properties of the spreading sequences. 'In synchronous DS-CDMA system, the code sequence in the receiver is exactly same with that in the

Table 22.2 Construction of 3-level, 6-length gray inverse gray (GIG) codes

3-git ternary gray code			3-git ternary inverse gray code			3-git ternary GIG code						{0,1,2} mapped to {−1,0,1}					
0	0	0	0	0	0	0	0	0	0	0	0	−1	−1	−1	−1	−1	−1
1	0	0	0	1	1	1	0	0	0	1	1	0	−1	−1	−1	0	0
2	0	0	0	2	2	2	0	0	0	2	2	1	−1	−1	−1	1	1
2	1	0	1	2	0	2	1	0	1	2	0	1	0	−1	0	1	−1
0	0	1	1	0	1	0	1	0	1	0	1	−1	0	−1	0	−1	0
1	0	1	1	1	2	1	1	0	1	1	2	0	0	−1	0	0	1
1	0	2	2	1	0	1	2	0	2	1	0	0	1	−1	1	0	−1
2	0	2	2	2	1	2	2	0	2	2	1	1	1	−1	1	1	0
0	0	2	2	0	2	0	2	0	2	0	2	−1	1	−1	1	−1	1
0	2	1	0	1	2	0	2	1	0	1	2	−1	1	0	−1	0	1
1	2	1	0	2	0	1	2	1	0	2	0	0	1	0	−1	1	−1
2	2	1	0	0	1	2	2	1	0	0	1	1	1	0	−1	−1	0
2	0	1	1	0	2	2	0	1	1	0	1	1	−1	0	0	−1	1
0	0	1	1	1	0	0	0	1	1	1	0	−1	−1	0	0	0	−1
1	0	1	1	2	1	1	0	1	1	2	1	0	−1	0	0	1	0
1	1	1	2	2	2	1	1	1	2	2	2	0	0	0	1	1	1
2	1	1	2	0	0	2	1	1	2	0	0	1	0	0	1	−1	−1
0	1	1	2	1	1	0	1	1	2	1	1	−1	0	0	1	0	0
0	1	2	0	2	1	0	1	2	0	2	1	−1	0	1	−1	1	0
1	1	2	0	0	2	1	1	2	0	0	2	0	0	1	−1	−1	1
2	1	2	0	1	0	2	1	2	0	1	0	1	0	1	−1	0	−1
2	2	2	1	1	1	2	2	2	1	1	1	1	1	1	0	0	0
0	2	2	1	2	2	0	2	2	1	2	2	−1	1	1	0	1	1
1	2	2	1	0	0	1	2	2	1	0	0	0	1	1	0	−1	−1
1	0	2	2	0	1	1	0	2	2	0	1	0	−1	1	1	−1	0
2	0	2	2	1	2	2	0	2	2	1	2	1	−1	1	1	0	1
0	0	2	2	2	0	0	0	2	2	2	0	−1	−1	1	1	1	−1

transmitter. Orthogonal codes are most suitable for synchronous communication. Two sequences are said to be Orthogonal when the Cross-correlation (intercode correlation) between them is zero. Codes with high autocorrelation and low Cross-correlation are preferred in synchronous communication. Walsh Codes are perfectly orthogonal, fixed power, binary user codes. In this paper, the performance of different spreading sequences is evaluated by maximum, mean, and squared sum of correlation values for even, odd, and aperiodic cross-correlations, Mean Square Aperiodic AutoCorrelation (MSAAC), Mean Square Aperiodic Cross-Correlation (MSACC), and Figure of Merit (FOM) [12–14]. If x(k) and x(k + m) represent the nondelayed and delayed versions of a code word then "m" is the number of units by which a code word is delayed. "N" is the length of a code word x. Some of the correlation metrics and merit factor are defined as follows:

Table 22.3 Sample sets of proposed 3-level, 6-length and 4-level, 8-length spreading codes

3-level, 6-length orthogonal code set		4-level, 8-length orthogonal code set	
Decimal equivalent	Basis elements {−1, 0, 1} code words	Decimal equivalent	Basis elements {−3, −1, 1, 3} code words
0	−1 −1 −1 −1 −1 −1	0	−3 −3 −3 −3 −3 −3 −3 −3
494	1 −1 −1 −1 1 1	49,215	3 −3 −3 −3 −3 3 3 3
426	0 1 −1 1 0 −1	45,285	1 3 −3 −3 3 1 −1 −1
194	−1 1 0 −1 0 1	11,385	−3 1 3 −3 −1 3 1 −1
624	1 0 1 −1 0 −1	43,350	1 1 1 −1 −1 −1 −1 1
316	0 −1 1 1 −1 0	18,891	−1 −3 1 −1 3 −3 1 3
		39,525	1 −1 1 1 −1 1 −1 −1
		43,690	1 1 1 1 1 1 1 1
		27,285	−1 1 1 1 1 −1 −1 −1
		42,585	1 1 −1 1 −1 −1 1 −1

The discrete aperiodic Correlation is defined as

$$C_{x,y}(m) = \frac{1}{N}\sum_{m=1-N}^{N-1} x(n)y(n+m)$$

The Mean Square Aperiodic AutoCorrelation (MSAAC) for a Code set containing M sequences is given by

$$R_{AC} = \frac{1}{M}\sum_{x=1}^{M}\sum_{m=1-N,\, m\neq 0}^{N-1} \left|C_{x,x}(m)\right|^2$$

Mean Square Aperiodic Cross-Correlation (MSACC) is defined as

$$R_{CC} = \frac{1}{M(M-1)}\sum_{x=1}^{M}\sum_{y=1,\, y\neq x}^{M}\sum_{m=1-N}^{N-1} \left|C_{x,y}(m)\right|^2$$

Figure of Merit (FOM)

$$FOM = \frac{C_{x,x}^2(0)}{\sum\limits_{m\neq 0}\left|C_{x,x}(m)\right|^2} = \frac{N^2}{2\cdot\sum\limits_{m=1}^{N-1}\left|C_{x,x}(m)\right|^2}$$

Even Cross-Correlation $\theta_{x,y}(m) = C_{x,y}(m) + C_{x,y}(m-N)$

Odd Cross-Correlation $\widehat{\theta}_{x,y}(m) = C_{x,y}(m) - C_{x,y}(m-N)$

Multilevel integer-valued orthogonal codes proposed by Poluri [4] are considered for comparison along with Walsh and Gold codes. The auto- and cross-correlation properties of the proposed 3-level, 6-length and 4-level, 8-length user code sets and nearer length Walsh (length 8), Gold (length 7) and Poluri codes are displayed in Table 22.4. Codes with high autocorrelation and low cross-correlation

Table 22.4 Auto- and cross-correlation metrics for popular and proposed codes

Parameter	W_8	G_7	Multilevel integer-valued codes by Poluri		Proposed multilevel orthogonal codes	
			3-level, 6-length (RP6)	4-level, 8-length (RP8)	3-level, 6-length	4-level, 8-length
MSAAC	2.375	0.9388	0.9167	1.045	1.2731	2.1186
MSACC	0.6607	0.8503	0.8167	0.8507	0.7648	0.8228
MAXEC	1	0.4286	0.75	0.7	1	1
MAXOC	1	0.7143	0.75	0.75	0.8165	0.85
MAXAC	0.875	0.5714	0.75	0.675	0.75	0.8944
MEANEC	0.0977	0.3024	0.2616	0.2467	0.2322	0.1829
MEANOC	0.2129	0.2641	0.2778	0.2512	0.2605	0.2535
MEANAC	0.138	0.1991	0.1692	0.1807	0.1888	0.1671
SSEC	14	19	11	11	12	19
SSOC	23	17	12	12	12	19
SSAC	18.5	18	12	12	12	19
FOM	0.4211	1.0652	1.0909	0.9569	0.7854	0.472

are preferable in synchronous multi-user communications. From Table 22.4, it can be observed that the proposed codes have correlation properties better than Gold and very close in their values to nearer length Walsh codes.

22.4 Bit Error Rate Performance Analysis

In DS-CDMA systems, every data bit is directly multiplied with a spreading code. The resulting signal modulates the radio frequency (RF) carrier. Spreading signal consists of a number of code bits called chips. Ratio of the spreading signal chip rate to the original data bit rate is called the spreading gain or processing gain. In DS-CDMA systems, all users use the same transmitting carrier and can simultaneously transmit their data. Receiver employs coherent demodulator to de-spread the received data using a locally generated code sequence. To be able to perform the despreading, receiver needs to know not only the code sequence that was used in the transmitter to spread the data, but also to synchronize the received signal with the locally generated code. After despreading and detection, original data bits are recovered.

Due to simultaneous transmissions of multiple users over the same channel, signal detection at the receiver for each user is limited by the interference caused by number of users, cross-correlation values among the user codes, delayed versions of the code of interest due to multipath, transmitted power levels of other users as well as their timing synchronization. Utilization of user codes for data spreading in a synchronized system eliminates Multiple Access Interference. Bit Error Rate (BER) performance is carried out for synchronous communication over Additive White Gaussian Noise (AWGN) channel for 2-user DS-CDMA system. BER performances of proposed 3-level, 6-length and 4-level, 8-length user code

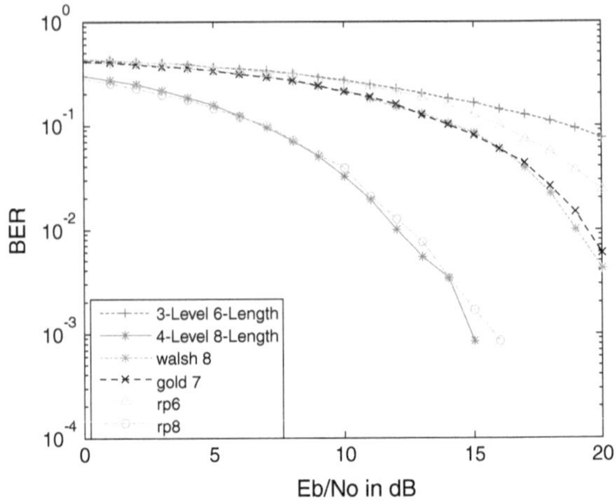

Fig. 22.1 BER performance of Walsh, Gold, Poluri codes and proposed codes for synchronous DS-CDMA communication over multipath AWGN channel

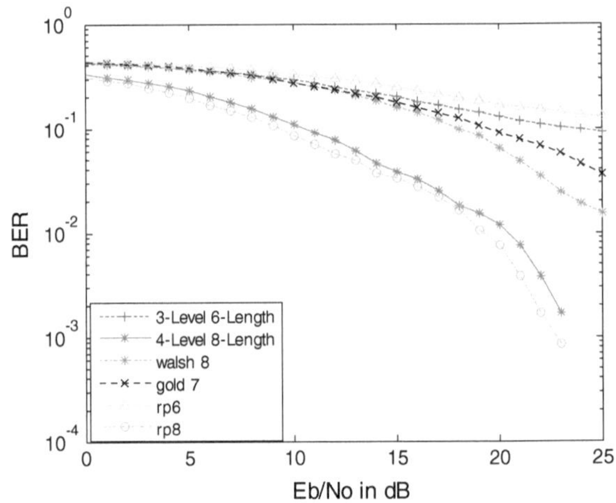

Fig. 22.2 BER performance of the Walsh, Gold, Poluri codes and proposed codes for two user synchronous DS-CDMA communication over AWGN channel

sets along with Walsh, Gold, and Multilevel Poluri codes is displayed in Fig. 22.1. BER performance of these codes for DS-CDMA Communication over Multiple (three) paths is displayed in Fig. 22.2. And it can be concluded from the figures that 4-level, 8-length user codes have better performance over all other codes.

22.5 Conclusions

Performance analysis of ML user codes constructed using ternary and quaternary GIG codes for direct sequence CDMA communication is proposed in this work. The design methodology proposed allows to construct r-level, 2n-length user code sets. Using ternary (r = 3) Gray and Inverse Gray codes 3-level, 6-length user code sets are obtained. Quaternary (r = 4) GIG codes result in 4-level, 8-length code sets. In this paper, construction of ML spreading codes is limited to 3- and 4-levels only. The higher length code sets (12, 16, 24, 32 … etc.) can be constructed recursively using these basic sets. Spreading code sets of higher levels (r = 5, 6, …) also can be obtained using the discussed construction procedure. Auto- and cross-correlation properties reveal that the proposed codes are superior to Gold codes and are competent with Walsh codes. Performance analysis of the Walsh, Gold and the proposed codes for 2-user DS-CDMA system over AWGN channel for synchronous communication is carried out. From the Bit Error Rate performance, it can be observed that the proposed codes outperform the existing binary user codes and are more suitable for synchronous communication and are superior in their performance compared to binary spreading codes.

References

1. Walsh JL (1923) A closed set of normal orthogonal functions. Am J Math 45(1):5–24
2. Liang CP et al (1999) Nonlinear amplifier effects in communications systems. IEEE Trans Microw Theory Tech 47(8):1461
3. Trinh QK, Fan P, Gabidulin EM (2006) Multi level Hadamard matrices and zero correlation zone sequences. IEE Electron Lett 42(13):748
4. Poluri R (2007) Design and performance of CDMA codes for multiuser communications. Ph.D thesis, New Jersey's Science and Technology University
5. Zhang J, Matolak DW (2009) Multiple level orthogonal codes and their application in MC-CDMA systems. J Comput Commun 32:492–500
6. Qiu X, Sha X, Ning X (2010) A multi-level orthogonal sequences design and implementation method. In: Proceedings of IEEE ICSP 2010, pp 1793–179
7. Gray F (1953) Pulse code communication. US patent #2632058, 17th Mar 1953
8. Sankar KJ, Pandharipande VM, Moharir PS (2004) Generalized gray codes, intelligent signal processing and communication systems. In: Proceedings of ISPACS'04, pp 654–659
9. Pham DM, Premkumar AB, Madhukumar AS (2011) Error detection and correction in communication channels using inverse gray RSNS codes. IEEE Trans Commun 59(4):975–986
10. Usha K, Jaya sankar K (2012) A technique for the construction of inverse gray codes. Int J Emerg Trends Eng Dev 2(2):337–341
11. Usha K (2012) Binary orthogonal code generation for multi user communication using n-bit gray and inverse gray codes. Int J Electron Commun Eng 5(2):165–174
12. Sarwate DV, Pursley MB (1980) Correlation properties of pseudo random and related sequences. Proc IEEE 68(5):593–619
13. Abu-Rgheff MA (2005) Introduction to CDMA wireless communications. Elsevier
14. Anil Kumar V, Mitra A, Prasanna SRM (2008) On the Effectivity of different pseudo-noise and orthogonal sequences for speech encryption from correlation properties. World Acad Sci Eng Technol 48:684–691

Chapter 23
A Novel Method for Construction of Structured Regular LDPC Codes with Girth Twelve Using Gray Code Representations

Vibha Kulkarni and K. Jaya Sankar

Abstract The Low-Density Parity-Check (LDPC) code is a linear block code specified by a parity-check matrix H. The construction process of LDPC codes considers the parameters such as row and column-weights, rate, girth, and code length. Regularity of the codes provides the advantage like simplicity of hardware implementation and fast encoding. Large girth speeds the convergence of iterative decoding and improves performance. To increase girth of a code and avoid short cycles, the parity-check matrix H must be sufficiently sparse, and hence, the block length must be large. Hence, there is a need to develop method of constructing LDPC codes over a wide range of lengths, rates, and girths. In this paper, we introduce a novel method for constructing structured regular LDPC codes with different densities, rates, lengths, and girths using Gray-code representations. The advantage of this algorithm compared to other methods is its flexibility in terms of rates, lengths, and girths. LDPC codes with column-weight two have low computational complexity and are promising for data storage and partial response channels. Hence, a modified version of proposed algorithm, to construct a column-weight two LDPC codes with girth 8 and 12, is also presented in this paper.

Keywords LDPC codes · Tanner graph · Girth · Gray codes

V. Kulkarni (✉) · K. Jaya Sankar
Department of ECE, Vasavi College of Engineering, Hyderabad, India
e-mail: vibhadk@gmail.com

K. Jaya Sankar
e-mail: kottareddyjs@gmail.com

V. Chakravarthi et al. (eds.), *Proceedings of International Conference on VLSI, Communication, Advanced Devices, Signals & Systems and Networking (VCASAN-2013)*, Lecture Notes in Electrical Engineering 258, DOI: 10.1007/978-81-322-1524-0_23,

23.1 Introduction

LDPC codes are originally developed by Gallager in 1960s [1, 2]. They were largely ignored for a long time. In last decade, researchers began to investigate on these codes. These can be applied to wireless, wired and optical communication systems, storage applications such as magnetic disks and compact disks. Wireless applications require low-power implementations with rates in Mbps. Storage applications require about 1 Gbps and high rate codes [3], while optical communication throughput can be above 10 Gbps [4]. LDPC codes are recommended in some communication systems such as digital video broadcasting (DVB-2) [5].

LDPC code construction requires the definite pattern in connecting rows and columns of a parity-check matrix. Construction methods can be either random (unstructured row-column connections) or structured (row-column connections predefined in some way). Random constructions have flexibility in design and construction, but lack in regularity of row-column connections, which increases decoder interconnection complexity. Structured constructions have regular interconnection patterns, but produce codes limited in rate, length, and girth. Structured connections reduce hardware complexity and the cost of encoders and decoders [6, 7, 8]. There is a need to develop methods that can produce a wide range (rate, length, and girth) of LDPC codes. In this paper, we propose a novel method of constructing structured LDPC codes with flexibility in rates, lengths, and girths.

Rest of the paper is organized as follows. Gallager's LDPC code construction method is discussed in Sect. 23.2. The proposed algorithm to construct LDPC codes, based on Gray-code representations, is presented in Sect. 23.3. Column-weight two LDPC codes, their applications and method of constructing these codes with girth eight are discussed in Sect. 23.4. Section 23.5 gives a construction method of column-weight two LDPC codes with girth 12. Section 23.6 concludes the paper.

23.2 Preliminaries

23.2.1 Gallager's Method of Constructing LDPC Codes

An LDPC code is defined as the null space of a parity-check matrix H that has structural properties, (1) each row consisting of "ρ" number of 1's (2) each column consisting of "γ" number of 1's (3) number of 1's in common between any two columns, denoted by "λ", is no greater than 1, i.e., $\lambda = 0$ or $\lambda = 1$ (4) both "ρ" and "γ" are small compared with length of the code and the number of rows in H [1, 2]. The density of H-matrix "r" is ρ / n (or) γ / v, where "n" is number of columns and "v" is the number of rows in H. The (γ, ρ) LDPC code is regular if row-weight and column-weight distributions are uniform for each row and column, respectively. If all the columns or all the rows of H do not have the same weight then LDPC code is said to be irregular.

Considering "k" to be a positive integer greater than 1, for a given choice of ρ and γ, Gallager's construction method is to form $[k\gamma \times k\rho]$ matrix H that consists of γ number of $[k \times k\rho]$ submatrices, $H_1, H_2, \ldots, H_\gamma$. Each submatrix has a total of "$k \times \rho$" number of 1's. For $1 \leq i \leq k$, the ith row of H_1 contains all its "ρ" number of 1's in columns $((i-1)\rho + 1)$ to $i \times \rho$. The other submatrices are merely column permutations of H_1. Then, the transpose of H matrix is, $H' = [H_1, H_2, H_3, \ldots, H_\gamma]$. The total number of 1's in H is "$k\rho\gamma$" and the total number of entries in H is "$k^2\rho\gamma$". Hence, the density of H is given by, $r = k\rho\gamma/k^2\rho\gamma = 1/k$. If $k \gg 1$, then H is sparse matrix.

The null space of H gives a linear block code of length $n = k\rho$. Though, Gallager proved that the minimum distance of the code is at least $\gamma + 1$, there is no information about the method for choosing column permutations of the submatrix H_1 to form the other submatrices, $H_2, \ldots, H_\gamma$ [9]. Computer searches are needed to find good LDPC codes for this method. In the present work, a structured way of generating submatrices is proposed which is based on the Gray-code representations.

23.2.2 LDPC Code Representation

LDPC codes can be graphically represented by bipartite graphs called Tanner graphs [10], as shown in Fig. 23.1. The Fig. 23.1a shows H matrix of size 4×8. Its Tanner graph is shown in Fig. 23.1b. The rows of H correspond to the check nodes c_1, c_2, c_3, and c_4 and the columns correspond to the bit nodes $v_1, v_2, \ldots, v_8$ of the Tanner graph. Set of edges, corresponding to the 1's in H matrix, connect these nodes. A chain of nodes, where initial and terminal nodes are the same, and that does not use the same edge more than once is a cycle. The shortest length cycle in the graph is called girth "g" of a code. Short cycles lead to inefficient decoding. Hence, LDPC codes with large girth are preferred.

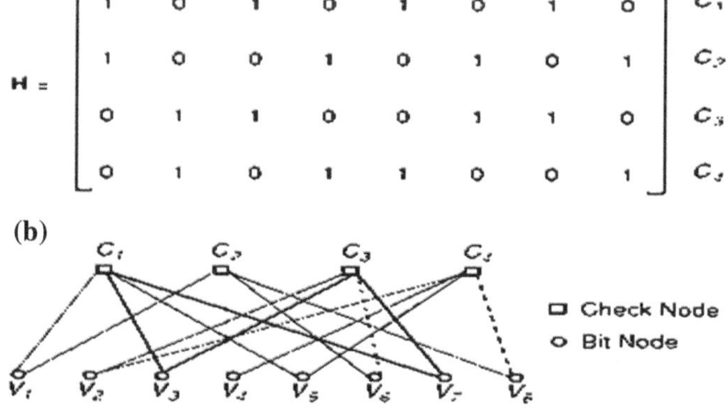

Fig. 23.1 **a** The parity-check matrix H of an LDPC code. **b** Corresponding tanner graph

23.3 Code Construction Using Proposed Algorithm

23.3.1 Selection of Point Set

Let "H" be the parity-check matrix of an LDPC code with "v" parity-check equations, i.e., H is v × n matrix, where "n" is the code length. We represent these parity-check equations by the elements of a point set "X." Let the elements of point set "X" be denoted as $X = \{X_0, X_1, X_2, X_3,\ldots X_\rho\}$, where "$\rho$" is the row-weight of H-matrix. The set of elements can be selected using either of the following two equations. The first element X_0 is zero in both the cases.

$$X_{i+1} = 2^i + X_i \tag{23.1}$$

$$X_{i+1} = 2X_i + 1 \quad \text{for i} = 0, 1, 2, 3, \ldots, \rho \tag{23.2}$$

23.3.2 Construction of SubMatrices

The proposed algorithm gives the structured way of constructing the submatrices as explained below: Let R_1H_1, R_1H_2, $R_1H_3,\ldots,R_1H_\gamma$ be the first rows of H_1, H_2, $H_3,\ldots$, H_γ submatrices, respectively.

- Construction of H1:
  The point set "X" is selected according to the required row-weight as $X = \{X_0, X_1, X_2,\ldots X_i\}$ where $i = \rho$. The elements of "X" form the first row of H_1, i.e., R_1H_1. The subsequent rows of H_1, i.e., R_2H_1, $R_3H_1,\ldots\ldots,R_{\rho+1}H_1$ are obtained by circularly left shifting the elements of their preceding rows, until the first row, R_1H_1, repeats.
- Construction of H_2:
  The first row of H_2, i.e., R_1H_2, is obtained by exchanging the first and second elements of the point set "X", keeping the rest elements in the same positions, i.e., set $X = \{X_1, X_0, X_2,\ldots X_i\}$ forms the first row of H_2. The subsequent rows of H_2 are obtained by circularly left shifting the elements, similar to that of submatrix H_1.
- Construction of H_3:
  Similarly, the first row of the third submatrix, i.e., R_1H_3, is obtained by exchanging the first and third elements of the point set "X," retaining the rest as they are. Hence, $X = \{X_2, X_1, X_0,\ldots X_i\}$ forms the first row of H_3. Subsequent rows are obtained as described for H_1 and H_2.

This procedure is repeated for all the "γ" submatrices. The minimum distance of the code is at least one more than the γ. Finally, these matrix elements are represented in Gray codes. The decimal numbers selected are such that, when converted into Gray-code representations, they have more number of 0's compared to number of 1's, so that the generated H-matrix is sparse. The proposed algorithm gives the systematic way of selecting the decimal numbers using one of the design equations. The sparsity of the matrix can be increased by increasing the number of bits used to represent the decimals into Gray code. Table 23.1 shows the various code rates, lengths, and densities of H-matrix with column-weight two and column-weight three codes, for different row-weights, without considering the girth of a code.

23.4 Column-Weight Two LDPC Codes

Gallager [2] has shown that column-weight two codes have minimum distance increasing only logarithmically with code length compared to a linear increase when the column weight is at least three. Despite the slow increase in minimum distance, column-weight two codes have shown potential in magnetic storage applications [3, 11, 12]. Their performance has been shown to be good enough for partial response channels and inter-symbol-interference (ISI) signals. They also have low computational complexity since there are only two column connections per row. Their encoders and decoders are simpler to implement, since they have lower computational complexity and storage complexity. They have better block error statistics properties as pointed out in [13]. When concatenating them with error correcting codes such as Reed-Solomon codes, these properties make LDPC codes with column-weight two promising for data storage and other applications [14].

Construction of Quasi-Cyclic LDPC codes with girth 6 based on finite geometries of lines and points of Euclidean and projective geometries over finite fields is proposed in [15]. Construction methods for column-weight two codes are found in [3, 11, 12, 16]. Girth eight column-weight two codes can be constructed from combinatorial designs [7]. Figure 23.2 shows an alternative graphical method of

Table 23.1 Code rates, densities and lengths under different ρ and γ

Row-weight (ρ)	Density of H (r)	Code length (n)	Column-weight(γ)	
			$\gamma = 2$ Code- rate $(1 - \gamma/\rho)$	$\gamma = 3$ Code-rate $(1 - \gamma/\rho)$
4	0.200	20	0.500	0.250
5	0.166	30	0.600	0.400
6	0.142	42	0.666	0.500
7	0.125	56	0.714	0.571
8	0.111	72	0.750	0.625
9	0.100	90	0.777	0.666
10	0.090	110	0.800	0.700

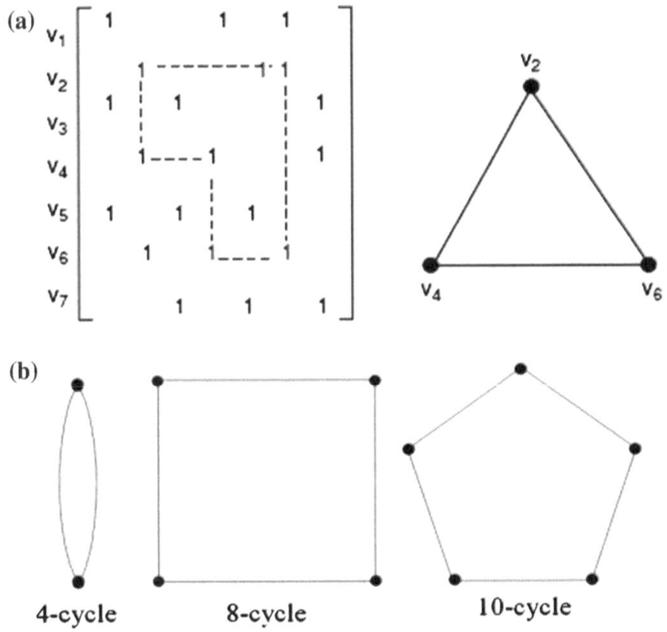

Fig. 23.2 **a** A 6-cycle in an H matrix and its structure graph. **b** Examples of cycles with length 4, 8, and 10

representing H-matrix using structure graph [15]. Structure graph is used to identify cycles in LDPC H-matrices. Two distinct edges between two nodes in a structure graph stand for 4-cycle. A 6-cycle is a triangle comprising three points and three edges between any two points. An 8-cycle is a loop composed of four points and four edges. Figure 23.2 shows examples of cycles with length 4, 8, and 10. In this paper, a method to construct column-weight two codes with girth 8 and 12 is proposed. We used structure graph to find the girth of H-matrix.

Girth of H-matrix affects the decoding performance of a code. Large girth improves code performance, whereas small ones, especially of length four, degrade performance. The Message Passing Algorithm (MPA) performance depends on the girth of a code. Sullivan [17], using bit error rate simulations, showed that large girth codes perform better than those with lower girths. If the number of 1's that are in common between any two columns is greater than 1, then 4-length cycle exists. Hence, the devised algorithm must provide a method so that no two rows share "1" in more than one column. Proposed algorithm provides a method of constructing H-matrices, which are free from 4-length cycles.

23.4.1 Construction of Column-Weight Two Codes with Girth of Eight

The H-matrix of column-weight two codes is constructed with two submatrices H_1 and H_2. We present a method for construction of H, by modifying an algorithm given in Sect. 23.3, to achieve a girth of eight.

Construction of H1

The point set "X" can be selected according to the required row-weight, i.e., $i = \rho$. Thus, the set is $X = \{X_0, X_1, X_2, X_3, \ldots, X_\rho\}$. This forms the first row of H_1, i.e., R_1H_1. The subsequent rows of H_1 are obtained by cyclically shifting the elements of first row, either to left or to right, until the first row R_1H_1 repeats.

Construction of H2

The elements of the first row of H_1, in reverse order, form the first row of H_2, i.e., R_1H_2. The subsequent rows are obtained by cyclically shifting the elements of first row, either to the left or to the right, until R_1H_2 repeats.

The girth obtained with this construction method is eight, which we have verified with the corresponding structure graph. Table 23.2 shows code-sizes for (n,2,ρ) codes with different ρ values.

Code expansion

With the construction method given above, each submatrix consists of ρ number of $\rho \times \rho$ matrices which we call blocks. The H-matrix constructed for any selected value of ρ, is called a "base matrix" for weight ρ and it is a matrix with minimum code size. For example, 6×9 matrix for $\rho = 3$ is a base matrix for row-weight of 3. Base matrix can be expanded to different levels as shown in the table. In expansion level 1, each "1" in $\rho \times \rho$ block in base matrix, is replaced by a corresponding $\rho \times \rho$ matrix in which it appears and a "0" is replaced by a $\rho \times \rho$ zero matrix. In expansion level 2, each "1" in the expanded matrix of level 1, is replaced by a $\rho \times \rho$ block of a base matrix and a "0" is replaced by $\rho \times \rho$ zero matrix. Same procedure is followed for expansion level 3 and so on.

Table 23.2 Code size and code rates for (n, 2, ρ) code with girth eight

Code size					
ρ	Min. size	Level 1	Level 2	Level 3	Code-rate
3	6 × 9	18 × 27	54 × 81	162 × 243	1/3
4	8 × 16	32 × 64	48 × 96	192 × 384	1/2
5	10 × 25	50 × 125	250 × 625	1,250 × 3,125	3/5
6	12 × 36	72 × 216	432 × 1,296	2,592 × 7,776	2/3
7	14 × 49	98 × 343	686 × 2,401	4,802 × 16,807	5/7

23.5 Construction of Column-Weight Two Codes with Girth 12

It is reported in [16] that decoding performance improves with higher girth. Hence, we present a modified algorithm to achieve a girth of 12. The proposed method is based on blockwise construction. The elementary block B1 is constructed by arranging the set elements in the form of a column. The other blocks B_2 and B_3 are obtained with minor modification on the elementary block B_1. The following steps indicate the construction method:

- The elements of point set "X" form column of B_1. The first element X_0 is excluded from the set.
- B_2 is obtained by shifting the bottom element of B_1 to the top and shifting rest all elements down to their succeeding rows.
- Similarly, B_3 is obtained by shifting the top two elements to the bottom and rest of the elements are shifted up, to their preceding rows.
- Now the submatrix H_1 is formed by concatenating three B_1 blocks.
- The submatrix H_2 is obtained by concatenating B_1, B_2, and B_3.
- Finally, the elements of the blocks are represented in Gray code.

As an illustration, let $\rho = 3$ and $i = 7$ for a $(n, 2, \rho)$ code. Block B_1 is constructed by selecting the set elements as $X = \{X_1, X_2, X_3, \ldots, X_7\}$. Representing these elements in 7-bit Gray code, the size of block B_1 is 7×7. The elements of block B_2 are $X = \{X_7, X_1, X_2, \ldots X6\}$. Similarly, block B_3 is constructed using set $X = \{X_3, X_4, \ldots X_1, X_2\}$. Now, using these blocks, H-matrix is constructed as below:

$$H = \begin{bmatrix} B_1 & B_1 & B_1 \\ B_1 & B_2 & B_3 \end{bmatrix}$$

This construction method generates H-matrix of size (14×21) with column-weight 2 and row-weight 3. This matrix we call it as a "base matrix". The code can be expanded as explained in Sect. 23.4. The code size is (98×147) in expansion level 1. This can be expanded to the second level to get the code size of $(686 \times 1{,}029)$. Thus, the procedure can be repeated, for different expansion levels, to increase the code length.

23.6 Conclusion

A Gray-code-based novel method for constructing LDPC codes with flexibility in rates, lengths and girths is presented in this paper. This work presents algorithms for construction of column-weight two structured regular LDPC codes with girth 8 and girth 12. The developed algorithm is flexible in rates, lengths, and girths. These codes are systematically constructed and their H submatrices are generated using a set of decimals selected from one of the defined equations.

References

1. Gallager RG (1962) Low density parity check codes. IRE Trans Inf Theory IT-8:21–28
2. Gallager RG (1963) Low density parity check codes, MIT Press, Cambridge
3. Song H, Liu J, Vijaya Kumar BVK (2002) Low complexity LDPC codes for partial response channels. IEEE Globecom 2002(2):1294–1299
4. Berrou C, Glavieux A, Thitimajshima P (1993) Near Shannon limit error correcting coding and decoding: turbo codes. In: proceedings of international conference on communications, pp 1064–1070
5. Eroz M, Sun F, Lee L (2004) DVB-S2 Low-density parity-check codes with near Shannon limit performance. Int J Satell Commun Netw 22:269–279
6. Yeo E, Pakzad P, Nikolic B, Anantharam V (2001) High throughput low-density parity-check decoder architectures. In: proceedings of GLOBECOM'2001, vol 5. San Antonio, TX, pp 3019–3024
7. Karkooti M, Cavallaro J (2004) Semi-parallel reconfigurable architectures for realtime LDPC decoding. Int Conf Inf Technol (ITCC) 1:579–585
8. Lee JKS, Lee B, Hamkins J, Thorpe J, Andrews K, Dolinar S (2004) A scalable architecture for structured LDPC decoder. IEEE Int Symp Inf Theory 292–299
9. Lin S, Costello DJ (2011) Error control coding, 2nd edn. Pearson
10. Tanner RM (2001) A recursive approach to low complexity codes. IEEE Trans Inf Theory 47(2):498–519
11. Song H, Liu J, Kumar B (2004) Large girth cycle codes for partial response channels. IEEE Trans Magn 40(4):2362/2371
12. Li Jing, Kurtas Erozon (2004) A class of high-rate, low-complexity, well-structured LDPC codes from combinatorial designs and their applications on ISI channels. IEEE Trans Commun 52(7):1038–1042
13. Song H, Liu J, Vijaya Kumar BVK (2002) Low complexity LDPC codes for magnetic recording. IEEE Globecom 2002, Taipei, Taiwan, R.O.C
14. Zhang H, Moura JMF (2003) The design of structured regular LDPC codes with large girth IEEE Globecom 4022–4027
15. Kou Y, Lin S, Fossorier MPC (2001) Low-density parity-check codes based on finite geometries: a rediscovery and new results. IEEE Trans Inf Theory 47(7):2711–2736
16. LuJ, Moura JMF, Niesen U (2004) Grouping-and-shifting designs for structured LDPC codes with large girth. IEEE Int Symp Inf Theory 236–236
17. O'Sullivan M (2006) Algebraic construction of sparse matrices with large girth. IEEE Trans Inf Theory 52(2):719–727

Chapter 24
Neuro-Curvelet Model for Efficient Image Compression Using Vector Quantization

Arun Vikas Singh and K. Srikanta Murthy

Abstract In many multimedia applications, such as image storage and transmission image, compression plays a major role. The fundamental objective of image compression is to represent an image with least number of bits of an acceptable image quality. A technique based on second-generation curvelet transform and Back-Propagation Neural Network (BPNN) has been proposed. The image compression is accomplished by approximating curvelet coefficients using BPNN. By applying BPNN into compressing curvelet coefficients, we have proposed a new compression algorithm derived from characteristic of curvelet transform. Initially, the image is translated by fast discrete curvelet transform and then based on their statistical properties; different coding and quantization schemes are employed. Differential Pulse Code Modulation (DPCM) is employed to compress low-frequency band coefficients and BPNN is used to compress high-frequency band coefficients. Subsequently, vector quantization is performed on BPNN hidden layer coefficients, thereby resulting in a reconstructed image with less degradation at higher compression ratios. For a given bits per pixel (bpp), the Curvelet Transform with Back-Propagation Neural Network (BPNN) gives better performance in terms of Peak Signal-to-Noise Ratio (PSNR) and Computation Time (CT) when compared to Wavelet Transform with BPNN and JPEG.

Keywords Image compression · Wavelet transform · Curvelet transform · Back-propagation neural network · Vector quantization

A. V. Singh (✉)
Department of ECE, PESIT, 100 Feet Ring Road, BSK 3rd Stage, Bangalore, Karnataka 560085, India
e-mail: arunvikas@pes.edu

K. S. Murthy
Department of CSE, PESSE, Bangalore, Karnataka, India
e-mail: srikantamurthy@pes.edu

V. Chakravarthi et al. (eds.), *Proceedings of International Conference on VLSI, Communication, Advanced Devices, Signals & Systems and Networking (VCASAN-2013)*, Lecture Notes in Electrical Engineering 258, DOI: 10.1007/978-81-322-1524-0_24,

24.1 Introduction

In multimedia applications, data compression has recently become more vital, as compression methods are being rapidly developed to compress large data files such as images, while maintaining high-quality image. There are many compression techniques, some of them are differential encoding quantization techniques like scalar and vector quantization, predictive coding, transform coding. Transform coding is most efficient technique among all these [1].

Wavelet transforms is an important mathematical tool for hierarchically decomposing functions and is used to compress images at higher compression ratios with higher PSNR values [2]. But from the observations, it is clear that wavelets may not be the best selection for presenting natural images because the wavelets are blind to the smoothness along the edges usually found in images. Therefore, wavelets cannot provide the "sparse" representation for an image due to the intrinsic drawback of it. The natural images are represented sparsely with the help of new two transforms, that is, ridgelet and curvelet transforms [3] which are represented by basis elements which demonstrate very high directional sensitivity and are exceedingly anisotropic. The first generation of curvelet is enhanced giving way to the fast discrete curvelet transform (FDCT) [4, 5] which are faster, far less redundant and simpler.

The neural-network-based approaches [6] used for data processing have inherent parallel processing capabilities will result in excellent results. The image compression can be achieved when the number of units (neurons) in input and output layers is same in a two layered neural network and has reduced the number of units (neurons) in hidden layers. This approach is proposed by Sonhera et. al [7].

24.2 Digital Curvelet Transforms

From the continuous-time definition of curvelet transform [8], frequencies near the angle and near the dyadic corona are extracted smoothly by the window U_j. The Cartesian arrays are not adaptable to rotations and coronae. As an alternative, these concepts are suitably substituted by Cartesian counterparts, known as "Cartesian coronae". They are based on concentric squares instead of concentric circles and shears, as shown in Fig. 24.1 [8].

The "Cartesian" window is defined as [8]

$$\widetilde{U}_j(\omega) := \widetilde{W}_j(\omega)V_j(\omega) \tag{24.1}$$

A window $\widetilde{W}_j(\omega)$ is defined as [8]

$$\widetilde{W}_j(\omega) = \sqrt{\varphi_{j+1}^2(\omega) - \varphi_j^2(\omega)}, \quad j \geq 0 \tag{24.2}$$

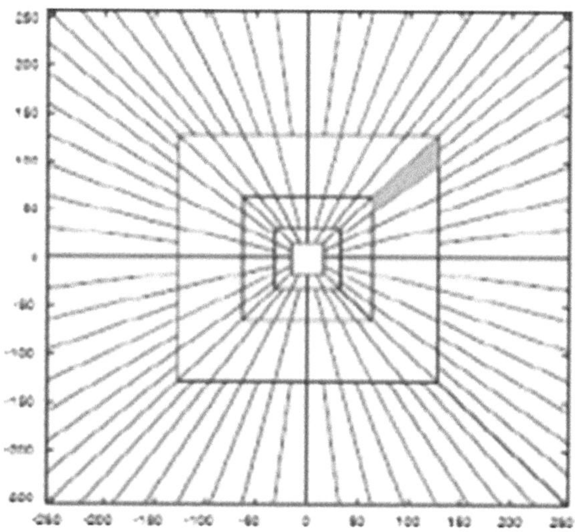

Fig. 24.1 Digital curvelet tiling of space and frequency

where ϕ is the product of low-pass one-dimensional windows

$$\varphi_j(\omega_1, \omega_2) = \varphi(2^{-j}\omega_1)\varphi(2^{-j}\omega_2) \tag{24.3}$$

The function ϕ is varied in the range, $0 \leq \varphi \leq 1$. The value of φ may be equal to 1 on [−1/2, 1/2] and dies out outside the range [−2, 2]. The digital curvelet transform coefficient is given by [8].

$$c(j, l, k) = \int \widehat{f}(\omega)\widetilde{U_j}\left(s_{\theta_l}^{-1}\omega\right)e^{j\left\langle s_{\theta_l}^{-T}b, \alpha x\right\rangle} d\omega \tag{24.4}$$

24.3 Neural-Network for Data Compression

Image compression coding is one of the applications where Back-propagation neural-network can be applied directly. Figure 24.2 shows the multilayered neural-network structure. Here, three-layered Back-propagation neural-network is designed which consists of one input layer, one output layer, and one hidden layer. To achieve image compression, the number of neurons at the hidden layer represented by k has to be designed and is less than that of the neurons at the input layer and output layer.

Here, the n-dimensional input vector is referred to as n pixels from one sub band. Each neuron at the hidden layer is connected to all weights and can be described by $\{w_{ij}$, j = 1, 2, ..., k and i = 1, 2, ..., n$\}$, which can also be represented in matrix form of size k × n.

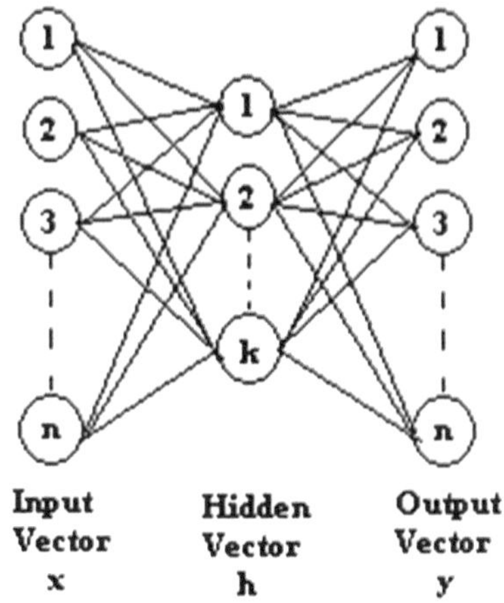

Fig. 24.2 A multilayered neural network

The connections from the hidden layer to the output layer are represented by $\{w_{ij}\}$, is the weight matrix of size n × k, which is set equal to the transposition of matrix [w]. To achieve image compression, the network is trained in such a way that the weights w_{ij} scales the n-dimensional input vector into a vector of k-dimension ($k < n$) at the hidden layer. Thus, the optimum output value is produced which makes the quadratic error minimum between input and output layer.

24.4 Implementation

The original image is first decomposed by the second-generation Curvelet Transform to obtain curvelet coefficients. Based on the statistical properties of different frequency bands, different coding and quantization schemes are practiced. Since the lowest sub-band have a large amount of image energy, the Differential Pulse Code Modulation (DPCM) is used to encode these coefficients. The finer scale sub-bands coefficients are compressed using BPNN. Furthermore, without much degradation of the reconstructed image, compression ratio can be increased when the hidden layer coefficients of BPNN are vector quantized [9]. Using entropy coding, all these coefficients is encoded. Here, Huffman encoding technique is used to encode these coefficients. As the finest scale sub-bands has very little amount of energy, these coefficients discarded directly with little perceptible effect on the image compression.

24.5 Experimental Results and Discussion

The performance of the proposed technique is illustrated by the experimental results. The various 256 × 256 gray-scale images (Lena, Girl, Girl (tiffany), House, Mandrill, Peppers, Saturn, Girl1, and Tree) are assessed by the proposed technique. The reconstructed image quality is evaluated using an objective measure known as PSNR (in decibel) and is defined as

$$\mathrm{PSNR} = 10\log_{10}\frac{255^2}{MSE} \text{ where } \mathrm{MSE} = \frac{\sum_{i=1}^{W}\sum_{j=1}^{H}\left(x_{ij} - \widehat{x_{ij}}\right)^2}{W \times H}$$

where the original and reconstructed pixel are denoted by x_{ij} and $\widehat{x_{ij}}$, respectively, and the size of the image is W × H.

The experimental results of the conventional and proposed algorithm are listed in Table 24.1 in terms of PSNR and Computation Time (CT) for different bpp using BPNN. The reconstructed images are shown in Figs. 24.3 and 24.4. From the Table 24.1, it is clear that for a given bpp, the curvelet transform with BPNN yields better PSNR and Computation Time (CT) is lesser when compared to wavelet

Table 24.1 PSNR (dB), bpp and Computation Time (CT) result of different test images

Image	With wavelet, BPNN and VQ			With curvelet, BPNN and VQ		
	PSNR	bpp	C T in seconds	PSNR	bpp	C T in seconds
Lena	26.9246	0.1850	75.554907	26.9436	0.1838	72.767328
Girl	28.6206	0.1913	71.091093	28.6288	0.1912	70.546348
Girl (tiffany)	28.7292	0.1914	74.392028	28.7342	0.1900	69.682748
House	27.7189	0.1905	73.088866	27.7260	0.1903	72.900415
Mandrill	23.5239	0.2023	84.540647	23.5315	0.2015	76.084689
Peppers	26.8401	0.1902	75.827844	26.8691	0.1902	72.430759
Saturn	33.4971	0.1593	78.693808	33.5246	0.1586	70.172824
Girl 1	26.9216	0.1839	84.242040	26.9414	0.1833	67.020727
Tree	22.7727	0.2004	74.997895	22.7941	0.2003	74.512500

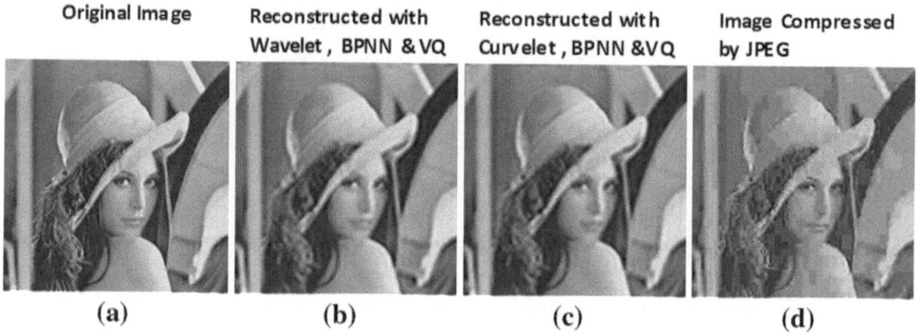

Fig. 24.3 Reconstructed images. **a** Lena. **b** PSNR is 26.9246 and bpp is 0.1850. **c** PSNR is 26.9436 and bpp is 0.1838. **d** PSNR is 25.39 and bpp is 0.20

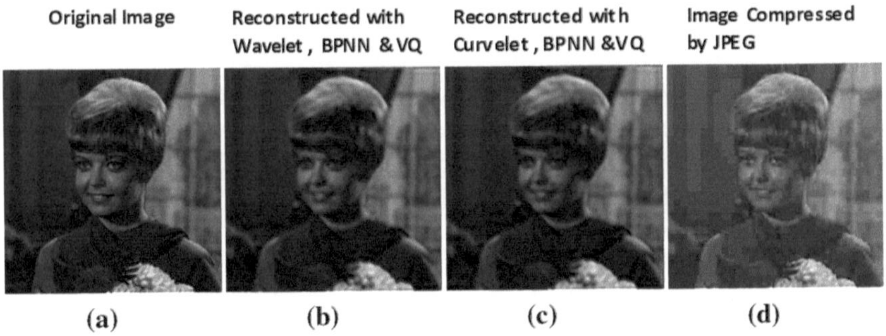

Fig. 24.4 Reconstructed images. **a** Girl. **b** PSNR is 28.6206 and bpp is 0.1913. **c** PSNR is 28.6288 and bpp is 0.1912. **d** PSNR is 28.48 and bpp is 0.19

transform with BPNN technique and JPEG compression. This is because the Back-Propagation Neural-Network (BPNN) has major advantage that their parameters are adaptable, which gives better compression rates after training the images. Furthermore, wavelets have the restrictions in handling the curve and line singularities of the image. This drawback is surmounted by employing second-generation curvelet transforms. When the wavelet transform is used, the serious problem like the block effect exists. But with the use of curvelet transform, this block effect is completely eliminated, and in image quality, there is remarkable improvement.

24.6 Conclusion

A compression algorithm based on second generation of curvelet transform and BPNN has been proposed. It is evident from the experimental results that the compression performance of our method gains much improvement in comparison to compression method based on wavelet transform with BPNN. The algorithm works reasonably well for declining block effect at higher compression ratios. The proposed image compression algorithm is more efficient when compared to the wavelet transform with BPNN and JPEG compression.

References

1. Iqbal MA, Javed MY, Qayyum U (2007) Curvelet-based image compression with SPIHT. In: International conference on convergence information technology. IEEE Computer Society
2. Talukder KH, Harada K (2007) Haar wavelet based approach for image compression and quality assessment of compressed image. IAENG Int J Appl Math
3. Candès EJ, Donoho DL (2000) Curvelets: a surprisingly effective nonadaptive representation for objects with edges. In: Rabut C, Cohen A, Schumaker LL (eds) Curves and surfaces, Vanderbilt University Press, Nashville, TN, pp 105–120

4. Candès EJ, Donoho DL (2004) New tight frames of curvelets and optimal representations of objects with piecewise C2 singularities. Commun Pure Appl Math 57(2):219–266
5. Candès EJ, Demanet L, Donoho DL, Ying L (2005) Fast discrete curvelet transforms. In: Applied and computational mathematics, pp 1–43. California Institute of Technology
6. Mougeot M, Azeneott R, Angeniol B (1991) Image compression with back propagation: improvement of the visual restoration using different cost functions. Neural networks 4(4):467–476
7. Sonehara N, Kawato M, Miyake S, Nakane K (1989) Image data compression using a neural network model. IJCNN con. IEEE cat. No. 89CH2765-6, vol 2, pp 35–41
8. Li Y, Yang Q, Jiao R (2010) Image compression scheme based on curvelet transform and support vector machine. Expert Syst Appl 37:3063–3069
9. Wang H, Lu L, Que DS, Luo X (2002) Image compression based on wavelet transform and vector quantization. In: International conference on machine learning and cybernetics, vol 4, pp 1778–1780 (4–5 Nov 2002)

Part IV
Advanced Devices

Chapter 25
A Novel Design of Narrowband Bandpass Filters on PTFE Laminate Using Radial Stubs

K. B. Pramod, H. V. Kumaraswamy, K. B. Praveen and S. Shyam Sundar

Abstract This paper proposes a novel building block for designing Narrowband Bandpass Hairpin filter on PTFE laminate using the Radial stubs. The proposed circuit block mainly consists of coupled lines, microstrip lines, and the Radial stubs. The process starts with the theoretical design procedure of the filter. Tuning and optimization of the design is achieved using AWR Microwave office tool. Finally, the result of the optimized design is suitably presented for implementation. Insertion loss (IL) less than 5 dB and the Return loss (RL) better than 12 dB is achieved with the 11.65 GHz center frequency. Electro Magnetic (EM) simulation results show good agreement with the linear schematic model results.

Keywords Advanced wireless revolution · Microstrip filters · Advanced numerical models · Radial stubs

25.1 Introduction

Bandpass filters are essential building blocks in communication design. It can reduce the harmonic and spurious emission for the transmitters and may improve the rejection of interfering signals for receivers. The hairpin resonator filter is one

K. B. Pramod (✉) · S. Shyam Sundar
Department of Electronics Engineering, Jain University, Bangalore, India
e-mail: pramod63putta@gmail.com

S. Shyam Sundar
e-mail: shyamss.jain@gmail.com

H. V. Kumaraswamy
Department of Telecommunication, R. V. College of Engineering, Bangalore, India

K. B. Praveen
Department of Telecommunication, Dr. Ambedkar Institute Technology, VTU, Bangalore, India
e-mail: prvn.guru@gmail.com

V. Chakravarthi et al. (eds.), *Proceedings of International Conference on VLSI, Communication, Advanced Devices, Signals & Systems and Networking (VCASAN-2013)*, Lecture Notes in Electrical Engineering 258, DOI: 10.1007/978-81-322-1524-0_25,

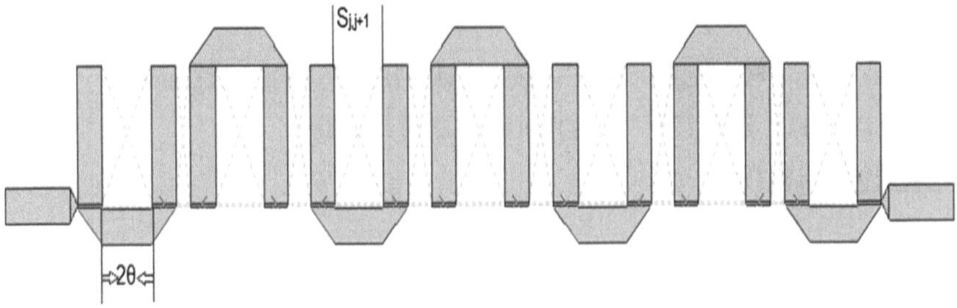

Fig. 25.1 Seventh order hairpin filter where θ is the slide factor and Sj, j + 1 is the spacing between the resonators

of the most popular micro strip filter configurations used in the higher microwave frequencies. It is easy to manufacture because it has open-circuited ends that does not require Direct to Ground System (DGS) [1].

The concept of hairpin filter is same as parallel coupled $\lambda/2$ resonator filters. The advantage of hairpin filter over end coupled and parallel coupled microstrip realizations is the optimal space utilization and susceptible to interelement coupling and radiation effects than a typical edge-coupled topology (Fig. 25.1).

The ceramic-filled PTFE laminate is used in this design because of its higher dielectric property which in turn responsible for the required narrowband response [2]. The ceramic-filled PTFE laminate has several advantages over the less expensive FR4 substrate, W and h are the width and height of the microstrip line.

$$\frac{W}{h} = \frac{8\exp(A)}{\exp(2A) - 2} \tag{25.1}$$

$$A = \frac{Z_0}{60}\sqrt{\frac{\varepsilon_r + 1}{2}} + \frac{\varepsilon_r - 1}{\varepsilon_r + 1}\left(0.23 + \frac{0.11}{\varepsilon_r}\right)$$

where, Z_0 = Characteristic impedance, ε_r = Dielectric constant.

25.2 Design Methodology

RT/Duroid 6010LM laminates feature ease of fabrication and stability in use. They have tight dielectric constant and thickness control, low moisture absorption, and good thermal mechanical stability.

Table 25.1 Odd and even mode impedance values and the width and spacing between resonators obtained using TLine software

Even–mode $(Zoe)_{j,\ j+1}$	Odd–mode $(Zoo)_{j,\ j+1}$	Spacing $S_{j,\ j+1}$	Width (in mils)
78.777	37.9177	33.565	15.0256
59.7244	43.0544	50.845	42.480
57.3072	44.3672	75.236	45.256
56.9688	44.5688	85.602	47.256

25.2.1 Initial Lumped Design

The filter design specification has the central frequency of 11.65 GHz. A bandwidth of 1.7 GHz and 40 dB attenuation desired at 10.05 GHz and 13.25 GHz. Using the insertion loss method, the order of the filter is found to be 7 by using Eq. (25.2).

$$N = \frac{\cos^{-1}\sqrt{(10^{0.1\times IL} - 1)/(10^{0.1\times\alpha} - 1)}}{\cos^{-1}\omega/\omega_c} \tag{25.2}$$

Using the Table 4.05-2(a) of [3], the prototype parameters for N = 7 can be obtained. For this design, the prototype parameters are as follows:

$$g_o = g_8 = 1,\ g_1 = g_7 = 1.3722,\ g_2 = g_6 = 1.3781,\ g_3 = g_5 = 202756,$$
$$g_4 = 1.5001.$$

25.2.2 Distributed Design

After the lumped design, it is transformed to distributed design using Kuroda's identity and the Richards Transformation. This uses the circuit models available in the tool. Using the TLINE of Eagleware Genesys [4], the values for the resonator spacing S and the width of the traces can be obtained and are shown in Table 25.1:

25.3 Implementation

25.3.1 Advanced Numerical Circuit Models

The linear elements incorporate closed form solution techniques that may be evaluated with a linear circuit simulator for high degree of accuracy with respect to electromagnetic simulation by conventional coupled line circuit and electromagnetic models. Resistive losses in the metal and dielectric losses in the substrate are characterized in these models [5] (Fig. 25.2).

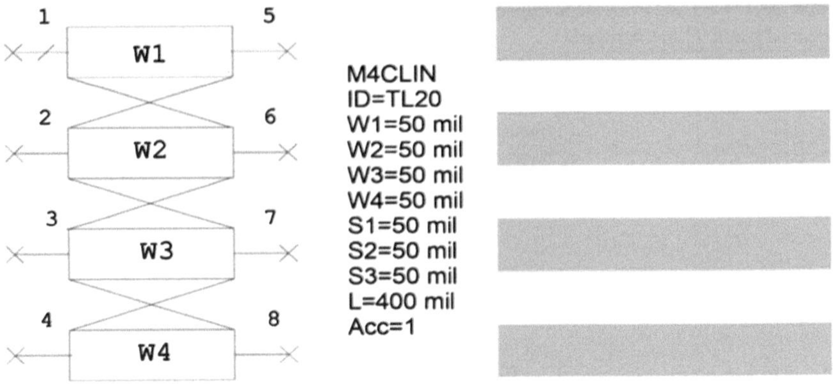

Fig. 25.2 AWR MXCLN advanced numerical element model where X = 4

25.3.2 Linear Schematic Model

The linear simulation schematic with advanced numerical models is presented in Fig. 25.3.

In the schematic, the M12CLIN element forms the middle section of the filter, and the two M14CLIN elements form the top and bottom sections. The MLIN element is connected to port three of the MTEE element, and ports 1 and 2 are connected to the first line segments of the M10CLIN elements.

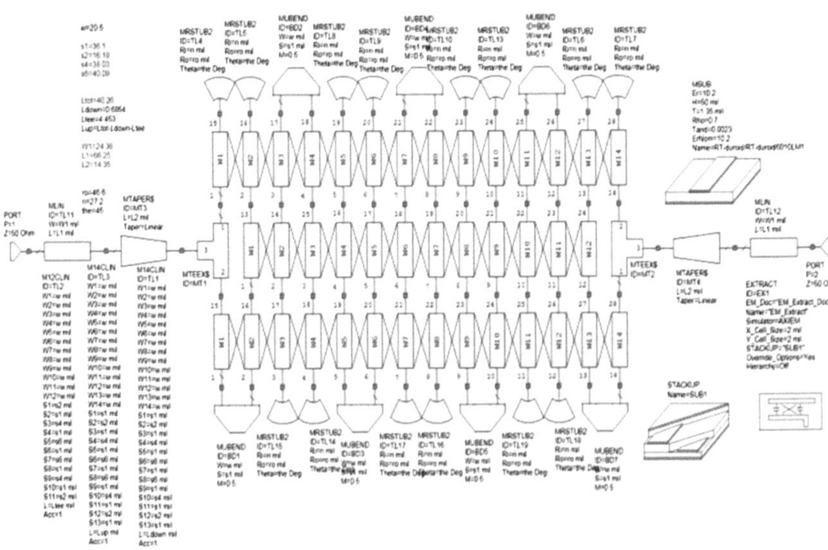

Fig. 25.3 Microwave office schematic of the hairpin filter

25.3.3 Seventh-Order Hairpin Bandpass Filter

The implementable structure of hairpin topology is chosen for taking advantage of accuracy of linear element simulation. Another advantage is that MXCLIN elements of simulation can be used, which can be tuned and optimized. The tapping length of the filter is calculated using the Eq. (25.3).

$$ltap = \frac{2L}{\pi}\sin^{-1}\left[\sqrt{\frac{\pi}{2}\frac{Z_c}{Q_s}}\right] \tag{25.3}$$

$$2L = \frac{\lambda_g}{\pi} = \frac{\lambda_0}{2\sqrt{\varepsilon_{\rm re}}} = \frac{\rm C}{2f_0\sqrt{\varepsilon_{\rm re}}}. \tag{25.4}$$

25.3.4 Electromagnetic Model

Creating an EM structure manually is to specify the enclosure parameters. In this analysis, the X direction was specified to be 957.4 mils and the Y direction 81.5 mils and, air 100 mils thick, Rogers 6,010, 50 mils thick with a relative dielectric constant of 10.2 and loss tangent of 0.0023 with specified dielectric layers. The enclosure top and bottom boundaries were set. After creating the stack up, the Material, Conductor, Dielectric layers, Thickness of the Substrate and the top layer, Material Thickness, Meshing, the Grid size have been defined and the Simulator is finally selected (Fig. 25.4).

25.4 Result and Analysis

The Insertion loss plot of a hairpin filter design is as shown in the Fig. 25.5. Here, the insertion loss at center frequency is −0.9693 dB. This is less than specification of −5 dB. The response is in good agreement with 1 dB bandwidth requirement. This is shown by the horizontal marker line at 1 dB. The return loss plot for the hairpin design is as shown in the Fig. 25.5. The return loss of −15 db is better than the specification of −12 db and the lumped element result.

The rejections plot of the hairpin filter design is as shown in the Fig. 25.6. The stop band rejections at the skirts of the filter response are found well in the linear simulation (Fig. 25.7).

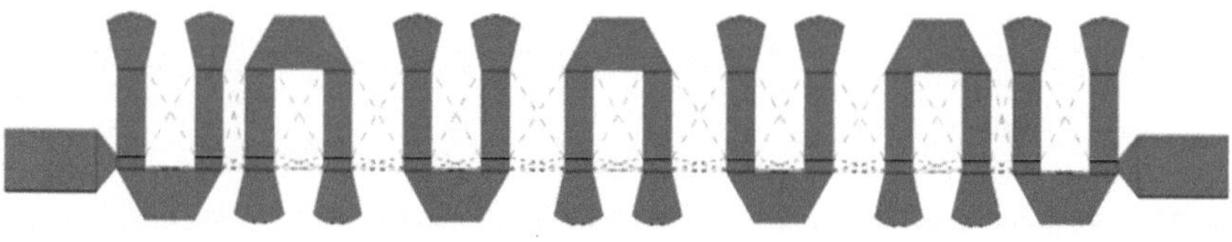

Fig. 25.4 Filter layout

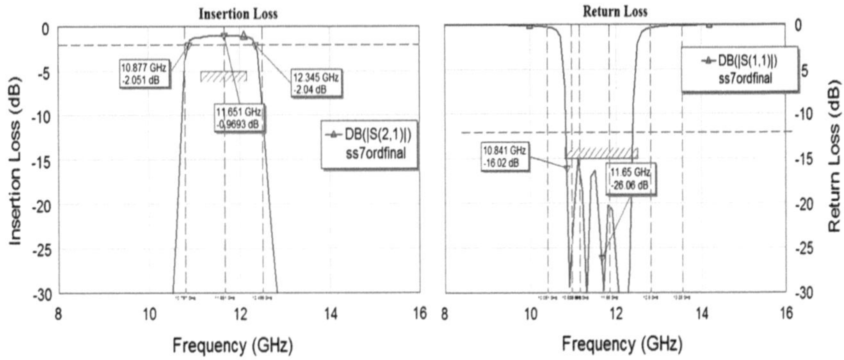

Fig. 25.5 Insertion loss and return loss of a linear schematic model of the hairpin filter

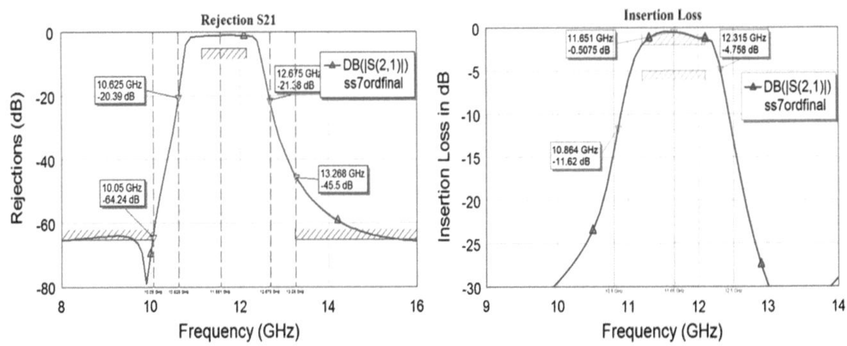

Fig. 25.6 Rejections of a linear schematic model of the hairpin filter and insertion loss of an electromagnetic model of the hairpin filter

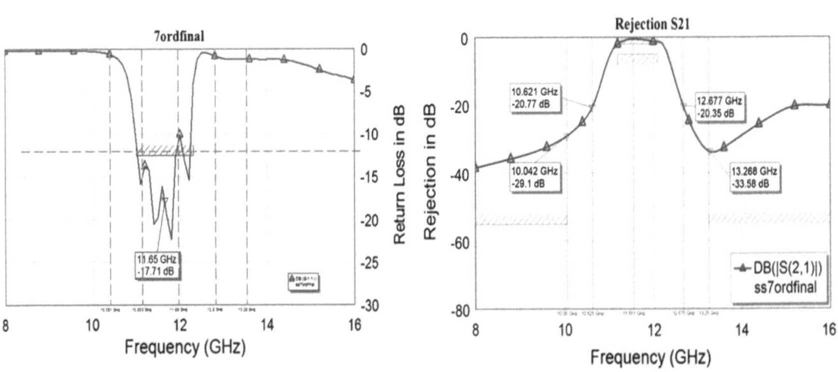

Fig. 25.7 Return loss and rejections of an electromagnetic model of the hairpin filter

Table 25.2 Comparison between lumped, hairpin, EM-simulated filters

Parameters	Lumped filter	Hairpin filter	EM simulation
Start Frequency	10.8 GHz	10.8 GHz	10.8 GHz
Stop Frequency	12.5 GHz	12.5 GHz	12.5 GHz
Insertion Loss @ (11.65 GHz)—−5 dB	−7.25 dB	−0.9856 dB	−0.5337
Return Loss @ (11.65 GHz)—−12 dB	−10.09 dB	−16 dB	−17.71 dB
Attenuation @ (10.625 and 12.675 GHz) −20 dB min.	−31.19 and −20.79 dB	−20.79 and −21.54 dB	−20.83 and −20.25 dB
Attenuation @ (10.05 and 13.25 GHz)—−40 dB min	−68.87 and 59.45 dB	−65.05 and −45.5 dB	−30.76 and −32.76 dB

25.5 Conclusion

The design of hairpin filter and its simulation was successfully done. Sequential procedure in designing hairpin filter was presented in this paper. After obtaining the appropriate order of the filter, the values of the odd and even impedances were computed using the admittance inverter parameters. Using a ceramic-filled substrate significantly decreased the final size of the filter due to its high dielectric constant. Another important factor that contributed to the decrease in size was the selection of the dielectric. The design of the 11.65 GHz Bandpass Hairpin filter on Ceramic-filled PTFE Laminate is designed and simulated using the AWR MWO tool. The results of lumped design, Hairpin design, and EM Simulation are briefly summarized and compared with each other in the Summary Table 25.2. The design shows the good agreement with the distributed Hairpin design and some of the parameters are eventually improved over the Lumped design.

25.6 Acknowledgments

The authors would like to thank Prof. K. T Srirangaraj CET, Jain University., Dr. TGS Moorthy, CET, Jain University., Prof. Chandrashekar K. Program Co-coordinator, CET, Jain University, Bangalore and Mr. Manjunatha Reddy H. V Technical Manager, RF and μW division, Icon Design Automation Pvt. Ltd. for their support and encouragement.

References

1. Boutejdarl A, Omar A A Miniature 3.1-GHz microstrip band-pass filter with a suppression of spurious harmonic using multilayer-technik and defected ground structure open-loop-ring (DGS)

2. Gupta K Micro strip Lines and Slotlines, Artech House
3. Matthaei GL, Young L, Jones EMT (1980) Microwave filters, impedance-matching networks, and coupling structures. Artech House, Boston, MA
4. Eagleware Corporation (2002) TLine Program. Genesys version 8.1, Norcross, GA
5. Applied Wave Research, Microwave Office 9.04 Online Reference Manual

Chapter 26
The Design and Simulation of 0.5 dB Noise-Figure RF Narrowband LNA

K. B. Pramod, H. V. Kumaraswamy and K. B. Praveen

Abstract This paper presents the design and simulation of 2-stage low-noise amplifier (LNA) for the application UHF range used for wireless communications and low-noise amplifier with bandwidth 800 M–1.2 GHz with optimization by using Enhancement Mode Pseudomorphic HEMT ATF34143 from Avago Technologies. The design and simulation uses lumped elements to implement the matching networks and proposed 2-stage is to achieve considerable gain. A 2-stage LNA has successfully designed and simulated with up to 33 dB forward gain, less than 0.58 dB noise figure and with good Voltage standing wave ratio (VSWR) from 1.5 to 1.6 at both input and output side by using advanced wireless revolution (AWR) Microwave office tool.

Keywords advanced wireless revolution · Low-noise amplifier · Radio frequency · Noise figure · Pseudomorphic high electron mobility transistor

26.1 Introduction

The low-noise amplifiers (LNA) is a electronic device used to suppress the noise of received input signals by improving gain without introducing much noise by itself at the front end of communication systems, by possibly very weak signals from the

---

K. B. Pramod (✉)
Department of Electronics Engineering, Jain University, Bangalore, India
e-mail: pramod63putta@gmail.com

H. V. Kumaraswamy
Department of Telecommunication, R. V. College of Engineering, Bangalore, India
e-mail: kumaraswamyhv@rvce.edu.in

K. B. Praveen
Department of Telecommunication, Dr.Ambedkar Institute Technology, Bangalore, India
e-mail: prvn.guru@gmail.com

V. Chakravarthi et al. (eds.), *Proceedings of International Conference on VLSI, Communication, Advanced Devices, Signals & Systems and Networking (VCASAN-2013)*, Lecture Notes in Electrical Engineering 258, DOI: 10.1007/978-81-322-1524-0_26,

antenna for the reduction of external as well as internal noise of the circuit [1]. Low-noise amplifier is used in a wide variety of applications in RF communication systems such as wireless computer networks, mobile phones, and satellite receiver. Its plays as a significant component in the receiving end of any communication system and its performance measured based on the Noise Figure, Gain in a Dynamic range, and stability.

26.2 Specification, Research, and Collected Data

All applications in wireless communication require and look for the parameters of the LNA such as Gain, Bandwidth, Noise Figure, Linearity, and Power Consumption. The goal of designer is to minimize noise figure, high gain with moderate linearity and establishing good impedance matching to other transceiver blocks. The additional constraint of low-power consumption is imposed in portable systems.

The Lorenzo [2], they have presented three low-noise amplifier topologies. The amplifiers were implemented in a standard 90 nm CMOS process and were operated with a 1 V supply voltage. The cascaded common-source low-noise amplifier achieved the best performance among the three with a simulated gain of 13.8 dB and noise figure of 1.7 dB, which makes it comparable to previously published works. Xu [3] presented a 50 MHz–1 GHz high linearity and low-noise amplifier MMIC with 0.15 um In GaAs PHEMT process. The ADS simulation results showed that it has good noise performance of 1.36 dB NF and excellent linearity of 17 dBm IIP3. Navaratne [4], in their paper LNA with a 0.8–1.8 GHz bandwidth is designed and Simulations show a noise figure between 2.8 and 1.3 dB, with gain 13 dB over the bandwidth.

26.3 Low-Noise Amplifier Transistor Selection

Selection of the transistor is the crucial stage in LNA design. Any transistor has its NF maximum available gain (MAG) and minimum intrinsic noise figure (NFmin). Therefore, after adding the matching and biasing sections, we cannot achieve gain more than MAG and Noise figure less than NFmin (Fig. 26.1).

K is the stability factor for a two port, defined as:

$$\mathrm{k} = \frac{1 - |S_{11}|^2 - |S_{22}|^2 + |\Delta|^2}{2|S_{12} \cdot S_{21}|} > 1 \tag{1}$$

B1 is the supplemental stability factor for a two port, defined as:

$$\mathrm{B1} = 1 + |\mathrm{S}_{11}|^2 - |\mathrm{S}_{22}|^2 - |\Delta|^2 > 0 \tag{2}$$

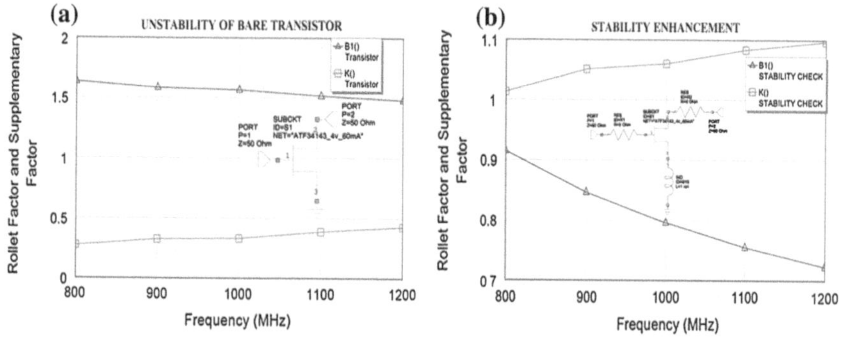

Fig. 26.1 **a** Stability check of the bare transistor with schematic **b** Stability enhanced result with schematic

where,

$$\Delta = S_{11}S_{22} - S_{12}S_{21}$$

A transistor is stabilized by adding small series/large shunt gate resistance or series output resistance. Now this circuit has satisfied conditions K > 1 and |·| < 1. B1 > 0 becomes unconditionally stable to calculate the Maximum available gain (MAG):-

$$G_{MAX} = \frac{|S_{21}|}{|S_{12}|}\left(K - \sqrt{K^2 - 1}\right) \quad (3)$$

where K is on the limit of unity, the above equation reduces down to:-

$$G_{MAX} = \frac{|S_{21}|}{|S_{12}|} \quad (4)$$

26.4 Noise Figure

Noise figure (NF) is a measure of degradation of the signal-to-noise ratio (SNR), caused by components in a radio frequency (RF) signal chain and defined as:

$$NF = 10\log\frac{SNRin}{SNRout} \text{ in dB} \quad (5)$$

SNRin and SNRout are Signal-to-Noise ratio of the circuit or system at input and output correspondingly. There are three key parameters that are needed for the noise figure analysis of an LNA with frequency, biasing condition are NFmin, Equivalent noise resistance Rn, and Optimum reflection coefficient Γ opt .

26.4.1 The Friis Formula for Noise Factor

Friis's formula for total noise factor of a cascade of stages is given as

$$F_{total} = F_1 + \frac{F_2 - 1}{G_1} + \frac{F_3 - 1}{G_1 G_2} + \cdots \quad (6)$$

where F_n and G_n are the noise factor and available power gain, respectively, of the nth stage.

$$F_{receiver} = F_{LNA} + \frac{(F_{rest} - 1)}{G_{LNA}} \quad (7)$$

where F_{rest} is the overall noise factor of the subsequent stages. According to the equation, $F_{receiver}$ the overall noise figure is dominated by the noise figure of the F_{LNA} and G_{LNA} the gain is sufficiently high. So in order to reduce $F_{receiver}$, we have to increase G_{LNA} and decrease F_{LNA} as much as possible.

26.4.2 The Friis Formula for Noise Temperature

Friis's formula can be equivalently expressed in terms of noise temperature:

$$T_{total} = T_1 + \frac{T_2}{G_1} + \frac{T_3}{G_1 G_2} + \cdots \quad (8)$$

26.5 Implementation of the Design

The design implementation requires adding microstrip lines between the lumped elements and placing of microstrip Tee at the junction. The junction arm has to be extended using MLINS. The grounding must be done using via. The design is implemented on FR4 substrate with the relative permittivity €r = 4.4 with a height of 1.6 mm. The substrate thickness is chosen to be 0.035 mm, Rho = 1. An inductor may be added in series with the transistor gate. As an ideal inductor has zero resistance, it generates no thermal noise. It improves stability by reducing the gain of the amplifier by a small factor (Fig. 26.2).

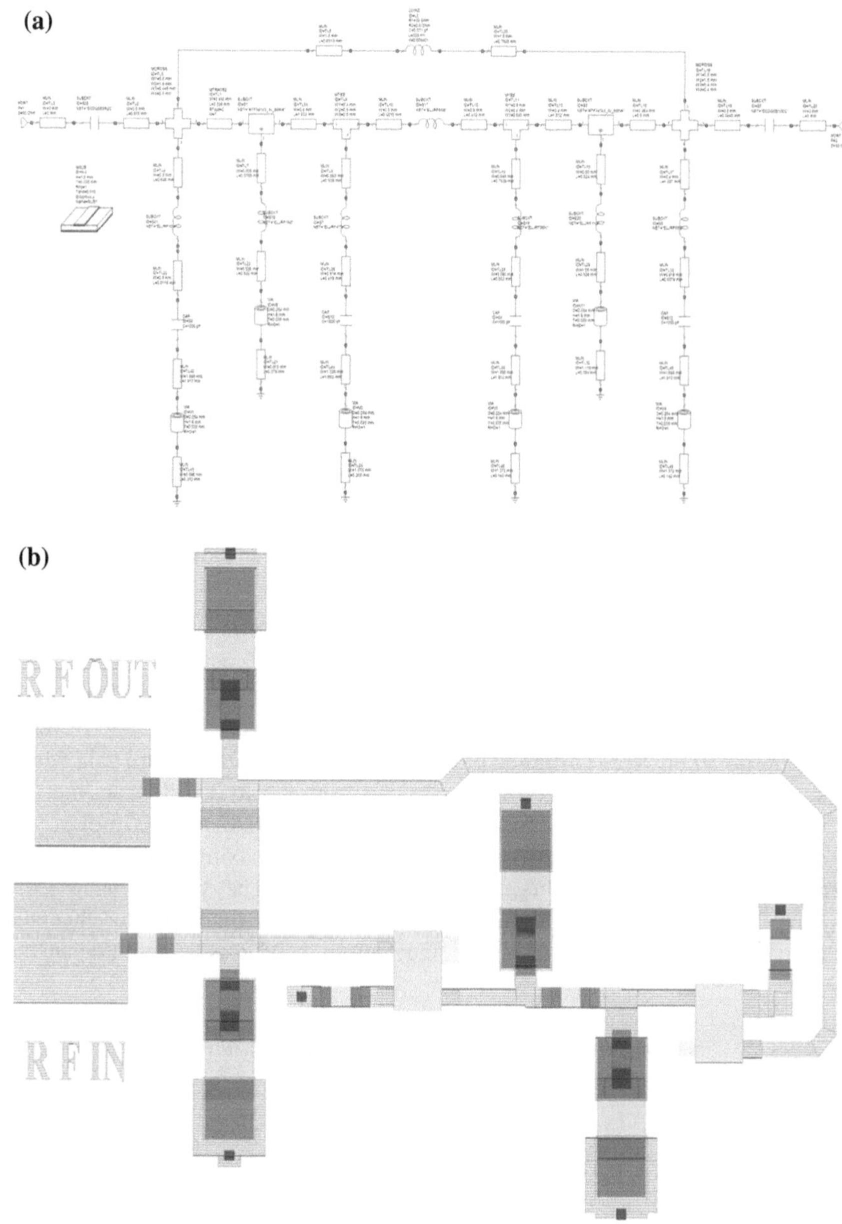

Fig. 26.2 **a** Complete circuit Schematic. **b** Corresponding layout

26.6 Results Analysis

The advanced wireless revolution design environment (AWRDE) features extensive post-processing capabilities, allowing the display of computed data known as "Measurements" on rectangular graphs, Smith Charts, tabular graphs, and 3D graphs.

The stability condition is satisfied which is shown in the Fig. 26.3 by Rollett's stability factor (K) > 1 and supplemental stability factor (B1) > 0 for the entire frequency range. Geometric Stability Factor: MU1 (Load) and MU2 (Source) computes the geometric stability factor of a 2-port which is the distance from the center of the Smith Chart to the nearest unstable point of the output load plane and input source plane [5]. The necessary and sufficient condition for unconditional stability of the two ports is that MU1 > 1 and MU2 > 1. The stability factor is computed from and result is shown in Fig. 26.3.

$$\mathrm{MU1} = \frac{1 - |S_{11}|^2}{|S_{22} - S_{11}^{*}\Delta| + |S_{21}S_{12}|} \tag{10}$$

$$\mathrm{MU2} = \frac{1 - |S_{22}|^2}{|S_{11} - S_{22}^{*}\Delta| + |S_{21}S_{12}|} \tag{11}$$

The maximum stable gain is the maximum gain that can be achieved by a potentially unstable device. Maximum stable gain is defined as the ratio of magnitude of S21 to the S12. The graph plotted in the Fig. 26.4 for forward gain which varies from 26 to 30 dB and NF_{Min} is varying from 0.3 to 0.4 dB from the frequency 800 MHz to 1.2 GHz. Noise Figure varies from 0.42 to 0.5746 dB.

SWR as VSWR is the ratio of the standing wave maximum voltage to the standing wave minimum voltage defines the VSWR. The VSWR value 1.6:1 denotes maximum standing wave amplitude that is 1.6 times greater than the minimum standing wave value. Input and Output VSWR is represented in the

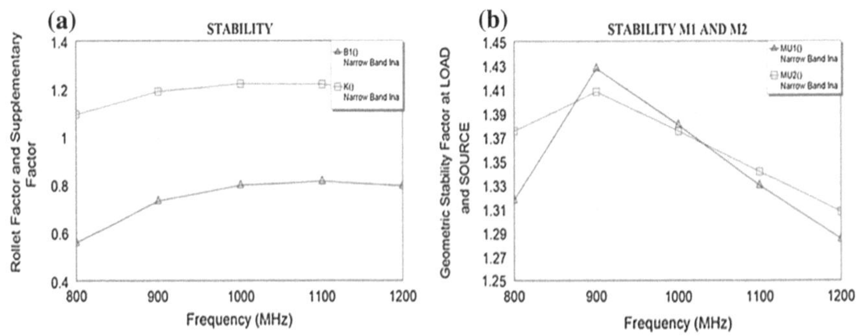

Fig. 26.3 **a** Shows Rollett's stability factor (K) and supplemental stability factor (B1) results, **b** Shows geometric stability factor MU1 (Load) and MU2 (Source)

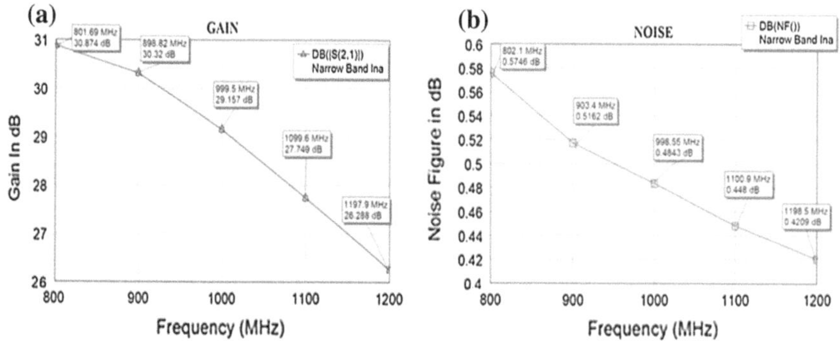

Fig. 26.4 **a** Shows forward gain mag S21 **b** Shows noise figure

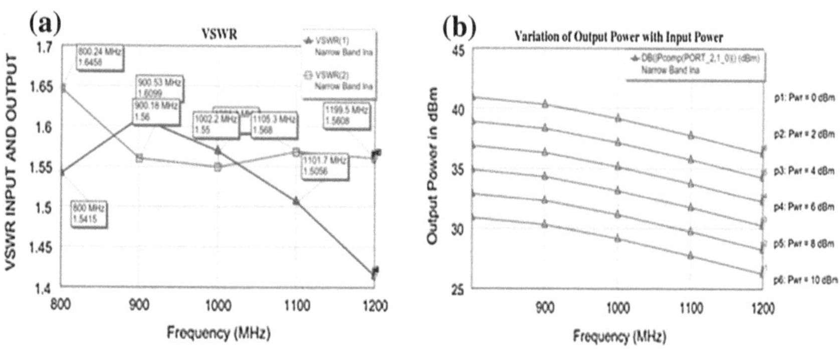

Fig. 26.5 **a** Shows VSWR at Input VSWR (1) and output VSWR (2), **b** Shows linear relationship between input and output power

above Fig. 26.5 as VSWR (1) and VSWR (1), Output VSWR relates to the magnitude of output port.

It is seen that the Output VSWR ranges from 1.5 to 1.6 and Input VSWR varies from 1.4 to 1.5 for the frequency 0.08 to 1.2 GHz. It relates to the magnitude of the voltage reflection coefficient and hence to the magnitude of S_{11} and S_{22} for the output port.

26.7 Conclusion

In this paper, a narrowband Low-Noise Amplifier (LNA) circuit is designed successfully for frequency bandwidth of 400 MHz (800 MHz to 1.2 GHz) with 30 dB gain and noise figure less than 0.57 dB throughout the frequency band with maximum VSWR of 1:1.6 at both input and output using E-PHEMT ATF34143 by Avago technologies. Circuit simulation is done in AWR Microwave Office 2010

Table 26.1 Comparison of reported LNA

References	Frequency (GHz)	Noise (dB)	Gain (dB)
[3]	0.05–1	1.36	16
[4]	0.8–1.8	2.8	13
[9]	1.9	0.9	16.3
This work	0.8–1.2	<0.57	26.2–30.8
This work	1.2	0.42	26.28

with very good overall performance apart from the ultra low-noise result (Table 26.1).

Earlier designs like Tao lianjuan [6], Loong, Hashim [7], Pramod [8] and many designs used resistors for stability and gain modifications. But in this design, we have achieved all the required specifications by using only inductors and capacitors which reduces the overall noise figure.

References

1. Design of a low noise amplifier with GaAs MESFET at ku_Band. In: Islam MR, Zahirul Alam AHM, Khan S, Arafat A (ed) A shabana department of electrical and computer engineering faculty of engineering international Islamic University, Malaysia, P.O. Box: 10, 50728 Kuala Lumpur, Malaysia. E-mail: rafiq@iiu.edu.my, doi:10.1109/ICCCE.2010.5556786 Publication Year: 2010, Page(s): 1–5
2. Comparison of LNA topologies for WiMAX applications in a standard 90 nm CMOS process. In: Lorenzo MAG, de Leon MTG (ed) Computer modelling and simulation (UKSim), 2010 12th International conference on doi:10.1109/UKSIM.2010.122 Publication Year: 2010, Page(s): 642–647 Cited by 2
3. Design of 50 MHz–1 GHz low noise and high linearity MMIC Amplifier. In: Xu J, Wang GZ, Zhang Y, Huang J (ed) Future computer and communication (ICFCC), 2010 2nd International conference on Volume: 3 doi:10.1109/ICFCC .2010.5497534 Publication year: 2010, Page(s): V3-489–V3-491
4. Noise figure optimization of a noise-cancelling wide-band CMOS LNA. In: Navaratne D, Beaulieu A, Belostotski L (2010) NEWCAS conference (NEWCAS), 2010 8th IEEE international doi:10.1109/NEWCAS.2010.5603720, Page(s): 125–128 Cited by 1
5. A 2.4 GHz Low-flicker-noise CMOS sub-harmonic receiver. In: Syu JS, Meng C, Wang CL (ed) Circuits and systems I: regular papers, IEEE transactions on vol 60(2) doi:10.1109/TCSI.2012.2215794 Publication Year: 2013, Page(s): 437–447
6. An 0.5–6 GHz Ultra-wideband low noise amplifier design Tao lianjuan University of electronic science and technology of China Chengdu, China taolianjuan@126.com Bao jingfu University of electronic science and technology of China Chengdu, China baojingfu@uestc.edu.cn, 978-1-4577-0321-8/11/$26.00 ©2011 IEEE, Page(s): 2516–2519
7. 1.575–2.48 GHz Multi-standard low noise amplifier using 0.18 μm CMOS with on-chip matching. In: Loong TT, Hashim A, Mustaffa MT, Noh NM (ed) School of electrical & electronic engineering, USM, engineering campus, Nibong Tebal, 14,300, Penang, Malaysia. 2011 IEEE symposium on industrial electronics and applications (ISIEA 2011), September 25–28, 2011, Langkawi, Malaysia, 978-1-4577-1417-7/11/$26.00 ©2011 IEEE, Page(s): 100–103

8. The design and simulation of 2.294 dB noise-figure RF wideband PHEMT LNA employing 2-stage cascade with single feedback. In: Pramod KB, Praveen KB (ed) Institute for research and development India (IRD INDIA), International conference on emerging trends in computer and image processing, ISBN Number: 978-93-81693-88-12, Bangalore, 24th February 2013, Page(s): 17–22
9. A design, analysis and optimization technique of 1.9 GHz CDMA low noise amplifier. In: Rahnama M, Kordalivand AM (ed) Signal processing and its applications (CSPA), 2010 6th international colloquium on doi:10.1109/CSPA.2010.5545336 Publication Year: 2010, Page(s): 1–5

Chapter 27
Design and Development of DC-DC Converter System to Drive a PZT Stack Actuator

Anand Kumar, P. Siva Subba Rao and K. Venkatesha

Abstract Basically, piezoelectric PZT, or lead zirconate titanate (Pb [Zr(x) Ti (1−x)] O3), is one of the world's most widely used piezoelectric ceramic materials as Advanced Devices. When the deformation occurs in PZT than that can be used as an actuator to drive or control the devices. The present work deals with the development and control of a converter to drive piezoelectric actuators. To generate the required force in the piezoelectric ceramics, high voltages are needed. That can be achieved by designing a bidirectional buck-boost converter. The converter is controlled by means of PWM control strategy. The output of the converter is given to the power amplifier to get high-power output and gain. The design is made to drive the PZT Stack actuator. The design is verified by PSpice simulations.

Keywords PZT · Buck-boost converter · PA93 · ANSYS

27.1 Introduction

The increasing number of applications where the piezoelectric actuators are used is motivating the development of specific converters to drive such actuators. Different switching converter topologies have been employed to drive such elements, including fly back [1], buck-boost [2], push–pull [3] and converters based on piezoelectric transformers [4]. The most common strategy to control the position and force of a piezoelectric actuator is the direct regulation of the voltage applied [3].

A. Kumar · K. Venkatesha
BNMIT, Bangalore, India

P. Siva. Subba. Rao (✉)
CSIR-NAL, Bangalore, India
e-mail: eenandu@gmail.com

V. Chakravarthi et al. (eds.), *Proceedings of International Conference on VLSI, Communication, Advanced Devices, Signals & Systems and Networking (VCASAN-2013)*, Lecture Notes in Electrical Engineering 258, DOI: 10.1007/978-81-322-1524-0_27,

The nonlinear relationship between displacement and voltage can be improved working with the electric charge instead of the voltage. This strategy has been used successfully by many authors [5–7]. The main drawback of the charge control is the necessity of sensing and integrating the current supplied to the load, with the corresponding problems involved. Piezoelectric loads differ from the conventional loads assumed in the conventional converters. They behave as voltage sources, and there is only current when the voltage is changing or when they are producing mechanical work. If the voltage is unchanged and the actuator is unloaded, there is no current. Hence, the converter can operate in either continuous or discontinuous conduction mode depending on the converter and load state. Furthermore, the voltage in the output capacitance (or load) cannot be considered constant, and therefore, the currents cannot be assumed linear. Such a complexity increase is to be introduced in the converter design in order to optimize the actuator drive performance.

The Fig. 27.1 shows the block diagram of the design, the input voltage is 28 V DC taken as the power supply in the aircraft. The DC-DC Converter converts the input 28 VDC voltage from one level to another DC level. This is accomplished by high switching frequency of 25 kHz. The converter is designed in such that it can produce adjustable output voltage of +200 V and −200 V. That is basically by PWM technique, we can obtain the adjustable voltage. This variable voltage will be the bias voltage for a power amplifier. The dual polarity output voltage (+Vs and −Vs) can be used as biasing voltage for the Power Amplifier (PA93, Cirrus Logic). The power amplifier generates required power to drive the PZT stack. By providing the control command voltage signal to the amplifier and by adjusting the gain of the amplifier, it is possible to obtain very high power output with required frequency that can be used to drive the PZT stacks actuators. Since the power amplifier operates continuously to drive the stack actuators is required to cool the

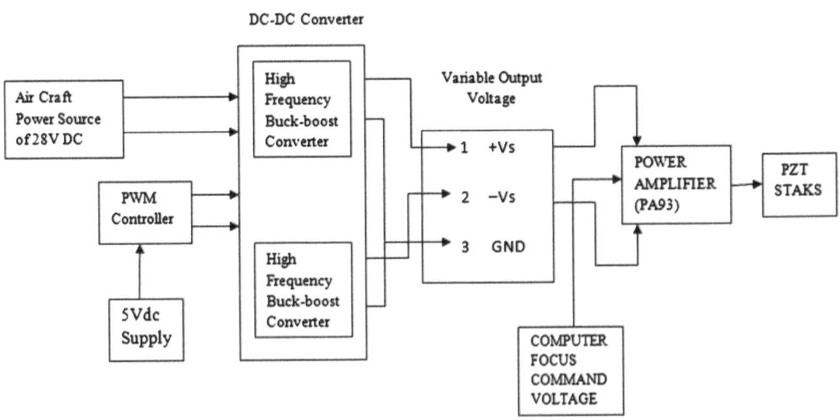

Fig. 27.1 Block diagram of DC-DC converter for PZT stack

PA93 IC using heat sinks. So heat sink is designed for effective cooling. The design verification of heat sink carried out by the simulation using ANSYS software tool.

27.2 Design of Buck-Boost Converter

The basic operational circuit of Buck-Boost converter is as shown in the Fig. 27.2. In the On-state (switch is closed), the input voltage source is directly connected to the inductor (L). This result in accumulating energy in L and the diode is reversed biased. In the Off-state (when switch is open), the inductor is connected to the output load, diode is forward biased, and the current will discharges through the load. It can be seen that the energy stored in L increases during On-time (as IL increases) and then decreases during the Off-state. L is used to transfer energy from the input to the output of the converter. The PWM pulses are given to the converter switches for the switching operations, which are provided by the PWM controller circuit. Basically, the PWM pulses are generated from the Astable mode and Monostable mode operations of circuit that will produce continuous PWM pulses there after that is given to the Driver circuit. That is because to provide isolation for the two converters switching pulses, finally the output from the driver is given to the converter switches appropriately.

The output of the converter is the adjustable DC voltage that can be given to the Power amplifier as Biasing voltage. The power amplifier generates required power to drive the PZT stack. By providing the control command voltage signal to the

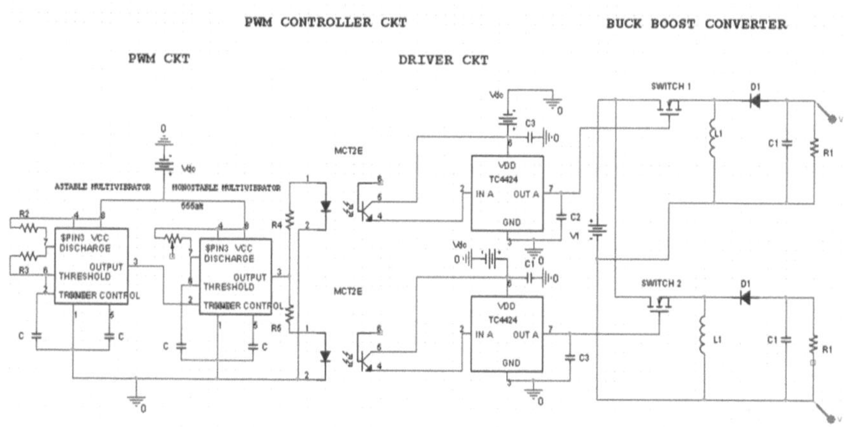

Fig. 27.2 Circuit diagram of Buck-Boost converter

amplifier and by adjusting the gain of the amplifier, it is possible to obtain very high power output with required frequency that can be used to drive the PZT stacks actuators.

27.3 Topology Selection for Converter Design

There are several methods of controlling SMPSs, e.g., pulse width modulation (PWM), frequency control, phase control, and cycle-by-cycle control. A PWM dc-dc converter is proposed for the investigated application because of its simple structure, well-known dynamic behavior, and possibility of a pulse-by-pulse current limiting and instantaneous shutdown. The output voltage in PWM converters is controlled against line and load variations by adjusting the duty ratio D of the switches

$$D = \frac{t_{on}}{t_{on} + t_{off}} = \frac{t_{on}}{T} \tag{27.1}$$

The buck-boost topology is selected as a basic power conversion cell. This topology has the ability to provide an Output voltage higher or lower than the input voltage can be easily implemented using a few circuit elements and is well-researched and established in its conventional unidirectional form. The simplified transfer function for the Buck-boost converter is given by

$$\frac{V_{out}}{V_{in}} = \frac{-D}{1 - D} \tag{27.2}$$

Although the topology of the conventional buck-boost converter is successfully used for constant-output-voltage Power supplies, it is not suitable for use with piezoelectric actuators.

27.4 Design and Modeling of Buck-Boost Converter

The design considerations are, Vs = 28 V (Aircraft supply), Load = 10 μF (PZT stack actuator), L = 10 mH, C = 10 μF, K = 0.87, Fr = 25 kHz as shown in the Fig. 27.3.

And the other components as, R (load) = 100 Ω and V_{OUT} = 200 V.

So the Output current I_{OUT} is given by, I_{OUT} (max) = $-V_{OUT}/R$ = 200/100 = 2A.

Ripple voltage, ΔV_C = 10 V (5 % of the output voltage).

Ripple current, ΔI_L = 0.02 A (1 % of the load current).

Calculation of inductance and capacitance:

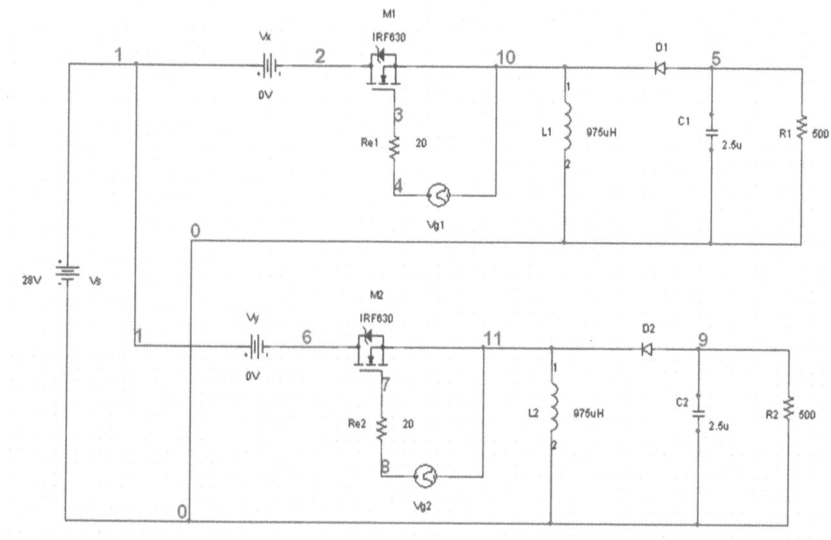

Fig. 27.3 PSpice ckt

$$L = \frac{\mathrm{I_L} * (1 - \mathrm{D})^2 * \mathrm{R} * \mathrm{T_S}}{\Delta \mathrm{I_L}}$$

$\mathrm{I_L}$ Load current
$\Delta\mathrm{I_L}$ Ripple current

$$L = \frac{2 * (1 - 0.87)^2 * 100 * 40 * 10^{-6}}{0.02}$$

L 6.7 mH

$$C = \frac{\mathrm{V_0} * \mathrm{D} * \mathrm{T_S}}{R * \Delta \mathrm{V_0}}$$

$\mathrm{V_0}$ Output voltage
$\Delta\mathrm{V_0}$ Output voltage ripple

$$C = \frac{200 * 0.87 * 40 * 10^{-6}}{100 * 10}$$

C 6.96 μF.

27.5 Simulation Results Using pSpice

The simulated results of voltage output and current are shown in the Fig. 27.4, and it is noted that the output 200 V is boosted from 28 V and the current across the inductor is 4.8 A.

27.6 Simulation of Heat Sink Using ANSYS

ANSYS is a general purpose finite element modeling package for numerically solving a wide variety of mechanical and other problems like static/dynamic structural analysis (both linear and nonlinear), heat transfer and fluid problems, as well as acoustic and electro-magnetic problems. So the heat sink design is essential here for the power amplifier because the operating temperature of the power amplifier is high, so cooling can be achieved by designing required heat sink. That can be designed by simulation using ANSYS. The thermal heat calculations are followed as per the model shown in Fig. 27.5 [8] and the thermal resistance calculations are shown in Table 27.1.

From the data sheet of PA93 [1].

Power dissipation, continuous @ Tc = 250 c is, 125 W Junction temperature, Tj = 1,500c. Ambient temperature, A = 25 °C.

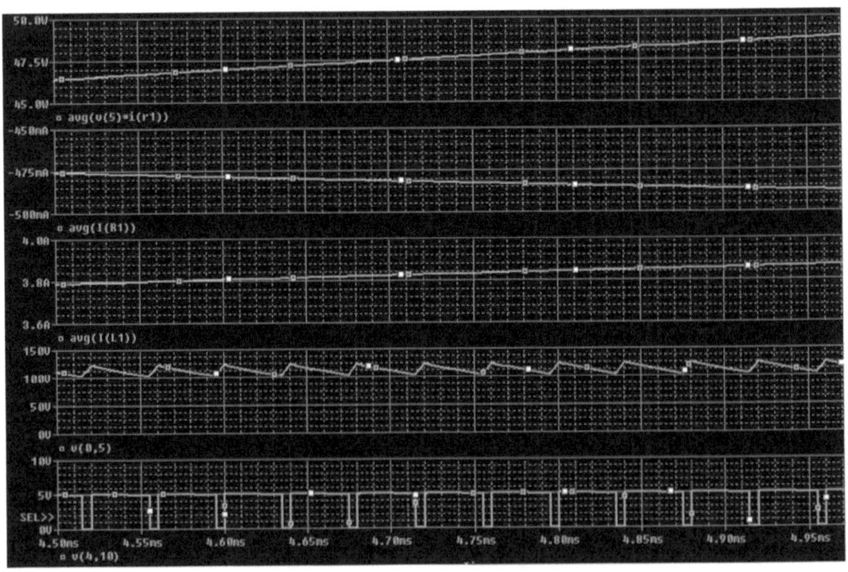

Fig. 27.4 Simulation result of Buck-Boost converter

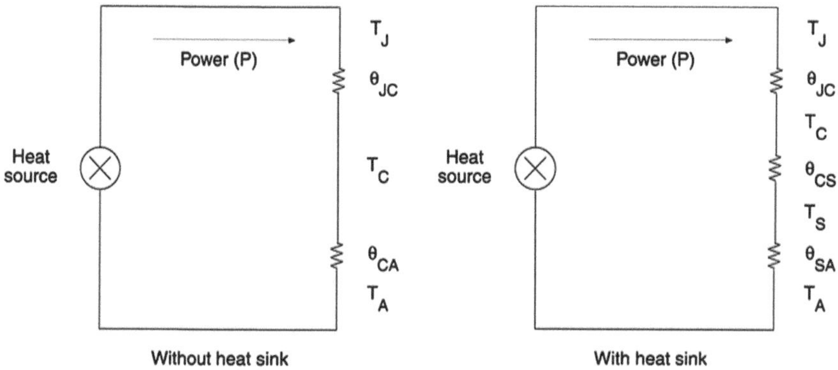

Fig. 27.5 Thermal Circuit

Table 27.1 Thermal Resistance Equation

Device	Equation
Without a heat sink	$\theta_{JATotal} = \theta_{JC} + \theta_{CA} = \frac{T_J - T_A}{P}$
With a heat sink	$\theta_{JATotal} = \theta_{JC} + \theta_{CS} + \theta_{SA} = \frac{T_J - T_A}{P}$

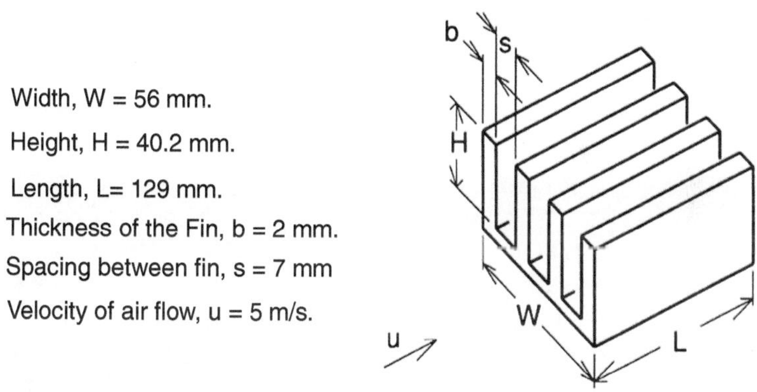

Fig. 27.6 Dimensions of heat sink

$$\theta_{JA\ Total} = \frac{150 - 25}{125} \tag{27.3}$$

Then $\theta_{\text{JA Total}} = 1^0$ C/W.

The dimensions of the heat sink is designed based on the junction surface area of the IC, for the element Aluminum (Flotran CFD 3D Flotran 142) are as follows as per the diagram as shown in Fig. 27.6:

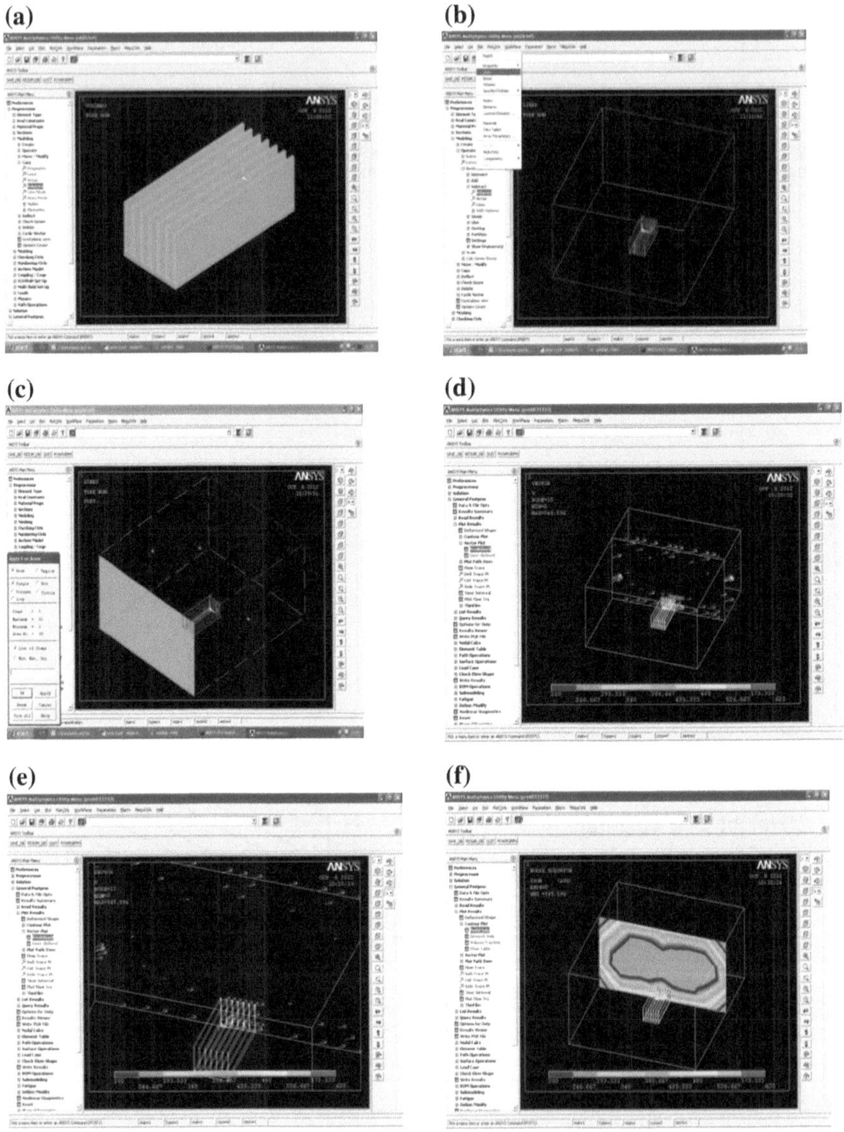

Fig. 27.7 Modeling of the heat sink in ANSYS and heat flow. **a** Metal plate model. **b** Heat sink inside the Box. **c** Air blows in to box using Fan. **d** Cross-sectional area showing heat flow. **e** Heat flow through heat sink. **f** Cross-sectional area showing the pressure

In the ANSYS, the heat sink modeled to know the heat distribution across the cross-sections as shown in the Fig. 27.7

27.7 Conclusion

The buck-boost converter is successfully designed and valuated by simulation method in this paper. This converter has been designed to deal with the drive of piezoelectric actuators and considers their particular requirements (capacitive load behavior). The control strategy is followed as PWM control. The power amplifier PA93 is used to drive the PZT Stack actuator. The designed converter is simulated, and the results obtained have shown to be reliable. The simulation of heat sink is done using ANSYS tool.

References

1. APEX Cirrus logic data sheet PA93 IC
2. Main JA, Newton DV, Massengill L, Garcia E (1996) Efficient power amplifiers for piezoelectric applications. Smart Mater Struct 5(766–775):282–284
3. Tseng SY, Chen YM, Huang YK, Hsieh HT, Wu TF (2004) Quasiresonant fly back converter for transdermal drug delivery applications. In applied power electronics conference and exposition. APEC'04. Nineteenth annual IEEE, 3:1653–1659
4. Silva J (2001) Control methods for power converters. Chapter in power electronics handbook-Rashid
5. Chandrasekaran S, Lindner DK (2000) Power flow through controlled piezoelectric actuators. J Intell Mater Syst Struct 11:469–481
6. Chandrasekaran S, Lindner DK, Leo DJ (2002) Effect of feedback control on the power consumption of induced-strain actuators. J Intell Mater Syst Struct
7. Chandrasekaran S, Lindner DK, Smith RC (2000) Optimized design of switching amplifiers for piezoelectric actuators. J Intell Mater Syst Struct 11:887–901
8. Stiebel C, Janocha H (2000) New concept of a hybrid amplifier for driving piezoelectric actuators

Chapter 28
Development of Technique for Making Ohmic Contacts to PEDOT-PSS Films

S. Bindu, R. Anil Kumar and M. S. Suresh

Abstract Conducting polymers (PEDOT-PSS) belong to a category of materials which are peizoresitive and are used in variety of applications like strain gauges, gas sensors and many more. Since these conducting polymers are semiconductors, four probe measurement of film resistance is necessary to avoid errors due to contact resistance and spreading resistances. In a device, it is cumbersome to make four probe measurements and convenient to make two probe measurements. Large differences in the work function of polymers and contacts metals lead to rectifying contacts rather than ohmic contacts. A two probe measurement technique of making ohmic contacts to PEDOT-PSS film pristine and doped with 5 % DMSO is presented. It has been shown that by coating the contact probes with carbon nano powder, ohmic contacts can be achieved. This method can be employed for all conducting polymer measurements.

Keywords Conducting polymers · PEDOT-PSS · Rectifying contacts · Ohmic contacts · Four probe measurements · Carbon nano powder

28.1 Introduction

Uses of conducting polymers are increasing day by day. With the synthesis of more and better conducting polymers, the range of applications to which they are put to is multiplying. Some of the devices possible with these materials are strain

S. Bindu (✉) · M. S. Suresh
Department of Electronics and Communication, B N M Institute of Technology, 12th main road, 27th cross, Banashankari 2nd stage, PB No: 7087 Bangalore, India
e-mail: bindu.ct@gmail.com

R. Anil Kumar
Honeywell Fellow, Honeywell Technology Solutions (P) Ltd, Bangalore, India

V. Chakravarthi et al. (eds.), *Proceedings of International Conference on VLSI, Communication, Advanced Devices, Signals & Systems and Networking (VCASAN-2013)*, Lecture Notes in Electrical Engineering 258, DOI: 10.1007/978-81-322-1524-0_28,

gauges, OLED, and gas sensors. The resistance of a conducting polymer varies when gasses are adsorbed on its surface which is used to build gas sensors while conducting polymers that show peizoresistive property are used in building strain gauges of high gauge factor [1–5]. In all these devices, the bulk resistance of the polymer film is measured which necessitates formation of a metal polymer contact. Measurement of resistance of conducting polymer films pose several challenges as they are semiconducting with band gaps in the range of 1.5 eV and work function in the same range [6–8]. Metals such as Au, Ag, Cu, and Al have work functions in the range 4.5 to 5 eV, and this poses a problem. Differences in the work function of polymers and metals lead to rectifying contacts rather than ohmic contacts. Many overcome this problem by making four probe measurements which is good enough for laboratory measurements but not for commercial devices such as strain gauges, where simple two probe measurement is essential.

The formation of rectifying contact at polymer—metal interface has been identified and studied by impedance spectroscopy by some workers. They have made detailed study using cole–cole plots identifying different aspects of the interface [9] including bulk resistance. However, they are cumbersome for application in devices which need measurement of just bulk resistance of a film as in gas sensors and strain gauges. This paper proposes a simple technique by which an ohmic contact can be made between PEDOT-PSS (poly 3,4ethylenedioxy-thiophene—polystyrenesulfonic acid) films pristine and doped with DMSO (Dimethyl sulfoxide) and copper using carbon nano powder. The efficacy of the contact has been measured.

28.2 Experimental Details

Films of PEDOT-PSS are drop cast on a kapton sheet of 120 μm thickness using PEDOT-PSS procured from H.C. Stark, as dispersion in water, with weight ratio of 1:6. Before casting of film the kapton substrate is etched in potassium hydroxide for 10 min to make it hydrophilic and then ultrasonicated in acetone, triple distilled water and isopropyl alcohol for 30 min each. The films cast are annealed at 50–55 °C for about 24 h so that water evaporates and a thin polymer film of about 10 μm thick, 15 mm length and 5 mm width is formed. Contact on to the films is made by roller pressure contacts which are used for making four probe measurements. Tinned copper rollers of 0.61 mm diameter spaced 5 mm apart are used to make measurements. Tinned copper rollers are used for all four contacts to show its rectifying nature. Tinned copper rollers coated with carbon nano powder as two outer contacts and uncoated tinned copper rollers as two inner contacts are used. For coating rollers with carbon nano powder, tinned copper rollers are dipped in carbon ink made from carbon nano powder of 25 nm size and dried for 4–5 h. Pressure on the contacts is applied using a C clamp which holds the polymer film between two rigid Perspex sheets. Measurement of sheet resistance is made by potentiometer technique shown in Fig. 28.1 using Fluke 179 digital multimeter by

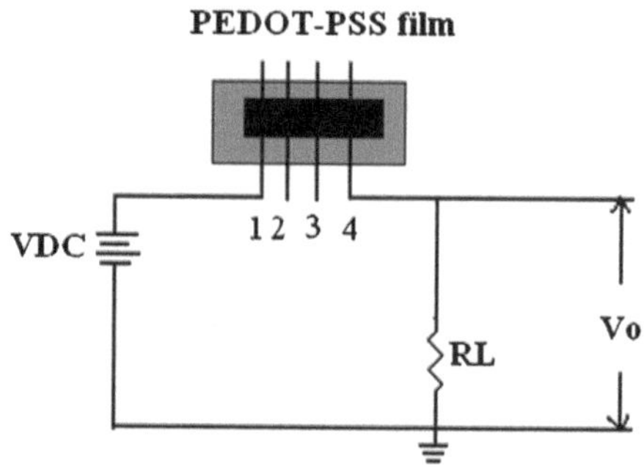

Fig. 28.1 Circuit used to measure film resistance using four probes

applying several DC voltages between contacts 1–4. The resistance of the film between 1–4 and 2–3 are determined by measuring the voltages at terminals V_1 (V_{DC}), V_2, V_3 and V_4 (V_O) and ground as below.

$$R_{14} = ((V_{DC}/V_O) - 1) * RL\ \Omega \tag{28.1}$$

$$R_{23} = (V_{23}/V_O) * RL\ \Omega \tag{28.2}$$

28.3 Results and Discussion

DC voltage is applied to the film between terminals 1 and ground and voltages between terminal 1, 2, 3, and 4 and ground are measured. From these voltages, potential across terminals 1–4 (V_{14}) and terminals 2–3 (V_{23}) are calculated. Resistance between 1–4 R_{14} may include contact resistances at those contacts which is very large if they are rectifying and small if ohmic. The resistance between 2–3 R_{23} does not include contact resistance, and it gives correct resistance between 2–3. If the contacts 1 and 4 are ohmic, the resistance R_{14} will be 3 times R_{23} as the contacts are equidistant. These can be determined by plotting voltage versus current which are linear if the resistance is constant. The plots of voltages between 1–4 and 2–3 versus current are shown in Fig. 28.2 for pristine PEDOT-PSS. The slopes of these plots give the resistance between these contacts which are fairly linear. The Fig. 28.2 also shows the equation that fits the results by regression and gives squared value of regression co-efficient which indicates the accuracy of fit. The resistance of pristine PEDOT-PSS film between contacts 1–4 is 238 KΩ while that between 2–3 is 30.75 KΩ. Correcting for the larger length of the film between 1–4 (three times that between 2–3) the film resistance between 1–4 would have been 92.5 KΩ but for the rectifying contacts. This shows that even though the V–I plot is linear the contact may be rectifying.

It is well known that the conductivity of PEDOT-PSS can be enhanced by several folds by doping with DMSO. In order to understand the behavior of doped PEDOT-PSS film, measurements were repeated with PEDOT-PSS doped with 5 %

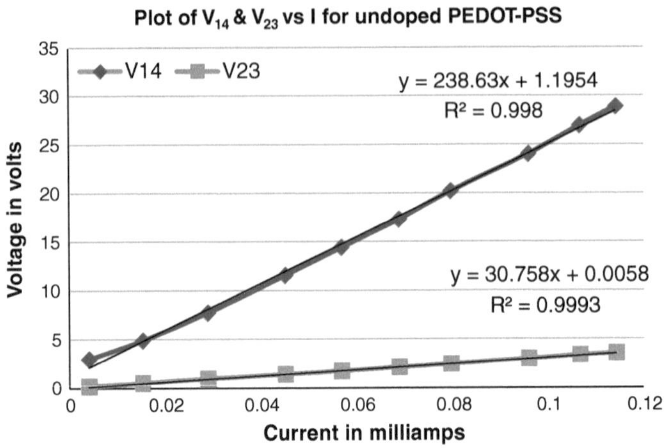

Fig. 28.2 V–I characteristics of undoped PEDOT-PSS film with tin-coated copper roller contacts

DMSO to investigate the change in the films resistance and also to study the rectifying behavior of the contacts. The results are shown in Fig. 28.3. It may be observed that V_{23} is linear with I indicating a constant resistance of 35.9 3Ω. However, the V–I plot of V_{14} with current is nonlinear similar to that of a diode with a break down voltage of 3.2 V. This indicates that tin-coated copper contacts make rectifying contact with PEDOT-PSS doped with 5 % DMSO.

Efforts were made to develop ohmic contact with PEDOT-PSS film. Carbon nano powder was selected as its work function is lower than that of PEDOT-PSS, and it is easy to prepare ink using powder. First, a layer of carbon nano powder was formed over the PEDOT-PSS film to make ohmic contact. Measurements show that this did make the contact ohmic but introduced large series resistance. Perhaps the ink reacts with the polymer film and increases resistance. Hence, a very thin layer of about 1 micrometer of carbon nano powder was formed over the tin-coated copper rollers. Measurements on doped PEDOT-PSS film were made

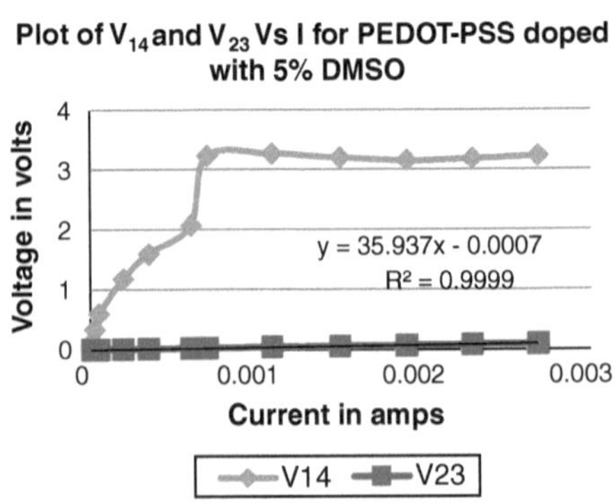

Fig. 28.3 V–I plot of doped PEDOT-PSS film with tin-coated copper contacts

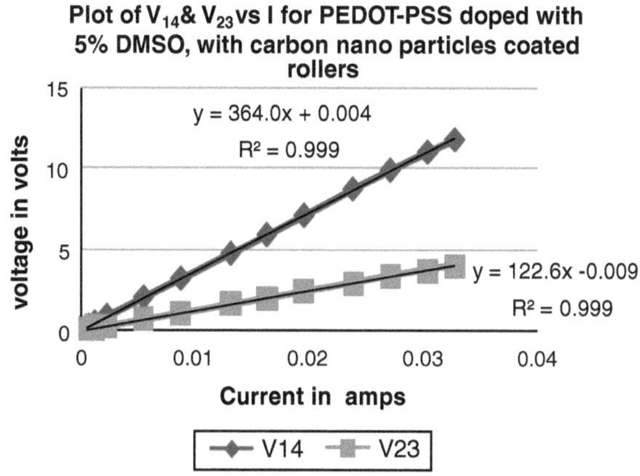

Fig. 28.4 V–I plot of doped PEDOT-PSS film with uncoated inner and Carbon nano powder coated outer electrodes

using carbon nano powder coated outer rollers which carry current and uncoated inner rollers. The results are shown in Fig. 28.4 which shows a linear relation between voltage and current between contacts 1–4 and 2–3. The measured resistance between 1–4 is 364 Ω which is about 3 times the resistance between contacts 2–3 of 122.6 Ω indicating formation of ohmic contact between PEDOT-PSS film and roller contacts. This also indicates that the series resistance added is negligible. The straight line fit is so good that the R^2 is 0.999.

More experiments are needed to ascertain the mechanism by which the contacts become ohmic. But one can guess that matching of work functions and tunneling could both be responsible.

28.4 Conclusion

It has been established that a tin-coated copper contact produces rectifying contact and cannot be used in devices. However, a thin coating of carbon nano powder on the tin produces an ohmic contact. The thickness of carbon layer should be very thin to reduce series resistance. A thin layer of carbon nano powder could be coated on the PEDOT-PSS film over which a metal (tin) could make ohmic contact in devices.

Acknowledgments We thank management of B N M Institute of Technology, Bangalore, for permission and support in carrying out the work. We also thank Dr Satish Amrutrao Patil from SSCU, IISc, Bangalore and Dr Praveen C R, Dept of Materials Engineering, IISc, Bangalore, for their help.

References

1. Cheng CH, Chao C, Cheung Y-N (2009) Piezoresistive strain gauge using doped polymeric fluid, United States Patent
2. Cheng C-H, Chao C, Cheung Y-N (2009) Fluidic piezoresistive strain gauge, United States Patent
3. Axelrod BW, Roukes ML (2008) Piezoresistive sensors, United States Patent
4. Mateiu R, Lillemose M, Hansen TS, Boisen A, Geschke O (2007) Reliability of Poly3,4ethylenedioxythiophene strain gauge. Micro Electron Eng 84:1270–1273
5. Lillemose M, Spieser M, Christiansen NO, Christensen A, Biosen A (2008) Intrinsically conductive polymer thin film. Piezoresistors 85:969–971
6. Pratt C (1996) Conducting polymers. D. Reidel Publishing Company, 1996
7. Petr A, Zhang F, Peisert H, Knupfer M, Dunsch L, Electrochemical adjustment ofthe work function of a conducting polymer, 207th ECS Meeting, Abstract #277, copyright ECS, Leibniz-Institut für Festkörper-und Werkstofforschung Dresden, Dresden, Germany
8. Schweizer TM (2005) Electrical characterization and investigation of the piezoresistive effect of PEDOT: PSS thin films. Thesis, Georgia Institute of Technology, April 2005
9. Lange U, Mirsky VM (2008) Separated analysis of bulk and contact resistance of conducting polymers: comparison of simultaneous two- and four-point measurements with impedance measurements

Chapter 29
A Comparative Study of Performance of Ring Resonator and Disk Resonator in Gas Sensing Applications

P. Rekha and M. S. Suresh

Abstract Upon exposure to gases, the surface properties of vapor-sensitive materials such as carbon nanotubes or conducting polymers change. These manifest as change in resistance or relative permittivity which are a measure of the gas concentration. A circular disk resonator using shift in resonant frequency as a measure to sense the gas has been reported for gas-sensing properties of carbon nanotubes. The sensor detected the presence of gases such as helium, oxygen, ammonia with a resonant frequency shift of 0.8, 2.3, and 3.55 MHz, respectively. Ring resonator is another promising resonating structure which can be explored for gas sensing applications. For gas sensing, based on resonant shift technique ring resonator offers more area for the gas to interact than the disk structure. Present research work aims at studying the resonating property of the ring resonator, analyze the performance, and compare with the existing results of the disk sensor using simulation tool HFSS. The gas adsorption is simulated using the change in relative permittivity (ε_r) of the substrate. Simulation results indicate that ring resonator exhibits higher selectivity compared to disk resonator. This improved selectivity makes the ring resonator detect more variety of gases compared to disk.

Keywords Gas sensing · Carbon nanotubes · Relative permittivity · Resonant frequency · Selectivity · HFSS · Disk resonator · Ring resonator

---

P. Rekha (✉) · M. S. Suresh
B N M Institute of Technology, Bangalore, India
e-mail: rekhap@bnmit.org

M. S. Suresh
e-mail: mskurke@scientist.com

V. Chakravarthi et al. (eds.), *Proceedings of International Conference on VLSI, Communication, Advanced Devices, Signals & Systems and Networking (VCASAN-2013)*, Lecture Notes in Electrical Engineering 258, DOI: 10.1007/978-81-322-1524-0_29,

29.1 Introduction

Gas sensing is an intensive research area. It is necessary to detect leak of explosive or toxic gases in industries. With increasing concern about global warming, it is essential to monitor and control the environment. High-performance gas sensors are in demand for identifying gases in the planetary atmosphere. Hence, highly sensitive, selective, fast, low-power and stable gas sensors are in demand.

Most common gas-sensing principle is the adsorption of gas molecules on sensing materials. Upon exposure to gases, the surface properties of vapor-sensitive materials change. Semiconductor metal oxide is a traditional material used in gas sensing. But it has the main drawback of high-operating temperature (200 to 500 °C) [1]. So, alternative materials are being investigated for gas sensing. Carbon nanotubes (CNT) is one such promising material. Inherent properties of CNT such as high specific surface area 1,580 m^2/g [2], porous surface are advantageous in gas sensing. High specific area of CNT offers ample sites for gas interaction resulting in more variation in surface properties of the gas sensor. These variations manifest as change in resistance or relative permittivity of the sensor. Change in permittivity of the sensor due to adsorption of gases manifest as change in magnitude of reflection coefficient and the resonance frequency of microstrip structures such as patch, ring.

Chopra and co-workers [2, 3] have reported gas-sensing property of CNT using a circular disk resonator. Single-wall CNT were coated on top of the disk using a conductive epoxy is shown in Fig. 29.1. The resonator detected gases such as helium(He), argon(Ar), nitrogen(N_2), oxygen(O_2), carbon monoxide(CO), and ammonia (NH_3) resulting in a resonant frequency shift of 0.8, 1.2, 1.91.2, 2.31.2, 3.41.2, 3.55 MHz, respectively. Figure 29.2 shows shift in resonant frequency for He, O_2, and NH_3 gases. The shift is measured using variation in return loss with frequency. The relation between resonant frequency and relative permittivity is

$$\mathrm{fr}_1 = 1.841\mathrm{c}/\left(2\pi \mathrm{a}\sqrt{(\varepsilon_r)}\right) \tag{29.1}$$

where fr_1 is the resonant frequency of the resonator, c is the speed of light in vacuum, "a" is the radius of the disk, and ε_r is the relative permittivity of the substrate.

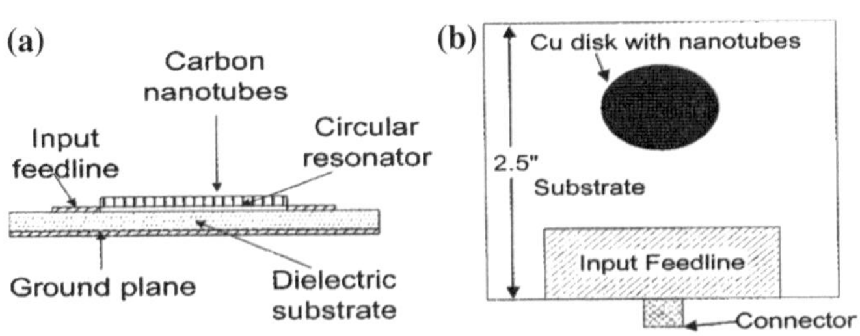

Fig. 29.1 **a** *Front* view of the disk resonator [3] disk. **b** *Top* view of the resonator [3]

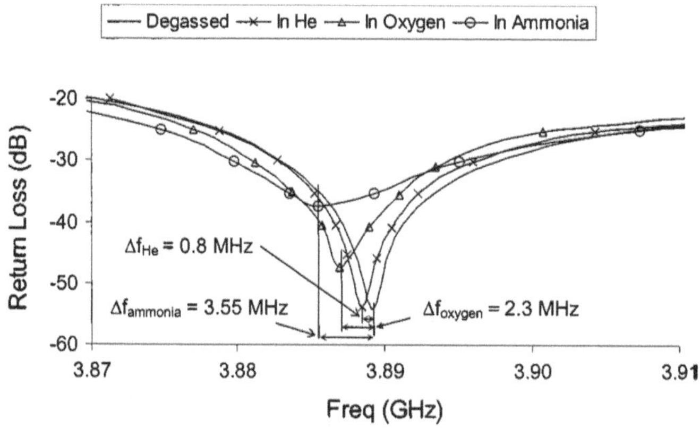

Fig. 29.2 Shift in resonant frequency when the sensor was exposed to He, O_2 and NH_3 [2]

In an effort to improve the performance of the gas sensor, alternate microstrip structure such as ring resonator was looked into. Ring resonator offers more dielectric space for interaction compared to disk. Present work compares the performance of disk resonator and ring resonator in gas sensing applications using simulations.

29.2 Modeling and Simulations

A disk resonator similar to the one reported in [2] was created using High-Frequency Structure Simulation (HFSS) tool. HFSS is a high-performance electromagnetic field simulator for 3D device modeling. HFSS can be used to calculate S-parameters, resonant frequency. Figure 29.3 shows the model of disk resonator using HFSS and the simulation results. Device specifications are: substrate: Rogers RO4003 (tm) with $\varepsilon_r = 3.37$ [2], thickness = 2 mm; Disk: copper with radius = 19.75 mm; Fig. 29.3 shows the results, and the resonant frequency is 3.8447 GHz.

Upon exposure to gases, ε_r of the CNT increases resulting in an increase in effective ε_r of the resonator [4]. This increase in ε_r, downshifts the resonant frequency. The disk modeled previously using HFSS was tested for its resonant performance with increasing ε_r, values. ε_r of the substrate was varied from 3.37 to 4, and the results are shown in Fig. 29.4.

Table 29.1 shows the variation of resonant frequency with ε_r of the substrate and selectivity of the disk resonator.

As second part of the simulations, a ring resonator was modeled using HFSS. Ring resonator offers more substrate space for interaction than to the disk. Figure 29.5 shows the 3-D model and simulation results. Device specifications are: Mean diameter of the ring = 19 mm with all other parameters same as disk.

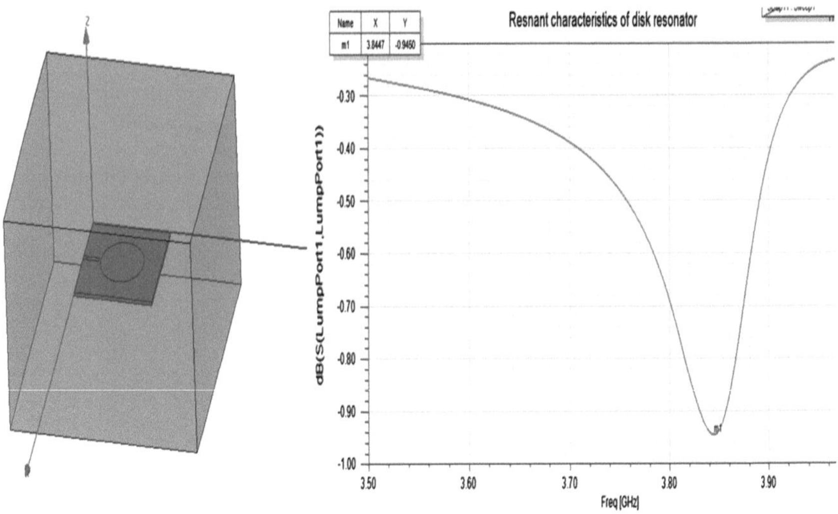

Fig. 29.3 3-D Model of the disk resonator using HFSS and the resonant characteristics (variation of return loss, S_{11} as a function of frequency)

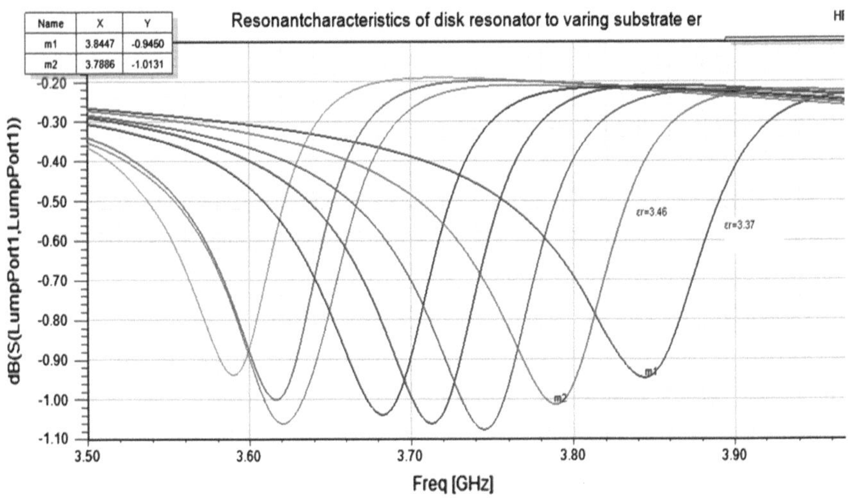

Fig. 29.4 Resonant characteristics of disk resonator to varying ε_r of the substrate

Relation between resonant frequency and the relative permittivity of the ring is given by

$$fr_2 = n * c / (2\pi a\sqrt{(\varepsilon_r)}) \tag{29.2}$$

Table 29.1 Change in resonant frequency with ε_r of the substrate and selectivity of the disk resonator

ε_r of the substrate	Resonant frequency of the disk resonator in GHz	Bandwidth (BW) in GHz/0.1 dB	Selectivity = f_r/BW
3.37	3.8447	0.0411	93.54
3.46	3.7885	0.0351	107.93
3.55	3.7454	0.0321	116.67
3.64	3.7124	0.0311	119.36
3.73	3.6823	0.034	108.30
3.82	3.6212	0.032	113.16
3.91	3.6162	0.03	120.54
4.00	3.5901	0.0291	123.37

where fr_2 is the resonant frequency of the resonator, n is number of half wavelengths around the ring, c is the speed of light in a vacuum, a is the mean radius of the ring, and ε_r is the relative permittivity of the substrate [5].

Simulation results in Fig. 29.5 shows a resonant frequency of 1.6052 GHz. Further, ring resonator was tested for resonant characteristics for increasing ε_r of the substrate. Figure 29.6 indicates simulation results for varying ε_r of the substrate from 3.37 to 3.46.

Table 29.2 shows the variation of resonant frequency with ε_r of the substrate and selectivity of the ring resonator.

Table 29.3 compares the selectivity of disk and ring resonator with variation in ε_r of the substrate.

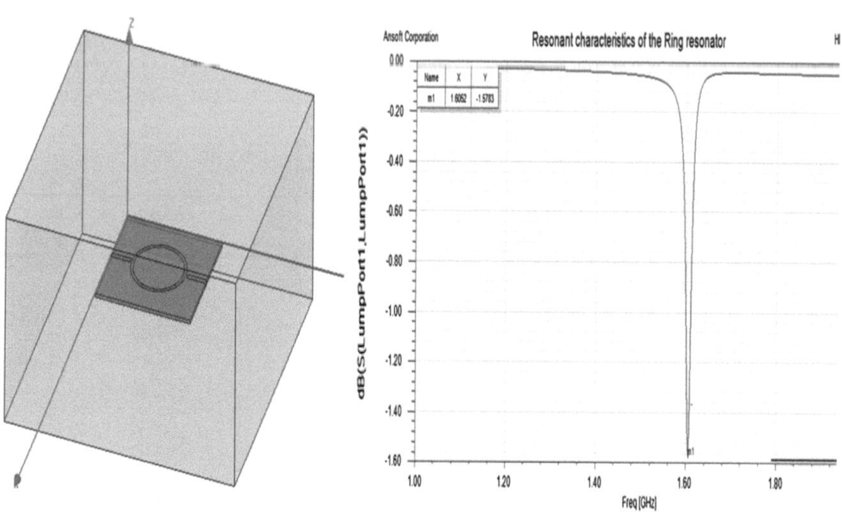

Fig. 29.5 3-D model of ring resonator in HFSS and the resonant characteristics

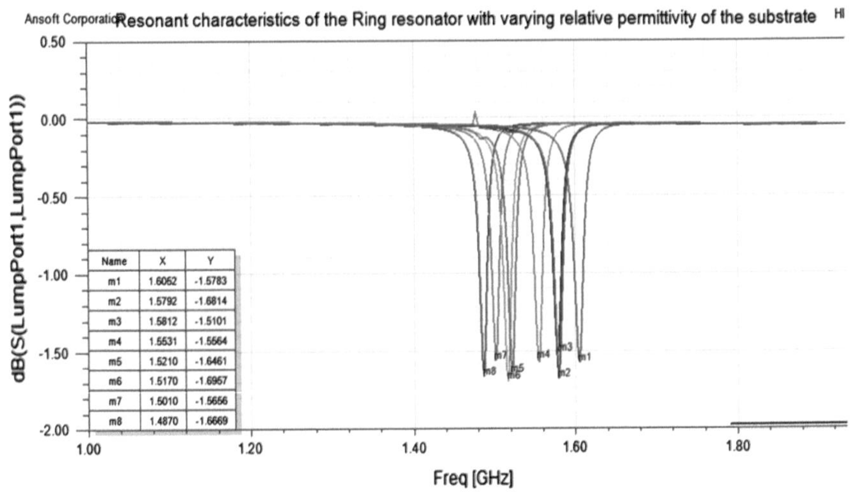

Fig. 29.6 Resonant characteristics of ring resonator to varying ε_r of the substrate

Table 29.2 Change in resonant frequency with ε_r of the substrate and selectivity of the ring resonator

ε_r of the substrate	Resonant frequency of the ring resonator in GHz	Bandwidth (BW) in GHz/0.1 dB	Selectivity = fr/BW
3.37	1.6052	6e-3	267.53
3.46	1.5791	4e-3	394.77
3.55	1.5771	6e-3	262.85
3.64	1.5551	6e-3	259.18
3.73	1.5210	6e-3	253.50
3.82	1.5170	6e-3	252.83
3.91	1.5010	6.1-3	246.06
4.00	1.4869	4e-3	371.72

Table 29.3 Comparison of selectivity of disk and ring resonator with variation in ε_r of the substrate

ε_r of the substrate	Selectivity of the disk resonator = f_r/BW	Selectivity of the ring resonator = f_r/BW
3.37	93.54	267.53
3.46	107.93	394.77
3.55	116.67	262.85
3.64	119.36	259.18
3.73	108.30	253.50
3.82	113.16	252.83
3.91	120.54	246.06
4.00	123.37	371.72

Values in Tables 29.1 and 29.2 indicate that the resonant frequency downshifts for increase in ε_r of the substrate from 3.37 to 4 for both disk and ring resonator. Table 29.3 compares the selectivity of disk resonator and the ring resonator. Results clearly indicate that ring has better selectivity compared to disk.

29.3 Conclusions

Table 29.3 indicates that ring has better selectivity compared to disk resonator. This shows that the same ring resonator gas sensor can detect more varieties of gases than disk sensor. Thus, ring resonator is a good alternate to disk resonator in gas-sensing applications.

References

1. Arshak K, Moore E, Lyons GM, Harris J, Clifford S (2004) A review of gas sensors employed in electronic nose applications. Sensor Review Emerald Group Publishing Limited. ISSN 0260-2288, 24(2):181–198
2. Chopra S, McGuire K, Gothard N, Rao AM, Pham A (2003) Selective gas detection using a carbon nanotube sensor. Appl Phys Lett 83(11):2280–2282
3. Chopra S, Pham A (2002) Carbon-nanotube-based resonant-circuit sensor for ammonia. Appl Phys Lett 80(24):4632–4634
4. Grujicic M, Cao G, Roy WN (2004) A computational analysis of the carbon nanotube based resonant circuit sensor. Appl Surf Sci 229:316–323
5. Amitec Mic10 Microstrip Trainer Experimental Manual Apss

Chapter 30
Simulation and Optimization of Channel Mobility in High-k/Metal Gate Nanoscale MOSFETs

Saptarshi Basak, Shashank Nagaraj and Rajendra K. Nahar

Abstract The performance of submicron MOSFET with a thin conventional SiO_2 gate dielectric degrades due to increasing gate leakage currents. High-k dielectric materials are used as alternate gate dielectric to overcome the problem. The introduction of high-k dielectric induces high interface charges which degrades both the mobility and threshold voltage of MOS device. Simulation of devices with suitable tools and models helps us to mimic the device performance. Often errors are possible during selecting the models and regions of probing during simulation of results. This paper investigates the role of interface charges on carrier mobility and other MOSFET parameters. The probing positions for extraction of mobility are optimized by simulating the mobility at various positions along the channel and at various depths in the channel. From simulation results, it is shown that higher mobility is obtained by probing in the middle of the channel, 1 nm below the HfO_2–Si interface. The performance of the high-k MOSFET with metal electrode and polysilicon electrode is also compared for various interface state charges.

Keywords Metal gate · Mobility · MOSFET · High-k · HfO_2

S. Basak
Birla Institute of Technology and Science (BITS), Pilani 333031, India
e-mail: writetosap001@gmail.com

S. Nagaraj (✉)
Uniphos Envirotronic Pvt. Ltd, Vapi 396195 Gujarat, India
e-mail: nshash@gmail.com

R. K. Nahar
Central Electronics Engineering Research Institute, Pilani 333031, India
e-mail: rknahar@gmail.com

V. Chakravarthi et al. (eds.), *Proceedings of International Conference on VLSI, Communication, Advanced Devices, Signals & Systems and Networking (VCASAN-2013)*, Lecture Notes in Electrical Engineering 258, DOI: 10.1007/978-81-322-1524-0_30,

30.1 Introduction

High-permittivity dielectric materials continue to be investigated for use as replacement dielectric for silicon dioxide in Metal Oxide Field effect Transistors (MOSFET) in the deep submicron regime [1, 2]. In devices with a channel length of 100 nm or below, the oxide thickness reduces to below 3 nm as per scaling laws. Dielectric thicknesses lower than 3 nm result in quantum mechanical tunneling through the device. This results in tunneling of charge carriers through the gate oxide into the channel and increase in leakage current through the device [3]. Passage of charge carriers through the thin oxide layer may result in creation of traps which further results in oxide degradation and breakdown [4, 5]. For this reason, high-k materials are used to physically increase the thickness of the gate dielectric without decreasing the effective gate capacitance.

The use of high-k material has its own issues as well. Hafnium-based materials are more promising and have been widely investigated. As-deposited HfO_2 layer have a number of defects which results in increased fixed oxide and interface charges [6, 7]. The magnitude of interface charge in Hf-based materials ($\sim 10^{12}$ cm^{-2}) is considerably higher than that in thermally grown SiO_2 [8]. These charges are formed during the high-k deposition and/or postdeposition processing which affect the device performance. Hence, the role of interface charges on device figures of merit (like threshold voltage (V_T), transconductance, output resistance (R_O), Drain current (I_D), subthreshold swing (1/S)) should be investigated. These charges also have a role to play in determining the effective mobility of the charge carriers in the channel as well as in the gate leakage current through the devices.

One of the unresolved issues that still affect high-k MOSFETs is the reduction of channel effective carrier mobility. Channel mobility degradation is one of the major issues of high-k/metal gate MOSFETs [9, 10]. Various sources of this degradation have been identified like Remote Coulomb Scattering (RCS), Remote soft-optical Phonon Scattering (RPS), and material parameter variations like interfacial layer thickness, permittivity and roughness. Yet there has been no consensus on which particular phenomenon dominates [11, 12].

In our work, we have designed and simulated 0.1 μm MOSFETs with HfO_2 gate dielectric. Two types of gate electrodes Poly-Si and TiN are chosen for our present study to understand the merits of high-k/metal gate technology. The high-k materials have a considerably larger number of interface charges as compared to silicon dioxide as reported in the literature. The simulated results show the variations in transistor figures of merit for various interface charge densities. Further, emphasis is given to understand the mobility aspects. The need for selecting the suitable mobility models during simulation is also explained. It has been observed that high-k/metal gate MOSFETs have high mobility degradation when the interface charges increases lending credence to the fact that coulomb scattering plays a major role at high charge concentrations.

30.2 Design and Simulation

In our study, we have process modeled 0.1 μm n-channel MOSFETs using HfO_2 as gate dielectric material. Two MOSFET structures, one with Poly-Si gate and the other with TiN metal gate electrode were designed using ATHENA 2D simulator of the SILVACO© suite. The simulated structures, which are based on fully scaled 100 nm gate length MOSFET's proposed in the International Technology Roadmap for Semiconductors (ITRS) [13], have gate length of 100 nm, with effective oxide thicknesses (EOT) of 3 nm. The dielectric constant of HfO_2 gate dielectric was considered to be 20 [14]. The substrate doping concentration is $N_{sub} = 3 \times 10^{17}\ cm^{-3}$. The source/drain extensions and deep source/drain junction depths are 45 and 75 nm, respectively. Titanium Silicide is used at source drain contacts to reduce the sheet resistance. The gate work-function for TiN is chosen to be 4.5 and 4.17 for Poly-Si. Once the process modeling is done in ATHENA, the device is simulated in ATLAS to determine the device characteristics. The device parameters were studied for three surface state densities viz. 5×10^{11} Charges/cm^2, 1×10^{12} Charges/cm^2, 5×10^{12} Charges/cm^2. The schematic of the design structure with TiN metal gate is as shown in Fig. 30.1.

30.3 Results and Discussion

We have investigated the effect of interface charges on the performance of both Poly-Si and metal gated MOSFETs. It is seen from Fig. 30.2 that threshold voltage of both the design structures decreases with increase in interface charges and the

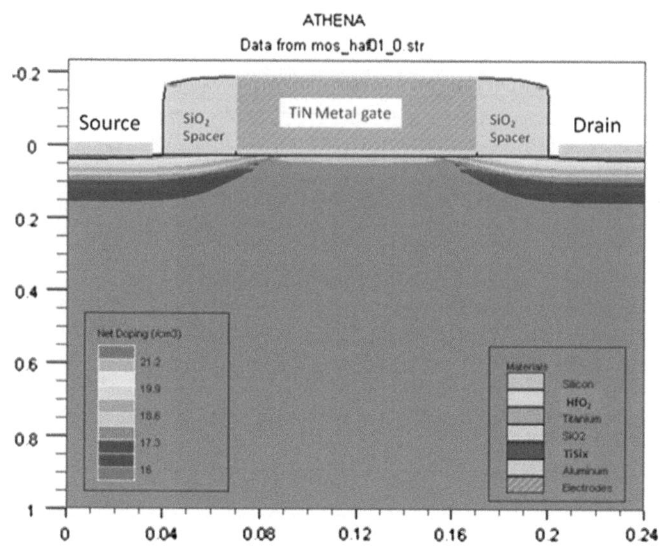

Fig. 30.1 Simulated structure of MOSFET with TiN gate

degradation is more in case of Poly-gated device structure than device with metal gate at higher interface charge densities. The threshold voltage degradation results in increased ON current (I_{ON}) in the devices with increase in interface charge density. The transconductance (g_m) also decreases for the same. Subthreshold swings (1/S) and drain current (I_D) increases. On comparison with poly-Si electrode, metal gate electrodes seem to perform better with a more I_D, transconductance and lower subthreshold swing. This proves that high-k dielectrics perform better in combination with metal gate electrodes. The simulated results are summarized in Table 30.1.

30.3.1 Selection of Appropriate Mobility Models

The total mobility in the inversion layer is given by Matthiessen's rule which is formulated as follows:

$$\frac{1}{\mu\text{total}} = \frac{1}{\mu\text{ph}} + \frac{1}{\mu\text{sr}} + \frac{1}{\mu\text{coul}} \tag{30.1}$$

where μtotal is the total mobility in the inversion region, μph is mobility due to phonon scattering, μsr is mobility due to surface scattering, and μcoul is mobility due to coulombic scattering. Matthiessen's rule summation is strictly correct when the scattering mechanisms that are being added are independent, but not mutually exclusive [15]. Numerous models have been included in the ATLAS viz. CVT (Lombardi's model) [16], Watt [17], modified Watt, SHI (Shirahata's model) [18], CONMOB (concentration dependent low field mobility), FLDMOB (transverse electric field dependent mobility), KLA (Klaassens' low field mobility) [19, 20],

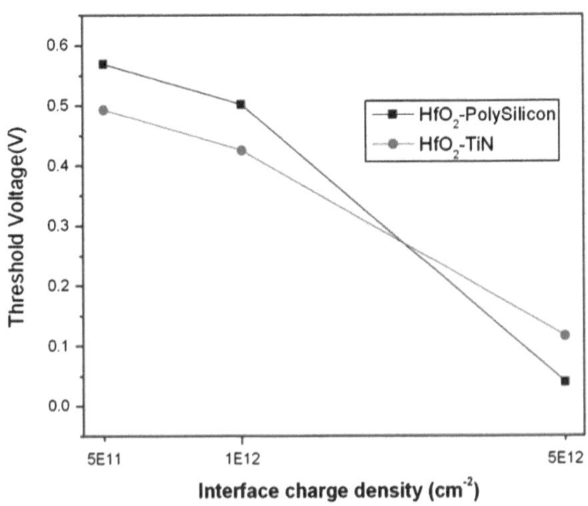

Fig. 30.2 Threshold voltage degradation of Poly-Si and TiN gate MOSFETs

Table 30.1 Simulated results of MOSFET structures for various interface charge densities

Interface Charges (cm^{-2})	Design Structure	V_T (V)	I_D (mA/ μm)	g_m ($\times 10^{-4}$ mho)	1/S (mV/ decade)	R_O (Ohms)
5×10^{11}	HfO_2–TiN	0.49	0.73	7.12	74.48	13159
	HfO_2–Poly-Si	0.56	0.31	5.86	85.66	20573
1×10^{12}	HfO_2–TiN	0.42	0.79	7.15	75.59	12768
	HfO_2–Poly-Si	0.50	0.36	5.92	85.62	19166
5×0^{12}	HfO_2–TiN	0.11	1.32	6.75	196.58	10488
	HfO_2–Poly-Si	0.04	0.81	5.59	162.5	13104

modified CVT, Yamaguchi [21], and TASCH model. Most of these models have the Matthiessen's rule built within them. However, all these models have some trade-off.

The Lombardi model [16] is physically based on phonon and surface roughness scattering and does not account for coulomb scattering due to interface charges. It includes all the effects of parallel as well as perpendicular electric field but cannot be used for channel doping in excess of 10^{17} cm^{-2} as it fails to follow the universal mobility curve [22]. The Shirahata model [18] takes Coulombic scattering into consideration but has been found that the approach is purely empirical with no physical basis [23]. Even though Yamaguchi [16] model includes the effect of all three scattering mechanisms, the Coulombic scattering term does not take into account the effect of interface trap charges. The Watt [21] model includes all the effects of phonon scattering, surface roughness scattering, and charge impurity scattering effects caused by the inversion charge carriers and the ions located in the oxide and interfaces. This was in line with our requirements. Hence, this model was selected specifically to show the scattering effects due to interface charges in high-k materials. The role of interface charges on the high field mobility of Poly-Si and TiN metal gated MOSFET structures are shown in Figs. 30.3 and 30.4, respectively. It can be observed that the electric field increases with the increase in interface charges while the mobility degrades.

The components affecting the effective mobility of the carriers in the transistor channel are the horizontal electric field, vertical electric field, and carrier velocity. As devices are reduced in size, the electric field typically also increases and the carriers in the channel have an increased velocity. However, at high fields, there is no longer a linear relation between the electric field and the velocity. When the horizontal electric field is high, the carriers in the transistor channel are accelerated to a maximum velocity. Above a critical field, the velocity is no longer related to electric field and reaches a constant level. This velocity saturation is caused by the increased scattering rate of highly energetic electrons, primarily due to optical phonon emission. This effect increases the transit time of carriers through the channel. When the vertical electric field is high the carriers in an n-channel device are strongly attracted to the silicon surface where they rebound. This results in surface scattering causing a reduction in the recombination time and the mobility

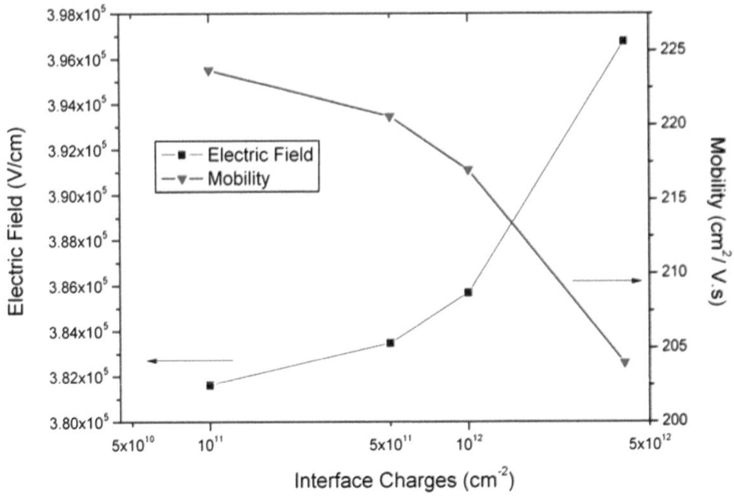

Fig. 30.3 Field dependent mobility of Poly-Si-gated MOSFET structures for various interface charge densities

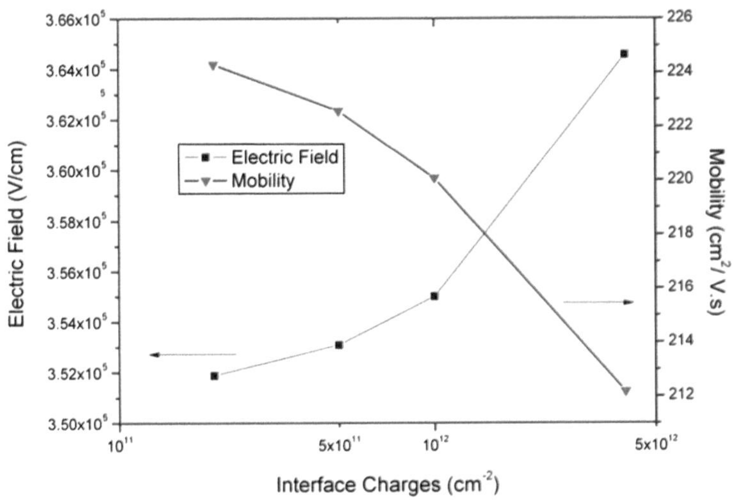

Fig. 30.4 Field dependent mobility of TiN-gated MOSFET structures for various interface charge densities

of the carriers [23]. Surface scattering due to irregularities of the surface near the interface and coulomb scattering due to presence of interface charges very near to the interface can be considered as the main reasons for mobility degradation.

The mobility of both the device structures (HfO_2-Poly and HfO_2-TiN) was probed at different locations along the channel-X1 near the source, X2 middle of

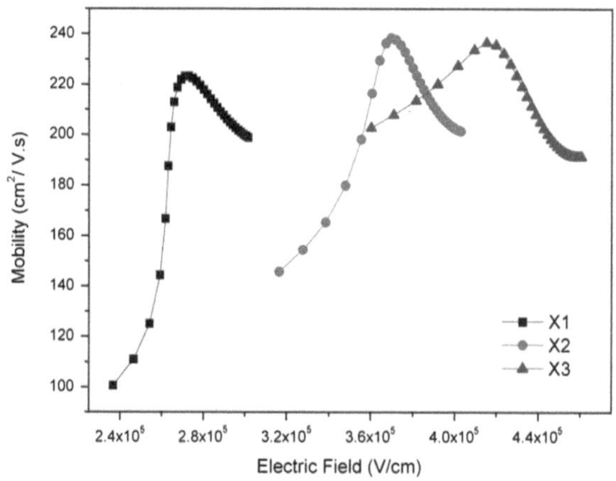

Fig. 30.5 Mobility of HfO_2-Poly gate devices at different locations along the channel

the channel and X3 near the drain. From Figures 30.5 and 30.6, it can be noticed that the mobility curves shift toward right along the electric field axis. This explains the presence of high electric field near the drain. Even though there is not much difference in the peak mobilities, probing the mobility at the middle of the channel seems to give better numbers than at the source or drain edges. It can also be observed that the mobility attains its peak at a lower electric field in metal gate structures compared to poly gate structures.

The MOSFET design structures were probed at different depths from the interface to extract the mobility. Three locations Y1 very near to the surface just 1 nm below the HfO_2-Si interface, Y2 which is 3 nm below the interface and Y3 still deeper which is 13 nm below the interface. All the three regions were considered to lie in the inversion layer of the channel. It can be observed from Figs. 30.7 and 30.8 that the mobility curves shift toward the left as we probe deeper from the interface. This shows that the electric field is decreasing as we probe deeper and thus the mobility seems to be following the universal mobility curves at low electric fields. Good numbers for mobility (fact that some of the mobility numbers were comparable to the ones published in literatures) were obtained when probed below 1 nm from the interface [24, 25]. Hence, optimizing the probing region during simulating the results is very important to get accurate values. As we probe still deeper into the bulk, the mobility will eventually equal the bulk mobility.

30.3.2 Graphs and Tables

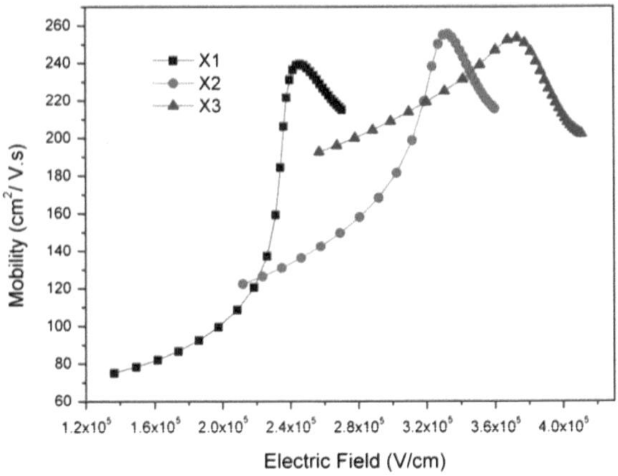

Fig. 30.6 Mobility of HfO_2–TiN gate devices at different locations along the channel

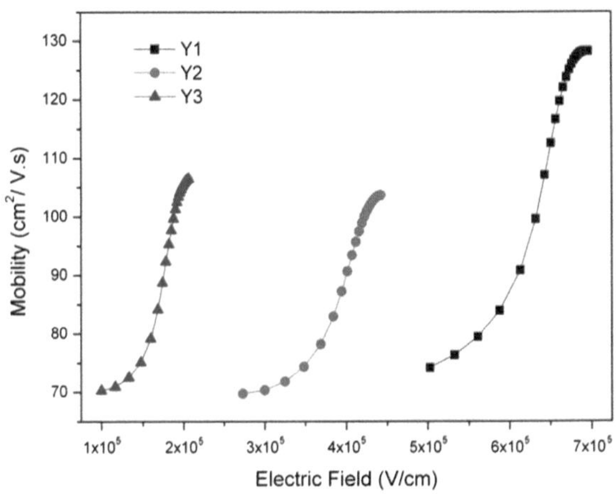

Fig. 30.7 Mobility of HfO_2-Poly gate devices at different depths from HfO_2 to Si the interface

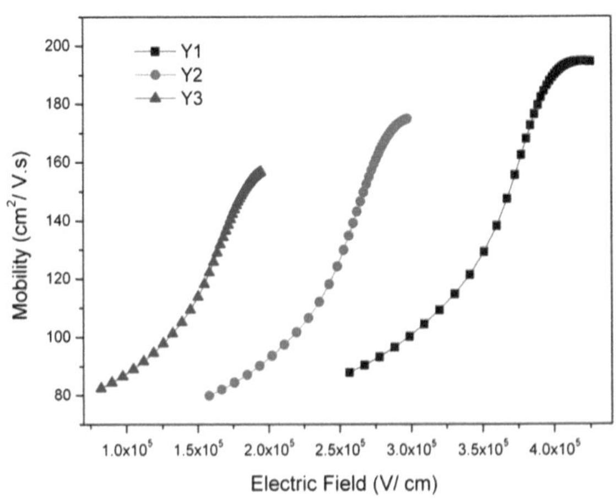

Fig. 30.8 Mobility of HfO_2–TiN gate devices at different depths from the HfO_2–Si interface

30.4 Conclusion

The role of interface charges on the high-k-based MOSFETs has been compared for Poly-Si and metal gate electrodes. At high interface charge densities, the MOSFET with Poly-Si gate are found to degrade more compared to metal gate structures. The replacement of polysilicon with metal gate seems to tackle the mobility degradation issue in high-k based devices to a greater extent. The non-linearity of mobility curves at high electric fields has been studied for various interface trap densities. The Watt model was found to be the most suitable model for mobility extraction which includes most of the scattering phenomena. The deviation in the mobility curves with various regions of probing shows the need for optimization during simulation of devices.

Acknowledgments The authors thank the Director, CEERI for his encouragement during the course of this work. Financial support from Council of Scientific and Industrial Research (CSIR), New Delhi, India, under Emeritus Scientist research scheme is greatly acknowledged. The authors would also like to thank members of Sensors and Nanotechnology Group, CEERI, Pilani for their insightful discussions.

References

1. Fishetti M (2003) Scaling MOSFETs to the limit: a physicist's perspective. J Comput Electron 2:73–79
2. Wilk GD, Wallace RM, Anthony JM (2001) High-κ dielectrics: current status and material considerations. J Appl Phys 89(10):5243–5275
3. Jeong DS, Hwang CS (2005) Tunneling-assisted Poole-Frenkel conduction mechanism in HfO_2 thin films. J Appl Phys 98(1137):113701
4. DiStefano TH, Shatzkes M (1974) Impact ionization model for dielectric instability and breakdown. Appl Phys Lett 25(12):685–687
5. Chen IC, Holland S, Hu C (1985) Electrical breakdown of thin gate and tunneling oxides. IEEE Trans Electron Dev ED-32:413
6. Gusev EP, Cartier E, Buchanan DA, Gribelyuk M, Copel M, Okorn-Schmidt H (2001) Ultrathin high-k metal gate oxide on silicon: processing, characterization and integration issues. Microelecton Engg 59:341–349
7. Zhu WJ, Ma TP, Zafar S, Tamagawa T (2002) Charge trapping in ultrathin hafnium oxide. IEEE Electron Dev Lett 23:597–599
8. Nahar RK, Singh V, Sharma A (2007) Study of electrical and microstructure properties of high dielectric hafnium oxide thin film for MOS devices. J Mat Sci Mater Electron 18:615–619
9. Fischetti M, Neumayer D, Carttier E (2001) Effective electron mobility in Si inversion layers in MOS systems with a high-k insulator: the role of remote phonon scattering. J Appl Phys 90:4587–4608
10. Kotlyar R, Giles MD, Matagne P (2004) Inversion mobility and gate leakage in high-k/metal gate MOSFETs. Electron Devices Meeting, 2004. IEDM Technical Digest. IEEE International, pp 391–394
11. Weber O, Andrieu F, Casse M (2004) Experimental determination of mobility scattering mechanism in Si/HfO_2/TiN and SiGe:C/HfO_2/TiN surface channel n- and p-MOSFETs. IEDM Tech Dig, pp 867–870

12. Rhee S (2004) Effects of varying interfacial oxide and high-*k* layer thicknesses for HfO_2 metal oxide semiconductor field effect transistor. Appl Phys Lett 85(7):1286–1288
13. ITRS Roadmap (2007) San Jose, CA: Semiconductor industry association
14. Sen B, Sarkar C, Wong H, Chan M, Kok CW (2006) Electrical characteristics of high-k dielectric film grown by direct sputtering method. Solid State Electron 50:237–240
15. Lundstrom M (1990) Fundamentals of carrier transport. Addison-Wesley, Reading
16. Lombardi C (1988) A physically based mobility model for numerical simulation of non-planar devices. IEEE Trans CAD 7:1164–1170
17. Watt JT (1989) Ph.D. Thesis, Stanford University
18. Shirahata M, Kusano H, Kotani N, Kusanoki S, Akasaka Y (1992) A mobility model including the screening effect in MOS inversion layer. IEEE Trans Comput Aided Des 11(9):1114–1119
19. Klaassen DBM (1992) A unified mobility model for device simulation—I. Model equations and concentration dependence. Solid State Electron 35(7):953–959
20. Klaassen DBM (1992) A unified mobility model for device simulation—II. Temperature dependence of carrier mobility and lifetime. Solid State Electron 35(7):961–967
21. Yamaguchi K (1983) A mobility model for carriers in the MOS inversion layer. IEEE Trans Electron Dev ED-30:658–663
22. Mujtaba SA (1995) Advanced mobility models for design and simulation of deep sub-micrometer MOSFETs. Ph.D thesis. Stanford University
23. Veendrick HJM (1992) MOS ICs-from basics to ASICs. Weinheim. New York. Basel. Cambridge.VCH Publishers Inc
24. Casse M, Thevenod L, Guillaumot B, Tosti L, Martin F, Mitard J, Weber O, Andrieu F, Ernst T, Reimbold G, Billon T, Mouis M, Boulanger F (2006) Carrier transport in HfO_2/Metal Gate MOSFETs: physical insight into critical parameters. IEEE Trans Electron Dev 53(4):759–768
25. Lime F, Oshima K, Casse M, Ghibaudo G, Cristoloveanu S, Guillaumot B, Iwai H (2003) Carrier mobility in advanced CMOS devices with metal gate and HfO_2 gate dielectric. Solid State Electron 47:1617–1621

Part V
Signals and Systems

Chapter 31
An Improved Artificial Neural Network Based Emotion Classification System for Expressive Facial Images

G. Shivakumar and P. A. Vijaya

Abstract Developing systems and devices that can recognize, interpret, and process human emotions are an interdisciplinary field involving computer science, psychology, and cognitive science. A system has been developed in order to formally categorize the emotions depending on facial expressions. The feature selection is done based on facial action coding system which is basically a contraction or relaxation of one or more face muscles. Our goal is to categorize the facial expression using image into six basic emotional states: Happy, Sad, Anger, Fear, Disgust, and Surprise. Extraction of facial features from eye, mouth, eyebrow, and nose is performed by employing an iterative search algorithm, on the edge information of the localized face region in binary scale. Finally, emotion class assignment is done by applying the extracted blocks as inputs to a feed-forward neural network trained by back-propagation algorithm.

Keywords Batch back propagation · Training · Stimulation · Face action unit

31.1 Introduction

Computer vision is an area of artificial intelligence that focuses on making computers to emulate human vision, including learning, making inferences, and performing cognitive actions based on visual inputs. Face is rich in information about individual identity, mood, and mental state. The expression on face is the

G. Shivakumar (✉)
Department of Information Technology, MCE, Hassan, India
e-mail: gskmce@gmail.com

P. A. Vijaya
Department of Electronics and Communication Engineering, BNMIT, Bangalore, India
e-mail: pavmkv@gmail.com

V. Chakravarthi et al. (eds.), *Proceedings of International Conference on VLSI, Communication, Advanced Devices, Signals & Systems and Networking (VCASAN-2013)*, Lecture Notes in Electrical Engineering 258, DOI: 10.1007/978-81-322-1524-0_31,

most basic form of nonverbal communication. In the field of computer vision, the term facial-expression recognition often refers to the classification of facial features in one of the six standard emotions: happy, sad, fear, disgust, surprise, and anger as introduced by Ekman [1]. This attempt of an interpretation is based on the assumption that the appearances of emotions are universal across individuals as well as human ethnics and cultures [2].

31.2 Proposed Methodology

The objective here is to analyze expressive cues within the face which mostly take place as dialogs in a sitting position. Hence, we focus on the expressiveness of the upper part of the body in our work. We first extract and analyze face action units (FAUs) [3, 4]. The general system framework for uni-modal facial-expression-based emotion recognition is depicted in Fig. 31.1. The steps involved in the complete implementation of the work are as follows.

- Face detection.
- Face feature extraction.
- Feature point selection.
- Emotion recognition.

Face detection is a process that aims to locate a human face in an image. The process is applied on stored images acquired from a camera. The variation in faces could be due to race, gender, age, and other physical characteristics of an individual. The face is initially located using Viola–Jones algorithm. However, the area of an image being analyzed for a facial feature needs to be regionalized to the location with the highest probability of containing the feature. By regionalizing the detection area, false positives are eliminated, and the speed of detection is increased due to

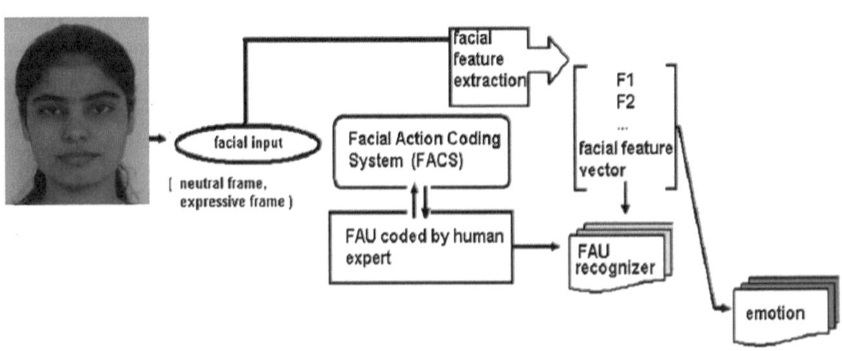

Fig. 31.1 Block diagram of the proposed system

the reduction of the area examined. The second step is using the isolated face to detect each feature [5].

Isolation of facial features, such as the mouth, eyes, and nose is done with a Haar classifier. The simple rectangular features of an image are calculated using an intermediate representation of an image, called the integral image (Fig. 31.2). The integral image is an array containing the sums of the pixels intensity values located directly to the left of a pixel and above the pixel at location (x, y) inclusive. These features, rather than using the intensity values of a pixel, use the change in contrast values between adjacent rectangular groups of pixels. The contrast variances between the pixel groups are used to determine relative light and dark areas. Two or three adjacent groups with a relative contrast variance form a Haar-like feature [6, 7].

To train the classifiers, two set of images are needed. One set contains an image that does not contain the object, which is going to be detected. These set of images are referred to as the negative images. The other set of images, the positive images, contain one or more instances of the object. The location of the objects within the positive images is specified by image name, the upper left pixel and the height, and width of the object. Only a tiny fraction of those features are needed to determine if a subimage potentially contains the desired object. In order to eliminate as many subimages as possible, only a few of the features that define an object are used when analyzing subimages. Cascading of the classifiers allows only the subimages with the highest probability to be analyzed for all Haar-features that distinguish an object [8]. Both the false alarm rate and positive hit rate can be increased by decreasing the number of stages.

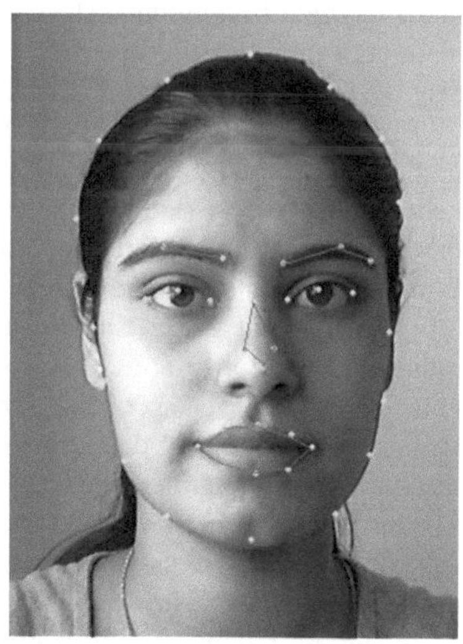

Fig. 31.2 Feature points in MCE sample database

31.3 Dataset for Experimentation

In spite of repeated calls for the need of a comprehensive, readily accessible reference set of face images that could provide a basis for benchmarks for all different efforts in research on machine analysis of facial expressions, no such database has been yet created that is shared by all diverse facial-expression-research communities. Cohn–Kanade face database [9] is one of the most commonly used datasets in the research on automatic facial-expression analysis. This is because all facial-expression sequences in the Cohn–Kanade face database start with a neutral expression. It helps us to differentiate facial activity (presence of AUs) from inactivity. The front-view face images were acquired using a CCD camera, under almost identical environments of distance, illumination, and background. The resolution of images is 128 $\times$ 128 pixels. There were 80 face images collected in our face image database, which includes eight different male persons and each person poses for several neutral, anger, and happiness expressions. Examples of expressive images collected in the database are shown below. In our database, we used 42 images as training data and 38 images for testing, both of which include neutral, anger, and happiness expressions. The Yale face database contains images of different facial expression (happy, sad, surprised, anger, and disgust) and configuration (with/without glasses). For every subject in a particular pose, an image with ambient (background) illumination was also captured. The availability of such standardized databases is important for scientific research as they provide a common testing ground to test the efficacy of different algorithms. We have developed our own data set called MCE database. The frontal face MCE database were acquired using Sony cyber shot 12 mp camera, with a certain distance of 50 cm under daylight background. The resolution of the images is 640 $\times$ 480 pixels. We have collected 180 persons face image and each person poses for happy, sad, disgust, fear, angry, and surprise in addition to neutral expressions.

The Face Action Units (FAUs), which represent the muscular activity that produces momentary changes in facial appearance. The exact appearance change varies from one person to another depending upon their bone structure, variations in the facial musculature, fatty deposits, permanent wrinkles, shape of features, etc. Common elements appear across people in the changes that occur in an action unit [10]. Ekman and Frisen studies suggested that anger, disgust, fear, happiness, sadness, and surprise are the six basic prototypical face expressions recognized universally. However, six universal emotion categories are not sufficient to describe all facial expressions in detail. In order to capture the subtlety of human emotion, recognition of fine grained changes and atomic movements of the face is needed. Ekman and Friesen developed the Facial Action Coding System (FACS) for describing face expressions by FAUs. In that work, 44 face action units have been defined. As FAUs can be classified either individually or in combination, among the 44, many are redundant, and hence, we have considered the dominant 26 FAUs (Table 31.1). For example, surprise is defined as a combination of four FAUs (the notation "+" refers to the linear combination of FAUs occurring together).

Table 31.1 Selected list of face action units

FAU	FAU description	FAU	FAU Description
1	Inner brow raised	14	Lip stretched
2	Outer brow raised	15	Lip tightened
3	Brow lowered	16	Lips pressed
4	Upper lid raised	17	Lips parted
5	Cheek raised	18	Jaw dropped
6	Lower lid tight	19	Mouth stretched
7	Nose wrinkle	20	Lips sucked in
8	Upper lip raised	21	Lid dropped
9	Lip corner pull	22	Eyes closed
10	Cheek puff	23	Eyes turned left
11	Dimpler	24	Eyes turned right
12	Lip corner depressed	25	Eyes turned up
13	Chin raised	26	Eyes turned down

Surprise = {FAU 1} + {FAU 2} + {FAU 5} + {FAU 25} database is yet to be created that is shared by all diverse facial-expression-research communities. Cohn–Kanade face database [11] is one of the most commonly used datasets. Here, all facial-expression sequences start with a neutral expression. It helps to differentiate the presence of Action Units from inactivity. The front-view face images were acquired using a CCD camera, under almost identical environments of distance, illumination, and background. The resolution of images is 128 × 128 pixels. There were 80 face images collected in our face image database, which includes eight different male persons and each person poses for several neutral, anger, and happiness expressions. In our database, 42 images were used for training and 38 for testing. The database developed by us is named as MCE database [11, 12]. The frontal face MCE database were acquired using Sony cyber shot 12 mp camera, with a certain distance of 50 cm under daylight background. The resolution of the images is 640 × 480 pixels. We have collected 180 persons face image, and each person poses for happy, sad, disgust, fear, angry, and surprise in addition to neutral expressions, however, six universal emotion categories are not sufficient to describe all facial expressions in detail. In order to capture the subtlety of human emotion, recognition of fine grained changes and atomic movements of the face is needed. Ekman and Friesen [1] developed the Facial Action Coding System (FACS) for describing face expressions by FAUs. As FAUs can be classified either individually or in combination, many of the 44 are redundant, and hence, we have considered the dominant 26 FAUs. For example, surprise is defined as a combination of four FAUs (the notation "+" refers to the linear combination of FAUs occurring together)

$$\text{Surprise} = \{\text{FAU } 1\} + \{\text{FAU } 2\} + \{\text{FAU } 5\} + \{\text{FAU } 25\}$$

31.4 Implementation

A multilayer, feed forward neural network with error back-propagation algorithm has been designed that functions as a classifier. Least mean squared errors in the output acts as a basis for adjustment of synaptic weights. Recalculation of the outputs is an iterative process carried on until the errors fall below a tolerance level. Learning rate parameters scale the adjustments from a previous iteration and adding to the adjustments in the current iteration. The neurons in the hidden and the output layer have biases which are connections from units whose output is always one. Back-propagation training takes place in three stages.

1. Feed forward of the input training pattern
2. Back propagation of the associated error
3. Weight adjustment.

Algorithm:
Let M1[i][j] be the weight on the connection from *i*th input neuron to *j*th neuron in the hidden layer, and M2[i][j] denotes the weight on the connection from *i*th neuron in the hidden layer to *j*th output neuron. X,Y, and Z denote the outputs of neurons in the input, hidden, and output layers, respectively, and θ represents the threshold value and β the learning rate parameter.

1. Output of *j*th hidden layer neuron is

$$Y_j = f\left(\left(\sum_i X_i M_1[i][j]\right) + \theta_j\right) \tag{31.1}$$

2. Response of *j*th output layer neuron is

$$Z_j = f\left(\left(\sum_i Y_i M_2[i][j]\right) + \tau_j\right) \tag{31.2}$$

3. *i*th component of vector of output differences is
   desired value − computed value = $P_i - Z_i$

4. *i*th component of output error at output layer is

$$E_i = (P_i - Z_i) \tag{31.3}$$

5. *i*th component of output error at the hidden layer is

$$t_i = Y_i(1 - Y_i)\left(\sum_j M_2[i][j]\ e_j\right) \tag{31.4}$$

6. Adjust the weight between *i*th neuron in hidden layer and *j*th output neuron.

$$\Delta M_2[i][j] = \beta_o Y_i e_j \tag{31.5}$$

7. Adjust the weight between *i*th input neuron and *j*th neuron in the hidden layer.

$$\Delta M_1[i][j] = \beta_h X_i t_j \tag{31.6}$$

8. Adjust the threshold value for the *j*th output neuron.

$$\Delta \tau_j = \beta_o e_j \tag{31.7}$$

9. Adjust the threshold value for the *j*th hidden layer neuron.

$$\Delta \theta_j = \beta_h e_j \tag{31.8}$$

10. Adjust the weights using the momentum parameter α.

$$\Delta M_2[i][j](t) = \beta_o Y_i e_j + \alpha \Delta M_2[i][j](t-1) \text{ and} \tag{31.9}$$

$$\Delta M_1[i][j](t) = \beta_h X_i t_j + \alpha \Delta M_1[i][j](t-1) \tag{31.10}$$

31.5 Experimental Results

The algorithm is tested on benchmark databases like Cohn–Kanade database, Yale database and also on MCE database. Our MCE face database contains 180 subjects with all six emotions. Certain images were selected for training set and few images for testing set, which are not used in training set as depicted in Table 31.2. Only five major features which are dominant have been considered in the analysis. They are the white pixel area of the eye, major and minor axis corresponding to the eye, orientation of eye brow, and major axis of nose. Table 31.3 shows these feature values for one sample subject. Figure 31.3 shows the distribution of white pixels in the eye for a sample subject, as a small window is moved from one extreme to another. The training error decreases as the epoch proceeds. However, nearer to 950 epochs, error starts increasing (Fig. 31.4). The synaptic weight change is maximum in the range of −1.5 to +1.5 for the synaptic connections between inputs and hidden layer. The confusion matrix is summarized in Table 31.4. Increasing number of hidden layers and number of neurons per hidden layer will not guarantee the better recognition. Figure 31.5 illustrates the behavior of the network for increasing the number of neurons in the hidden layer. Also, increased hidden layer neurons increases the synaptic connections, thereby decreasing the speed of training. Figures 31.6 and 31.7 exemplifies the distribution of emotion classes.

Table 31.2 Recognition rates using different ratios of training to testing images

Training sets	Testing sets	Validation sets	Percentage recognition
50	30	20	77.27
40	50	10	73.64
65	20	15	89.09

Table 31.3 Extracted features for a sample subject

Emotion	Features extracted				
	White pixel	Major axis	Minor axis	Orientation	Major axis (nose)
Normal	449	95	37	−8.92	70.43
Happy	229	95	46	−6.65	71.59
Surprise	333	119	71	−4.77	80.82
Disgust	486	98	43	−9.96	76.21
Sad	513	113	28	−1.04	71.59

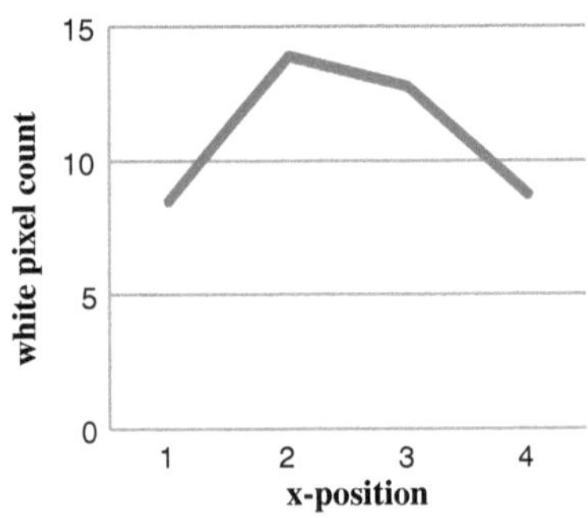

Fig. 31.3 Distribution of white pixels in eye

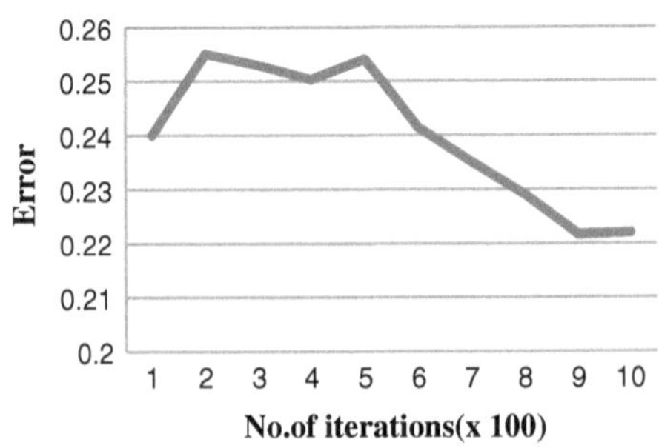

Fig. 31.4 Training curve

Table 31.4 Confusion matrix

Target output	Neutral	Angry	Surprise	Happy	Sad
Neutral	**84**	01	00	00	00
Angry	01	**69**	09	04	02
Surprise	00	05	**78**	00	02
Happy	00	04	06	**74**	01
Sad	05	18	01	02	**59**

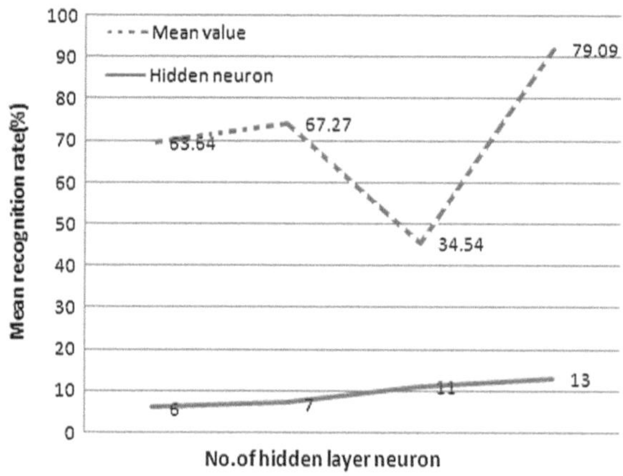

Fig. 31.5 Effect of hidden layer neurons

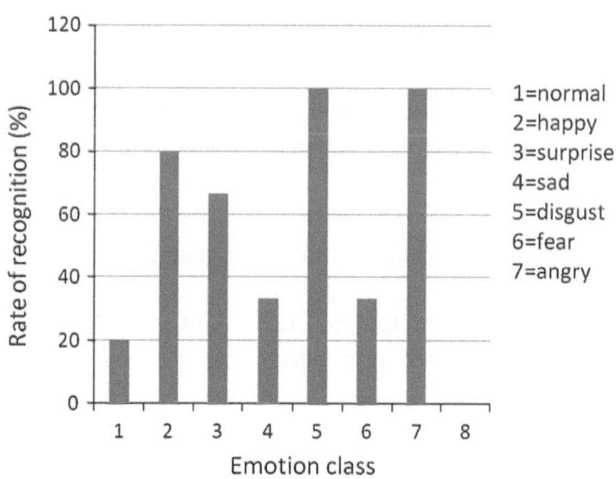

Fig. 31.6 Emotion class distribution

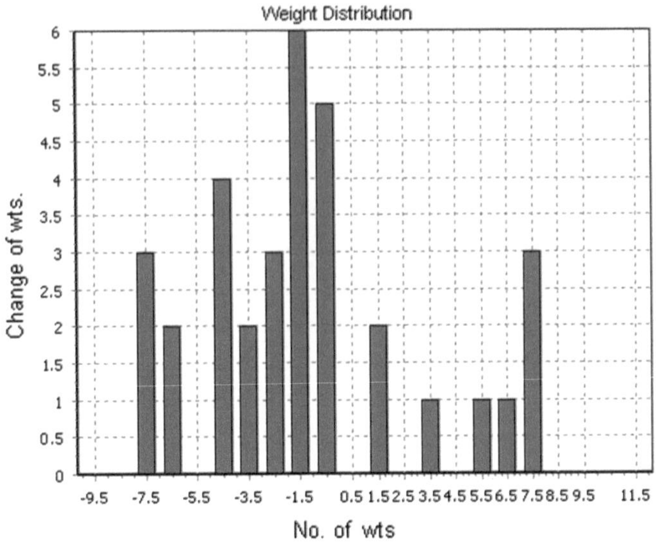

Fig. 31.7 Weight distribution in hidden layer

31.6 Conclusions

The proposed method has been implemented using Dev C++ compiler. The results obtained with our dataset are validated using the bench mark Cohen–Kanade database. The following conclusions can be drawn after successfully testing the network.

1. Best performance for batch back-propagation network is 89 %. Accuracy can be further increased with a still larger dataset.
2. This network is more stable with a good measure of repeatability.
3. The increase of training set increases the rate of recognition of the network. Hence, increasing number of samples in dataset, accuracy can be further increased.
4. By including additional facial and physiological features, we can improve the accuracy of emotion recognition.
5. Using neural network for emotion classification in real time is tedious and time-consuming, but it can be accurate. Whereas emotion classification without using neural network in real time is fast, but it is relatively less accurate.

References

1. Ekman P, Rosenberg EL (eds) (2005) What the face reveals: basic and applied studies of spontaneous expression using the facial action coding system (FACS), 2nd edn. Oxford University Press, Oxford, pp 271–286
2. Raheja JL, Kumar U (2002) Human facial-expression detection from detected in captured image using back-propagation neural network, digital systems group, central electronics engineering research institute (CEERI)/council of scientific and industrial research (CSIR), Rajasthan
3. Peter C, Herbon A (2006) Emotion representation and physiology assignments in digital systems. Interact Comput (Elsevier) 18:139–170
4. Taylor JG, Fragopanagos NF (2005) The interaction of attention and emotion. Neural Netw (Elsevier) 18:353–369
5. Gunes H, Piccardi M (2005) Fusing face and body gesture for machine recognition of emotions, 2005 IEEE international workshop on robots and human interactive communication, pp 94–99
6. Wilson PI, Dr Fernandez J (2006) Facial feature detection using haar classifiers, Texas A&M University, JCSC 21(4)
7. Shivakumar G, Vijaya PA, Anand RS (2007) Artificial neural network based cumulative scoring pattern method for ECG analysis. IEEE international conference on advances in computer vision and information technology, Aurangabad, pp 451–457
8. Shivakumar G, Vijaya PA (2011) Analysis of human emotions using galvanic skin response and finger tip temperature. Int J Synth Emotions (US) 2(1):15–25
9. Shivakumar G, Vijaya PA (2009) Face recognition system using back propagation artificial neural network. Int J Comput Sci Inf Technol 1(1):68–77
10. Ross A, Jain A (2003) Information fusion in biometrics. Pattern Recogn Lett (Elsevier) 24:2115–2125
11. Tian YL, Kanade T, Cohn JF (2001) Recognizing action units for facial expression analysis. IEEE Trans Pattern Anal Mach Intell 23(2):97–113
12. Shivakumar G, Vijaya PA (2012) Emotion recognition using finger tip temperature: first step towards an automatic system. Int J Comput Electr Eng (Singap) 4(3):252–255

Chapter 32
DSP–FPGA-Based Parallel Architecture for Acquisition and Compression of Instrumented Pipeline Inspection Gauge Data in Real Time

Sushil Kumar Bahuguna, Sangeeta Dhage, Siddhartha Mukhopadhyay and Y. K. Taly

Abstract The paper presents a DSP–FPGA-based parallel architecture for acquisition and compression of data in real time. The architecture is structured with high-performance DSP and acquisition hardware, implemented in FPGA. Hardware blocks for data acquisition with control logics, and FIFO are implemented in FPGA. FIFO interconnects DSP with acquisition hardware, running acquisition task in parallel for achieving maximum throughput. The data compression algorithms based on mean absolute deviation (μAD) is implemented on the DSP. The test results on field data show that the compression algorithm is very effectively implemented with the proposed architecture providing a very high compression ratio. The paper also presents the task management policy for implementing the scheme on DSP–FPGA hardware.

Keywords Data compression · DSP–FPGA · μAD · Real-time implementation

32.1 Introduction

High-end data acquisition systems often need to deal with large volumes of data and complex algorithms. Applications may need data compression in real time to reduce local memory and to achieve fast data transfer rates. One of such applications is in designing data acquisition system for instrumented pipeline inspection gauge (IPIG). An IPIG is an instrument for in-line inspection of pipelines. It acquires huge quantity of data from a single run [1] and simultaneously stores it after online compression. Such a computationally intensive and performance critical application requires more processing power than can be provided by

S. K. Bahuguna · S. Dhage · S. Mukhopadhyay (✉) · Y. K. Taly
Control Instrumentation Division, Bhabha Atomic Research Centre,
Mumbai 400085, India
e-mail: smukho@barc.gov.in

V. Chakravarthi et al. (eds.), *Proceedings of International Conference on VLSI, Communication, Advanced Devices, Signals & Systems and Networking (VCASAN-2013)*,
Lecture Notes in Electrical Engineering 258, DOI: 10.1007/978-81-322-1524-0_32,

common microprocessors. A number of software/hardware solutions may be considered for the system including fast DSP processors to process the acquired data in real time [2]. The task can be achieved by developing efficient algorithm, designing a customized architecture, and implementing the same on high-performance hardware.

Real-time data acquisition hardware performs three primary tasks. The first task is to acquire data, the second is to process the acquired data in real time, and third task is to store the processed data into onboard flash memory. The maximum throughput can be achieved by parallelizing these tasks. This needs appropriate design of system architectures with built in features that helps parallelizing. A great deal of work on parallel system architecture has been carried out over the past decades. An example is the DSP–FPGA architectures developed specifically for image processing application [3]. In the proposed architecture in this paper, the most time-consuming task, i.e., data compression and storage in onboard flash runs parallel to data acquisition task implemented on FPGA. A parallel architecture consisting of a floating point DSP and acquisition hardware implemented on FPGA is presented here. FIFO for interconnection network and specific data, control and status registers implemented within FPGA ensure high data transfer rate. Moreover, due to the presence of FPGA, the system architecture can be conveniently reconfigured to adopt various paralleling approaches.

The paper is organized as follows: In Sect. 32.2, the DSP–FPGA parallel architecture is discussed in details. Method used for the application of magnetic flux leakage (MFL) data compression and results using mean absolute deviation are described in Sect. 32.3. Implementation within DSP–FPGA hardware is described in Sect. 32.4. System performance justifying the proposed architecture is presented in Sect. 32.5. Section 32.6 concludes the paper indicating future scope of work.

32.2 The Proposed Architecture

The proposed DSP–FPGA parallel architecture is shown in Fig. 32.1. The signal acquisition and data compression plus storage tasks are made to run in parallel by implementing a FIFO on the FPGA.

Acquisition hardware on FPGA implements a counter, a decoder for multiplexer (mux-decoder), an address decoder circuit for system peripherals, control signals for ADC, and control and status registers. For the specific application of data acquisition system of the IPIG, for 18-inch nominal bore pipeline, 218 sensors are used to acquire data via a multiplexer card. The analog signals from the multiplexer card are digitized on board by a 16-bit ADC and the output is stored in FIFO. The data acquisition hardware is designed in such a way that in every acquisition, all 218 sensors' data are acquired sequentially without processor intervention. After acquisition and storage of one group of 218 signals in FIFO, an interrupt is generated for the DSP processor by the FPGA. The DSP copies the data

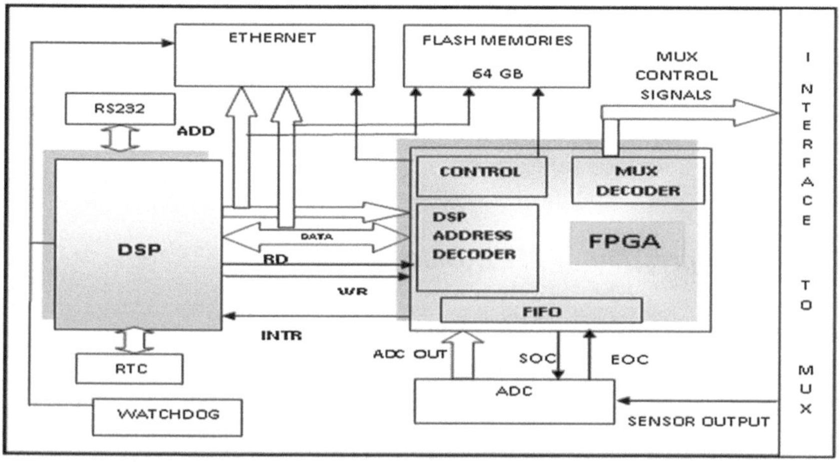

Fig. 32.1 Proposed architecture of DSP–FPGA parallel system

from FIFO and stores it in its internal RAM for further processing. As the FPGA controls the acquisition and fills the FIFO with next set of data, the DSP processor runs the compression algorithm on an earlier block of data. A flash memory provided on board is used to store the processed data. The DSP processor completes the compression and storage task before the new set of acquired data is ready in the FIFO. An Ethernet module is used to download the stored data off-line to the host computer for analysis and interpretation.

32.3 Online Compression of Magnetic Flux Leakage Data

In a real-time data acquisition system, the ultimate aim is to acquire, process, and store the data reliably within a fixed interval of time. An elegant and low-cost solution to the cumbersome data handling problem lies in employing efficient online data compression and storage technique. The aim is to reduce the volume of total data, so that it can be stored in an onboard flash or disk-on-chip. Onboard permanent memory substantially improves the reliability of the system by avoiding connectors and cables.

32.3.1 An Overview of Instrumented PIG Technology

Instrumented PIG tools have been indigenously developed at BARC and work on MFL principle. These tools are used for checking integrity and operability of cross-country pipelines. An IPIG magnetizes a section of carbon steel pipe wall using strong permanent magnets and senses leakage flux near inside wall. The developed tools have taken several commercial runs, and the MFL data from these runs have been analyzed for detection and characterization of metal loss defects [4]. Defects' size, estimated from the data acquired by the tools, matched to a good extent with the actual measurements of defects in dig site verifications.

32.3.2 Compression of MFL Data

An 18″ IPIG tool collects minimum 100 GB MFL data at 16-bit resolution from a run of 200 km. Quantity of the acquired data increases with the size of the tool. Handling such a large volume of data in real time is challenging and requires huge memory space and unacceptably long downloading, data analysis, and reporting time. An online data compression scheme could be employed to reduce the size of the stored data. A large volume of MFL signal does not contain any information. So, data compression algorithm implemented in the hardware is basically a data screening algorithm. Whenever there is any presence of anomaly in the pipeline, the level of magnetic flux leakage increases locally. Hence, there will be significant deviation of MFL signal from the baseline. Data screening algorithm utilizes this variability in the MFL signal, to differentiate between noisy block and informative block. The μAD is used as a robust measure of the variability of a MFL data [5]. For a univariate series, $\overrightarrow{x} = \{x[1], x[2] \ldots x[N]\}$ μAD is defined as the mean of absolute deviation from the data's median.

$$\mu AD = k * Mean(\overrightarrow{x} - Median(\overrightarrow{x})) \tag{32.1}$$

where k is a scaling factor and depends upon the probability distribution of the noise present in the signal. If for any block of data, the value of μAD is greater than the predetermined threshold, then the block of data is stored, otherwise rejected.

32.3.3 Data Compression Results Using μAD

The data screening algorithms work on a block of data. Steps involved in data compression/screening algorithm are as follows:

1. Divide the signal into blocks and calculate μAD for each block
2. If the μAD value is insignificant, reject that particular data block
3. Else retain the data block along with the index of the location.

Figure 32.2 shows raw and compressed MFL data in top and bottom panels, respectively. Parts of the signal in bottom panel show no variation indicating zero storage (total compression of the blocks). Results show that typically a

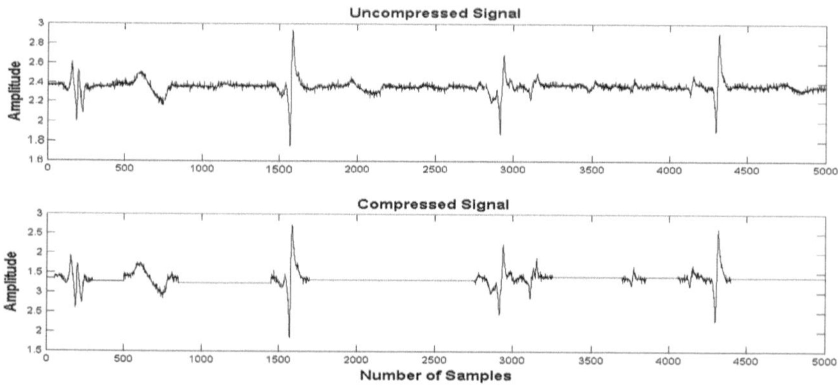

Fig. 32.2 Data compression on field data using μAD

compression ratio of 1:10 is achievable for MFL data acquired from the field. The compression ratio depends on pipe condition; if pipeline is cleaner then this ratio may further increase.

32.4 Implementation in DSP–FPGA Hardware

High-end DSP applications are computation intensive, which poses difficulty in real-time implementation. The efficient implementation of an algorithm on hardware can be done by identifying the functions/tasks, which are performance intensive and providing parallel hardware blocks for those tasks. In the present scheme, the processor exchanges data with FPGA through FIFO using interrupts and flags. The task management software has been coded in C although the data compression algorithm is implemented in the assembly (of DSP) for time optimization. The process diagram for DSP–FPGA architecture is shown in Fig. 32.3. The data acquisition process is continuously running in FPGA and is storing the data in FIFO without processor intervention. The FPGA generates an interrupt after acquisition of all 218 sensor/channel data. On receiving highest priority interrupt, DSP processor copies the data from FIFO to its internal RAM. The data buffer manager implemented in DSP distributes the acquired data into 218 different buffers. The μAD algorithm runs on a block of data on each channel; in this case, the block size is 50. The data buffer manager allocates the size of the buffer of each channel appropriately such that current data get collected in one buffer, while μAD code is run on the previous buffer. The compression code checks online; the presence of information and useful data is stored in onboard flash. Storage is again a task which runs in parallel. The hardware is provided with adequate buffer to match the overall throughput.

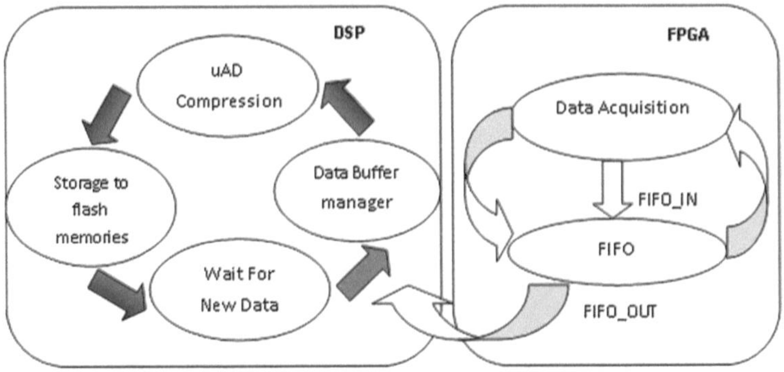

Fig. 32.3 Process diagram of DSP–FPGA parallel architecture

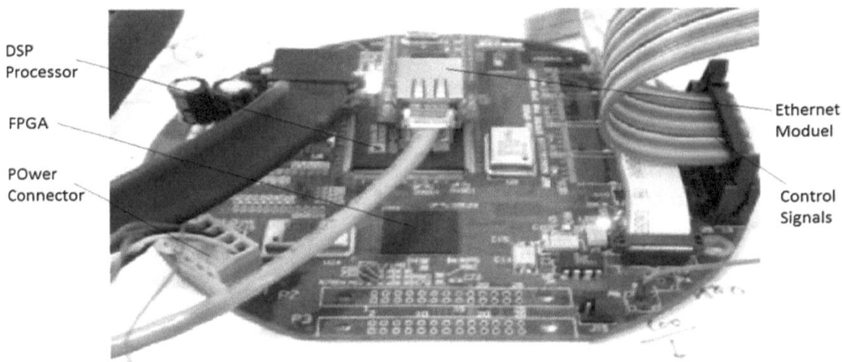

Fig. 32.4 DSP–FPGA-based data acquisition and processing prototype hardware

The proposed parallel architecture is implemented on a proprietary, very low power, DSP–FPGA hardware developed in-house (shown in Fig. 32.4).

32.5 System Performance Evaluation

The quantum of computations required to process a block of 50 samples (for all 218 channels) is shown in Table 32.1. Total numbers of floating point operations required for 218 channels are 218*31598 = 6888364. At a rate of 1 kHz sampling, acquisition system takes 50 ms time to collect a block of 50 sample data for all the channels. The DSP processor takes 10 ms for data storage into onboard flash, copy from FIFO, managing the data buffer and other housekeeping activities. So, the processor has 40 ms time to execute the compression algorithms. It actually takes

Table 32.1 Number of floating point operations for processing 1 channel

Operation	Signal length	
	50 point (in flops)	100 point (in flops)
Sorting of data (qsort)	31,498	1,25,362
Finding the deviation	50	100
Mean of deviation	50	100
Total	31,598	1,25,562

35 ms time to execute the compression algorithm for a 50 sample block, which justifies the need of DSP–FPGA parallel architecture for this application. The table also shows the flop needed for processing a 100 sample block.

32.6 Conclusions

The paper presents a DSP–FPGA-based parallel architecture for data acquisition and compression of MFL signal in real time, needed for instrumented pipeline inspection gauge. The article reports implementation of a real-time μAD-based compression scheme for MFL data on a dedicated hardware. The implementation has been tested on field data, and a very sizable compression is achieved. The scheme is used in the data acquisition system of 18" IPIG which has been tested satisfactorily in test rigs. In future, this architecture could be enhanced by putting more DSPs and high-end FPGAs which support DSP functions for computationally more extensive applications.

References

1. Bahuguna SK, Mukhopadhyay S (2009) Implementation of wavelet based MFL data compression scheme on DSP hardware for IPIG, SACET-09. VECC, Kolkata
2. Bahuguna SK, Mukhopadhyay S, Bhattacharya S, Patil MB, Das S, Biswas BB (2006) Development of a DSP based data acquisition system for IPIG Project. BARC Newsletter 264:26–29
3. Yan L, Zhang T, Zhong S (2006) A DSP/FPGA—based parallel architecture for real-time image processing. In: Proceedings of the 6th world congress on intelligent control and automation, Dalian, China, 21–23 June 2006
4. Saha S, Mukhopadhyay S (2010) Empirical structure for characterizing metal loss defects from radial magnetic flux leakage signal. NDT E Int 43:507–512
5. Kathirmani S, Tangirala AK, Saha S, Mukhopadhyay S (2012) Online data compression of MFL signals for pipeline inspection. NDT E Int 50:1–9

Chapter 33
A Contourlet Transform-Based Versatile Watermarking Algorithm for Medical Images

Saritha Chakrasali and Ramachandran Murugesan

Abstract In this paper, an attempt is made to analyze the potential of contourlet transform (CT) for medical image watermarking. The effect of the embedding strength on the fidelity of the image is investigated. The algorithm is tested for its robustness against a few selected attacks like adding noise, cropping, and filtering attacks. Texture analysis is applied to understand the effect of watermarking in CT. Performance criteria such as root mean square error (RMSE), standard deviation (SDE), peak signal-to-noise ratio (PSNR), and normalized correlation (NC) are used to evaluate the potential of the proposed algorithm. The results show that the CT-based watermarking algorithm is comparable with the watermarking in the wavelet domain in terms of imperceptibility and robustness.

Keywords Contourlet transform · Lifting wavelet transform · Telemedicine · Watermarking · Medical images

33.1 Introduction

Teleradiology, the most popular application of telemedicine, has a major requirement of image security. Medical image watermarking plays a vital role in providing this security. The requirements of multi-resolution, localization, and critical sampling required in medical image watermarking are provided by the

S. Chakrasali (✉)
Department of Information Science and Engineering, BNM Institute of Technology, Bangalore, India
e-mail: saritha.chakrasali@gmail.com

R. Murugesan
Faculty of Allied Health Sciences, Chettinad Academy of Research and Education, Chennai, India
e-mail: rammurugesan@yahoo.com

V. Chakravarthi et al. (eds.), *Proceedings of International Conference on VLSI, Communication, Advanced Devices, Signals & Systems and Networking (VCASAN-2013)*, Lecture Notes in Electrical Engineering 258, DOI: 10.1007/978-81-322-1524-0_33,

wavelet system [1]. DWT has been successfully applied for watermarking medical images due its efficiency in representing smooth signals with point discontinuity [1–4]. Unfortunately, the lack of directionality and anisotropy of DWT leads to its inefficiency in describing bi-dimensional smooth contours. The contourlet transform (CT) addresses the inherent limitations of the traditional multi-scale representations [5]. CT is capable of capturing the directional information in addition to multi-resolution, localization, and critical sampling characteristics. In addition, it can be easily implemented as a combination of the Laplacian pyramid (LP) and the directional filter bank (DFB) for obtaining sparse expansion of typical images having smooth contours.

CT has been applied on medical images for feature extraction required in texture classification of ultrasound images [6] and for the enhancement and denoising of mammograms [7]. But the application of CT on medical images is yet to be explored to its full potential. CT provides an improvement in approximation based on keeping the most significant coefficients. This will directly lead to improvement in watermarking application. Because of these advantages, there is current interest to evaluate the application of CT for medical image watermarking [8–10]. Medical images belonging to different modalities exhibit different textures. CT provides multi-directional representation of an image in order to conform to the multi-resolution nature of HVS. It is also easily adjustable for detecting fine texture detail in any orientation at various scale level, thus helping to enhance the watermark intensity in the components. Therefore, in this paper, application CT for watermarking medical images of different texture is explored.

33.2 Proposed Watermarking Algorithm

The problem faced in teleradiology is to minimize the perceptibility and increase the robustness of a watermark to authenticate the medical images. We propose to design a preprocessing watermarking function that adds the watermark in selected coefficients of subbands at a specific decomposition level with high imperceptibility and robustness. At the receiver's end, the retrieved watermark provides the authentication and medical image is diagnosed by the specialist without any visual distortion added by the watermark. The goal of this algorithm is to provide imperceptibility and high resolution as much as possible, but without necessarily sacrificing robustness. Several image quality metrics have been developed to aid in visual analysis, to predict the visible differences between a pair of images, the input image and the watermarked image. For quantitative evaluation, three common error measurements, viz. RMSE, SDE, and PSNR, are widely employed [11, 12]. In this work, a list of attacks are used to evaluate the watermarking algorithms. Differentiation between malicious and coincidental attacks is not made. After the attack, the retrieved watermark is compared to the original watermark and the normalized correlation (NC) is evaluated.

33.2.1 Watermark Embedding and Retrieval

The CT decomposition is performed on the original image. The outputs from the function are the horizontal and vertical coefficients at the required level. From the CT coefficients, the key K is generated. The equation used to calculate the upper threshold T_1 is given by

$$T_1 = 2^{\lfloor \log_2 C \rfloor - 1} \tag{33.1}$$

where C is the maximum coefficient in the subband, which is chosen for watermarking. The lower threshold T_2 is calculated as

$$T_2 = \left(\frac{L + T_1}{2}\right) \quad \text{and} \quad \text{if } T_2 < 0 \text{ then } T_2 = 0 \tag{33.2}$$

where L is the minimum coefficient in the subband. The coefficients C(i,j) are selected as follows:

$$T_2 < C(i,j) < T_1 \tag{33.3}$$

The positions of these selected coefficients are stored in the key. The watermark is a binary grayscale image. A one-dimensional binary array watermark is generated from this. This watermark is embedded in selected coefficients of the bands at a given level using the following Eqn:

$$C' = C(i,j) + alpha * W(1,k) \tag{33.4}$$

where C′(i,j) and C(i,j) are the watermarked and original coefficient, respectively, while W(1,k) is the watermark bit and alpha is the embedding strength.

CT reconstruction is performed to obtain the watermarked image.

The PSNR is computed from the original and watermarked images. The watermark retrieval process requires the original image, watermarked image, alpha, and the key. Both the original and watermarked images undergo CT decomposition and the required subbands of the given level are selected from both. The watermark retrieval is based on the following equation:

$$W(1,k) = \frac{C'(i,j) - C(i,j)}{alpha} \tag{33.5}$$

The watermark obtained is converted into grayscale image. This retrieved watermark is used to compute the NC. The pseudocode of the CT-based watermarking algorithm incorporating the embedding and retrieval schemes is as given below. The level of decomposition is fixed at 2 with 8 subbands (1–4 are horizontal subbands, and 5–8 are vertical subbands). The subband selected for watermarking are the horizontal subbands. This pseudocode was implemented in Matlab®.

```
Procedure CT_embed( input: I,level,subband,alpha,W; out-
put:IW, PSNR)

[(CTHcoeffs,CTVcoeffs)level, CTHcoeffs,CTVcoeffs)other_level)]
        ←CT_decomposition(I,level)

(CTHcoeffs,CTVcoeffs)'level,key) ← embed1((CTHcoeffs,
  CTVcoeffs)level, subband, W, alpha )

IW ← CT_reconstruction((CTHcoeffs,CTVcoeffs)level,
CTHcoeffs,CTVcoeffs)other_level)]
Compute PSNR

return IW, PSNR

end

Procedure CT_retrieval(input:
IW,level,subband,alpha,Watermarksize,key; output:W', NC)

[(CTHcoeffs,CTVcoeffs)level, CTHcoeffs,CTVcoeffs)other_level)]
  ← CT_decomposition(I,level)

[(CTHcoeffs,CTVcoeffs)'level, CTHcoeffs,CTVcoeffs)other_level)]
  ← CT_decomposition(IW,level)

W' ← retrieve1((CTHcoeffs,CTVcoeffs)'level,
(CTHcoeffs,CTVcoeffs)level,subband,key,alpha,watermarksize)

Compute NC

return W', NC

end
```

In the pseudocode, CTHcoeffs and CTVcoeffs refer to the horizontal and vertical CT coefficients, I and I_W refer to original and watermarked image, W and W′ refer to the watermark and retrieved watermark, alpha refers to the embedding strength, level refers to the level of decomposition, and NC and PSNR refer to normalized correlation and peak signal-to-noise ratio.

33.3 Input Data Sets

To test the performance of the proposed approach, the algorithm is at first validated using the standard peppers image (Fig. 33.1a). The size of the standard image is 512 × 512 and 34.6 kB. Following the validation of the proposed watermarking algorithm using peppers image, it was applied on the medical

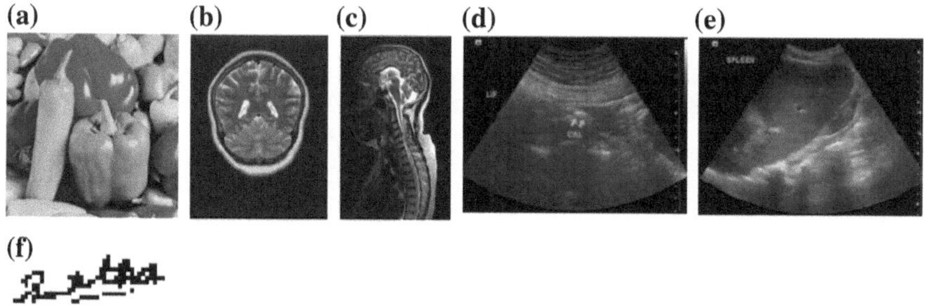

Fig. 33.1 Input data set: **a** standard peppers, **b** MRI1, **c** MRI2, **d** ultrasound1, **e** ultrasound2, **f** watermark

images. Medical images of different modalities were used in the watermarking experiments. The medical images were obtained from Dr.K.G.Srinivasan, Radiologist of KGS Advanced MRI and CT Scan Centre, Madurai.

The two MRI images of the brain, referred to as MRI1 and MRI2 with sizes 544 × 672(41.7 kB) and 432 × 672(29.6 kB), respectively, and two ultrasound images of kidney and spleen, referred to as ultrasound1 and ultrasound2 with sizes 672 × 512 (32.7 kB) and 656 × 512 (31.1 kB), respectively, are used in this work. Medical test images are shown in Fig. 33.1b–e. A binary image that contains a signature is used as the watermark (Fig. 33.1f). The first-order and second-order statistical texture analyses of the image indicated that among the test images, peppers have the highest entropy value signifying a coarser image, while ultrasound images have the low entropy values indicating smoother nature of the image. Further both the ultrasound images are fine-textured, smooth images with low contrast, while MRI1 and peppers images are coarser and high contrast images. MRI2 image is finer than MRI1.

33.4 Experimental Design

In this work, the effect of embedding strength on image fidelity was analyzed. For each of the images, the watermarking scheme was applied with gradually increasing embedding strengths. From the results obtained, the embedding value which yielded the optimal balance between PSNR and NC symbolizing imperceptibility and robustness, respectively, was selected as the optimal one. The decomposition level chosen here is 2. Further, the algorithm was first validated in an attack-free environment. The algorithm was also implemented in the LWT [13] for comparative evaluation. The level of decomposition and other parameters are set equivalently to that of CT. The various performance metrics and attacks are used to evaluate the image quality and robustness after applying the proposed CT-based watermarking algorithm.

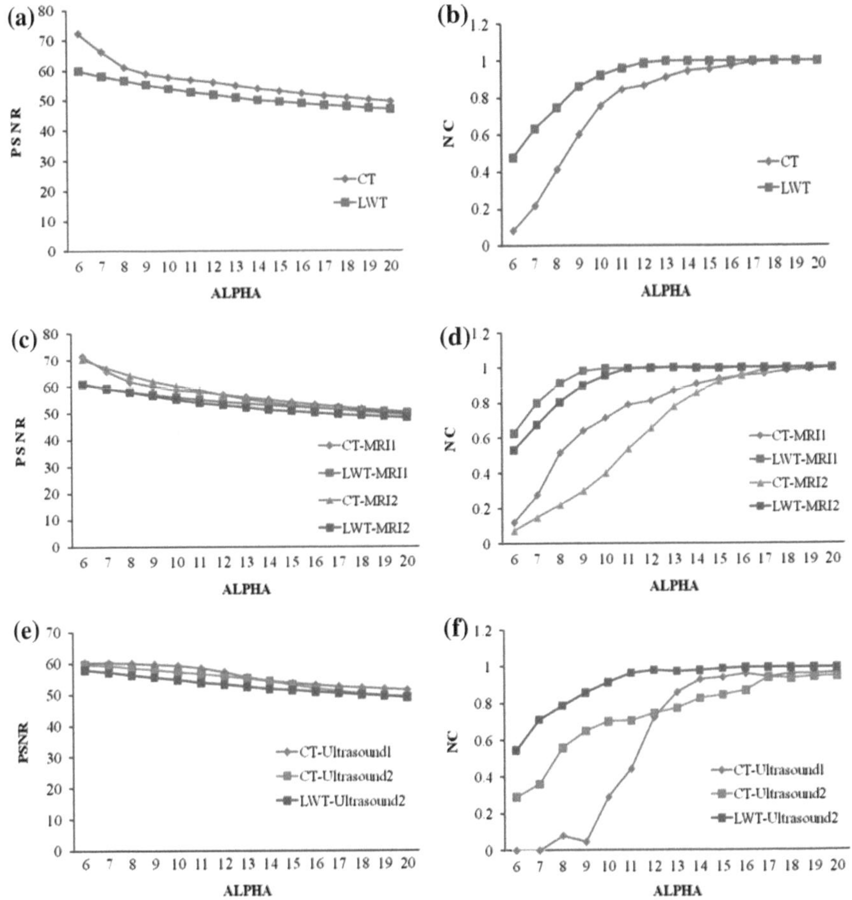

Fig. 33.2 Effect of alpha on imperceptibility and robustness of watermarked ultrasound1 and ultrasound2 images in CT and LWT domains (I) PSNR (y-axis) versus alpha (x-axis) (II) NC (y-axis) versus alpha (x-axis): **a–b** peppers, **c–d** MRI, **e–f** ultrasound

33.5 Performance Evaluation

The proposed watermarking algorithm was evaluated first by applying it on standard peppers image. Following this, it was applied on the medical images. Both MRI and ultrasound images are used as test images.

The embedding strength alpha was increased in steps of 1 from 6 to 20 in case of all the test images. The high values of embedding strength are attributed to the fact that both CT and LWT coefficient values are normalized between 0 and 255. Plots of imperceptibility (PSNR) versus alpha for peppers, MRI, and ultrasound images are given in Fig. 33.2a, c, e. From the graphs, it is clear that as the embedding strength is increased the PSNR decreases. Plots of robustness of watermark (NC) versus alpha for peppers, MRI, and ultrasound images are given in Fig. 33.2b, d, f, respectively.

The graphs indicate that as embedding strength is increased, the robustness of the watermark increases and the imperceptibility of the image decreases. For accurate diagnosis, medical images need to have PSNR > 38 dB. For reasonable robustness, the NC values should be greater than 0.8. These plots reveal that alpha values of 18, 20, 18, 20, and 20 for peppers, MRI1, MRI2, ultrasound1, and ultrasound2 provide robustness (NC > 0.8) with PSNR > 38 dB. The fine-textured and smooth ultrasound images show similar PSNR values for the same embedding strengths in case of both CT-based and LWT-based watermarking. In case of coarse-textured peppers, MRI1, and MRI2, the CT-based watermarked images show higher PSNR values than the LWT-based watermarked images, for the same embedding strengths. The CT-based approach shows lower robustness than the LWT-based approach for the same embedding strength in all the test images. Thus, the CT-based watermarking approach performs well on coarse-textured medical images. This observation correlates with the fact that CT is more suited for anisotropic images.

Both CT-based and LWT-based watermarking algorithms were performed using the alpha values selected on the basis of Fig. 33.2. Random noise will generate significant wavelet coefficients just like true edges, but is less likely to generate significant contourlet coefficients [5]. This resulted in the images requiring a higher alpha value for optimizing in the CT domain in comparison with LWT. A qualitative analysis of both the original and watermarked images given in Table 33.1 indicates the imperceptibility.

In addition to visual evaluation, a quantitative analysis of the watermarked images was also assessed. These results are presented in Table 33.2. Except for MRI2, the watermarked images show a decrease in SNR, indicating that the watermarking is akin to adding noise. Among all the four test images, MRI2 is the noisiest image as seen by the SNR of the input image.

All the watermarked test images exhibit high CNR and PSNR in both the domains. The improvement in approximation by contourlets, based on keeping the most significant coefficients, will directly lead to improvements in applications, like compression, denoising, and feature extraction [5]. But in case of watermarking fine-textured images like ultrasound leads to high RMSE, SDE, and F(I), because the significant contourlet coefficients are changed. While for the coarser-textured images like peppers, MRI1 and MRI2 watermarked in the CT domain show lower RMSE, SDE, and F(I). High RMSE and SDE indicate poor quality of image. The low value of F(I) indicates that there is no loss in the resolution of the MRI images in the contourlet domain.

33.5.1 Robustness to Attacks

Different types of attacks were performed to evaluate the performance of the proposed CT-based watermarking algorithm. After the attacks were applied, the

Table 33.1 Performance evaluation of the watermarking algorithm applied to test images in CT and LWT domains: **a** peppers, **b** MRI1, **c** MRI2, **d** ultrasound2, **e** ultrasound1

	Transform	CT	LWT
A	Retrieved watermark	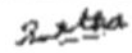	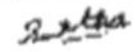
	Alpha	18	13
	PSNR	50.8	50.9
	NC	1	1
B	Retrieved watermark	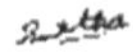	
	Alpha	20	13
	PSNR	50.4	53.8
	NC	1	1
C	Retrieved watermark	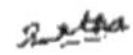	
	Alpha	18	12
	PSNR	50.6	53.1
	NC	0.99	1
D	Retrieved watermark	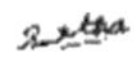	
	Alpha	20	16
	PSNR	48.8	50.7
	NC	0.95	1
E	Retrieved watermark	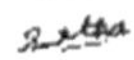	Cannot watermark because the number of coefficients selected in the LH region by this method was lesser than the watermark size
	Alpha	20	
	PSNR	51.2	
	NC	0.97	

Table 33.2 Quality metrics of test images watermarked in CT (*shaded*) and in LWT domains

Test images	SNR of input image	SNR of output image	CNR	RMSE	SDE	F(I) × 10¯7	PSNR (dB)	WPSNR (dB)
Peppers	6.74	6.73	1	0.82	0.82	18.67	50.8	62.8
		6.73	1	2.51	2.51	174.06	50.9	61.2
MRI1	5.75	5.75	1	0.06	0.06	0.21	50.3	62.2
		5.75	1	0.00	0.00	0.00	53.8	65.3
MRI2	4.70	4.70	1	1.57	1.56	153.08	50.6	62.3
		4.70	1	9.47	9.41	55542	53.0	62.3
Ultrasound2	6.62	6.60	1	2.78	2.76	531.27	48.8	60.9
		6.62	1	1.70	1.69	199.21	50.7	60.0

NC was computed for the recovered watermarks. A sample of the attacked images, the recovered watermarks and the NC values for MRI1 are shown in Fig. 33.3. The watermarked images in both CT and LWT were not affected by transmission on the public network. The tampering and cropping attacks depend on the localization of the coefficients selected for watermarking. Nevertheless, all the test images show good recoverability from these two attacks. Further, the results show that smoothening operation such as median filtering attack deteriorates the watermark considerably, but comparatively the images watermarked in CT domains are more robust to median filtering. Similarly, the watermark is robust against the

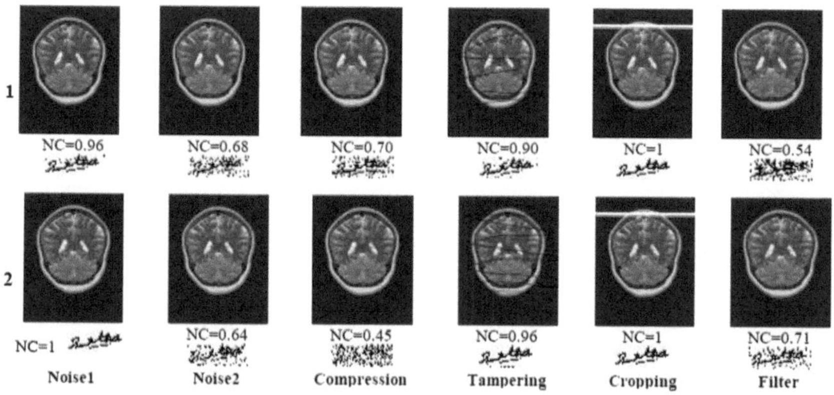

Fig. 33.3 Performance evaluation of watermarking in [1] CT and [2] LWT domains against various attacks using the MRI1 image

multiplicative speckle noise attack, while the additive Gaussian noise attack destroys the watermark almost completely as observed by the NC values.

The speckle noise commonly manifests itself as a fine-grained structure in an image, which leads to discontinuities at edge points. Speckle noise in medical images is less likely to generate significant coefficients as CT can combine the coefficients along the smooth curve. Hence, CT takes into account the edge direction without boosting the local strong variation caused by the noise. But Gaussian noise is capable of generating significant coefficients, and thus, in contourlet domain, the watermark is highly degraded. In the case of compression attack, the LWT watermarking scheme shows better performance than the CT approach for the ultrasound images. That is because compression degrades many fine-texture details that are salient in the ultrasound images. Therefore, it destroys some watermarked coefficients in the CT domains. For coarser image such as peppers and MRI however, the salient features survive compression.

33.6 Conclusion

In this paper, a versatile non-blind CT-based watermarking algorithm for medical image authentication and tamper detection is presented. This work is based on private key derived from the input image, thus embedding the watermark in selected coefficients. Both subjective and objective methods are used to evaluate the watermarked image quality and the robustness of the watermark. Experimental results demonstrate that the watermarked images are robust to attack and show PSNR $>$ 37 dB, thus providing invisible watermark. The NC value of $\sim$0.85 indicates effective recovery. Both MRI and ultrasound images can be successfully watermarked in the CT domain. CT is capable of capturing the directional information. The CT-based watermarking algorithm performs well for coarse MRI images. Thus, the CT-based watermarking algorithm is more suitable for coarse-textured medical images.

Acknowledgments The authors thank BNM Institute of Technology and Madurai Kamaraj University for the facilities and resources provided.

References

1. Podilchuk C, Zeng W (1998) Image-adaptive watermarking using visual models. IEEE J Sel Areas Commun 16:525–539
2. Cox I, Kilian J, Leighton F, Shamoon T (1997) Secure spread spectrum watermarking for multimedia. IEEE Trans Image Process 6:1673–1687
3. Nikolaidas A, Pitas I (2001) Region-based image watermarking. IEEE Trans Image Process 10:1726–1740
4. Hartung F, Kutter M (1999) Multimedia watermarking techniques. Proc IEEE 87:1079–1107
5. Do M, Vitterli M (2005) The contourlet transform: an efficient directional multiresolutional image representation. IEEE Trans Image Process 14:2091–2106
6. Katsigiannis S, Keramidas EG, Maroulis D (2010) A contourlet transform feature extraction scheme for ultrasound thyroid texture classification. Eng Intell Syst Spec Issue Artif Intell Appl Innovations 18(3/4):171
7. Zhang X, Xie H (2009) Mammograms enhancement and denoising using generalized Gaussian mixture model in nonsubsampled contourlet transform domain. J Multimedia 4:389–396
8. Duong DA (2012) Robust and high capacity watermarking scheme for medical images based on contourlet transform. In: proceeding international conference on systems and informatics, pp 2183–2187
9. Das S, Kundu MK (2011) Hybrid contourlet-DCT based robust image watermarking technique applied to medical data management. In: proceeding 4th international conference on pattern recognition and machine intelligence, pp 286–292
10. Rahimi F, Rabbani H (2011) A dual adaptive watermarking scheme in contourlet domain for DICOM images. BioMed Eng OnLine
11. Kanjanawanishkul K, Uyyanonvara B (2003) Fast adaptive algorithm for time-critical color quantization application. In: Proceeding VIIth digital image computing: techniques and applications, pp 781–785
12. McGibney G, Smith M (1993) An unbiased signal-to-noise ratio measure for magnetic resonance images. Med Phys 20:1077–1078
13. Sweldens W (1997) The lifting scheme: a construction of second generation wavelets. SIAM J Math Anal 29:511–546

Chapter 34
Application of Wavelet Shrinkage Denoising Method Based on Genetic Algorithm to Partial Discharge Signals

Bindiya Tyagi and H. A. Vidya

Abstract This paper presents the technique for denoising of partial discharge (PD) signals using genetic algorithm (GA) based on wavelet shrinkage. For the high voltage (HV) equipment in services, it is desired to perform PD tests when the equipment is in operation. The measurements, however, are seriously affected by the interference signals resulting from different sources. In order to enhance the sensitivity of a PD online monitoring system, many digital signal processing methods have been put forward for removal of noises. But to obtain the optimum denoising GA is applied. Here, first, optimum signal-to-noise ratio was obtained using modified wavelet technique (WT), and then, the same was obtained using GA and compared.

Keywords Genetic algorithm · Optimization · Denoising · Signal-to-noise ratio · Mean square error (MSE)

34.1 Introduction

The failure of solid insulating systems of HV equipment can lead to catastrophic failure of equipments. It is therefore imperative that the degradation of the insulation in such systems be detected and quantified at an early stage in its development so that appropriate replacement/refurbishment can be arranged at the appropriate time. When degradation occurs in such systems, irrespective of the causative mechanism, it generally results in generation of PD. Once prevalent, PDs are one of the dominant mechanisms of degradation. For this reason, detection and

B. Tyagi (✉) · H. A. Vidya
Department of EEE, BNMIT, Bangalore, India
e-mail: bindiya_tyagi@redifmail.com

V. Chakravarthi et al. (eds.), *Proceedings of International Conference on VLSI, Communication, Advanced Devices, Signals & Systems and Networking (VCASAN-2013)*, Lecture Notes in Electrical Engineering 258, DOI: 10.1007/978-81-322-1524-0_34,

characterization of partial discharge activity is a key approach in insulation system condition monitoring [1].

PD measurement is a very powerful tool to detect defects in the insulation structure during manufacture and to ascertain inadequacies. PD measurements are done using commercially available PD detectors, which satisfy requirements stipulated in national and international standards [2]. PD measurements give wide variety of information about the degradation of insulation system. The PD events are generally transient phenomena diminishing in few cycles. The data need to be sampled at relatively high rates to capture the useful information. At the same time, the PD signals being recorded or transmitted at such frequencies are often contaminated with noise from the communication systems or surrounding electrical equipments. One of the problems that primarily arise in the analysis of the data is the noise content in the signals and its removal without sacrificing signal features which is achieved by wavelets.

The denoising process can technically be viewed as performing a linear or non linear filtering of the input signal. One of the conventional denoising methods is the Wiener filter and its variants. These methods rely on the statistical characteristics of the input signal and produce a minimum mean error output [3, 4]. These methods simultaneously remove high frequency components of the signal while denoising. The denoising is also performed using wavelet approach giving better results. WT is proposed as an alternative to Fourier transform (FT) for PD denoising to overcome the limitations of FT. The method suppresses noises within the signal more effectively than FT. One drawback with WT is the poor frequency resolution at high frequencies, which is overcome by wavelet packet transform. To achieve good denoising, it is crucial to select the parameters optimally, such as mother wavelet decomposition-level and thresholding-related parameters. Methods are proposed for optimal selection of the parameters; however, there is no guarantee of optimal selection of the complete set of parameters as they are considered individually. Moreover, considering parameters individually tends to be time consuming as selection process is not often automatic. To overcome these drawbacks, method based on GA is proposed to automatically optimize entire set of parameters resulting in best denoising performance. Nowadays GA is used as an effective tool in analysing data and pattern recognition which are very difficult or even impossible with traditional methods.

34.2 Genetic Algorithm

GA was introduced by John Holland at University of Michigan to allow computers to evolve solutions to difficult search and combinatorial problems, such as function optimization and machine learning. The working principle of GA is illustrated in Fig. 34.1. Major steps are generation of population of solution, finding objective function and fitness function and application of genetic operators.

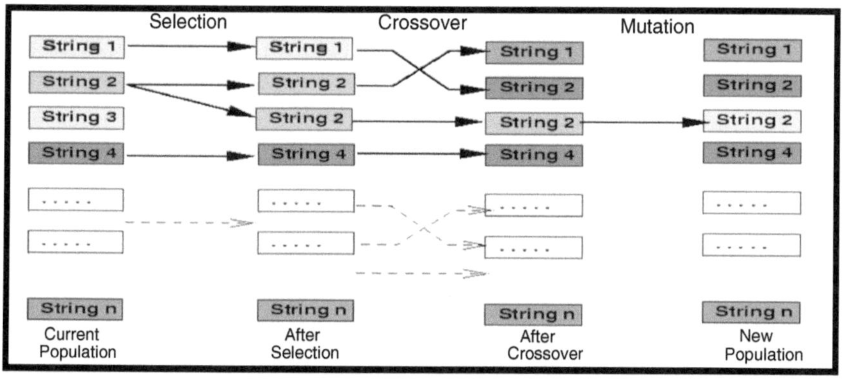

Fig. 34.1 The basic GA operation

Initially, the variables that describe the problem are coded. GA processes number of solutions simultaneously, and this collection of chromosomes is called a population, whose individual represents a feasible solution. This is called initial solution.

In second stage, individual members of population are evaluated to find objective function value. In the third step, objective function is mapped into a fitness function and then followed by the application of GA operators.

There are two GA operators, crossover and mutation, which generate new solutions from existing ones. In crossover, two chromosomes called parents having higher fitness function are combined together to form new chromosome called offspring. By iteratively applying crossover operator, genes of good chromosomes are expected to appear more frequently in the population, eventually leading to convergence to an overall good solution [5, 6].

In mutation, random changes are introduced in the characteristics of the chromosomes. Mutation is generally applied at the gene level. Mutation plays critical role in GA, where crossover leads the population to converge by making the chromosome in population alike, mutation reintroduces genetic diversity back into the population and assists the search escape from local minima. The working of GA can be as summarized in the Fig. 34.2.

34.3 Optimum Denoising

34.3.1 Description of Problem

The denoising process based on wavelet provided good results [7] but to obtain optimized denoising, selection of proper parameters like mother wavelet, number of levels of decomposistion, threshold has significant effect and GA can be applied for the same.

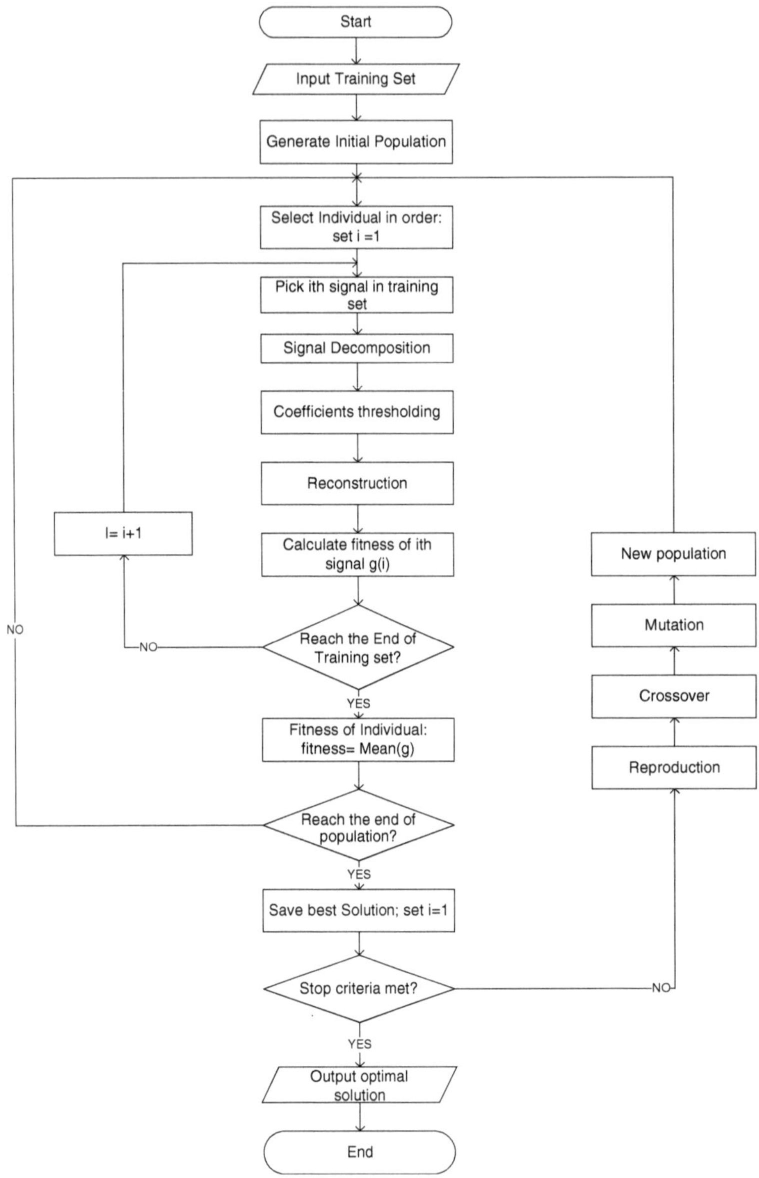

Fig. 34.2 Flow chart summarizing GA Technique

34.3.2 Denoising Based on Wavelet

The general wavelet-based denoising method proceeds in 3 steps:

1. Decomposing the noisy signal
2. Thresholding the wavelet coefficients.
3. Reconstructing the signal.

The decomposition of the noisy signal and reconstruction is performed in the same way as proposed in [7].

The second step, thresholding is very crucial in efficient noise cancellation. The key questions in thresholding are

- how to perform the thresholding
- how to find the appropriate threshold value for each scale given the noise model.

In this work, soft thresholding method is proposed as it results in a smooth signal. The success of the noise cancellation can be assessed using a number of measures. One of the measures involves the SNR defined by Eq. (34.1). The goal of denoising is to improve SNR. SNR is considered to be the objective function for GA as proper characteristic curve was obtained when SNR was plotted with respect to threshold provided noise model is known. The denoising action is also guided by a reasonable minimum % reduction in pulse amplitude (RPA) which is defined in Eq. (34.2).

$$SNR = 10 * \log \frac{\sum_{i=1}^{N} Y(i)^2}{\sum_{i=1}^{N} (X(i) - Y(i))^2} \tag{34.1}$$

$$\% \text{ Reduction } = \frac{X - Y}{X} * 100 \tag{34.2}$$

Where X(i) is the original signal, Y(i) is the denoised signal, N is the number of pulses, X is the peak amplitude (positive going peak pulse) of original or reference signal and Y is positive peak amplitude of recovered denoised signal [7].

34.3.3 Proposed Denoising Method

An advanced wavelet technique is proposed to find optimum SNR value and the threshold value. For the proposed GA technique, single variable problem is considered. With variable threshold and applying GA for different population, generation, crossover rate and mutation rate, SNR and RPA were determined, which were found to be same as obtained by wavelet technique in comparatively less time. Initially the technique is applied to the simulated noisy signal and later to practical PD signals obtained from corona discharge. The results obtained are compared with results obtained in [7].

34.4 Results and Discussions

34.4.1 Generation of Simulated Signal

In this work to generate simulated PD signals, methodology employed by Satish et al. [8] was used. A sampling interval of 25 ns ($F_s = 40$ MHz) was used. PD pulses were simulated by an exponentially decaying function. These pulses were convolved with the impulse response of a 2nd-order Butterworth filter to generate a signal corresponding to the quasi-integrated output of the PD detector. The authors of Ref. [8] clarified at this juncture that the bandpass filter forms an essential block in the simulation studies and is supposed to represent the cumulative frequency characteristics of the entire PD detector system as seen by a PD current pulse. Since quasi-integration of the PD pulses is the ultimate motive (achieved by bandpass filtering) and the frequency characteristics of most PD detector systems can in general be considered to be appropriate here, the output of this system corresponds to quasi-integrated signal whose peak value is proportional to the charge of input pulse. Quasi-integrated PD pulse for f_1/f_2 value of 80/300 kHz for 2nd-order Butterworth filter with pulse resolution time of 10.67 μs is obtained. In order to consider the effect of pulse resolution time into account, a train of pulses were analysed. PD signals were buried in the discrete spectral

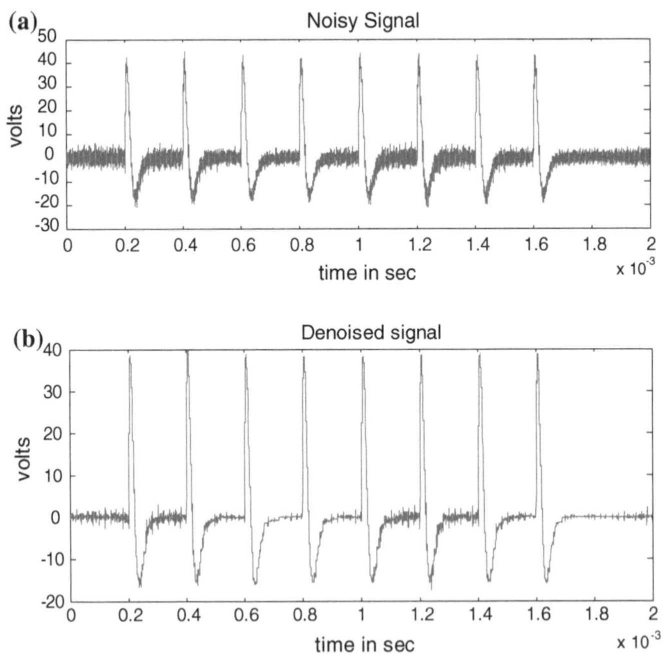

Fig. 34.3 **a** Noisy signal with minimum amplitude DSI and, **b** Denoised signal

interferences (DSI) of 8 different frequencies from 200 kHz to 1 MHz with modulating frequency of 1 kHz and random noise. DSI signals were generated following the procedure adopted by other researchers in past [8, 9]. Amplitude modulated sine waves with 40 % modulation were considered.

34.4.2 Denoising of Simulated Signal

The simulated signal is mixed with minimum amplitude DSI and random noises. When denoised with the proposed modified wavelet technique and after long manual mathematical calculations, the optimum value of SNR was obtained to be 46.72 dB at the threshold value of 3.2. The RPA is obtained to be 1.22 %. Next, the same signal was denoised with wavelet-based GA technique keeping population size as 20, generation count as 20 and crossover as 0.8, the optimum SNR 46.63 dB was obtained at the threshold value of 2.95, which is approximately the same as that obtained previously, but in comparatively less time. The same signal when denoised using the wavelet technique mentioned in [7], SNR obtained was 20.56 dB and RPA was 5.76 %. Figure 34.3 shows the PD signal with minimum amplitude of DSI and the denoised signal.

The optimum SNR and the threshold value obtained by applying GA are as shown in Fig. 34.4.

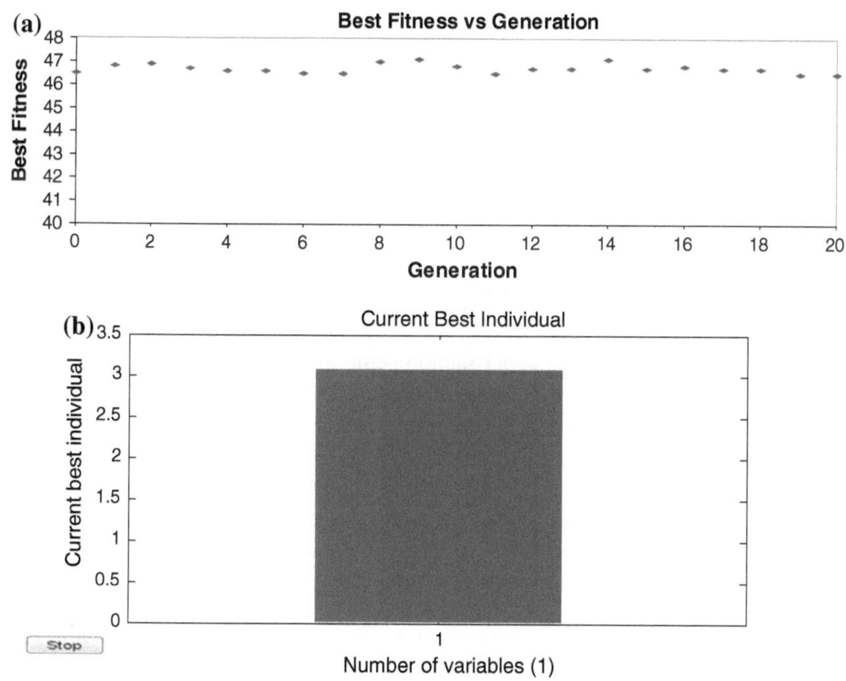

Fig. 34.4 **a** Best fitness vs Generation and, **b** Current best individual vs Number of variables

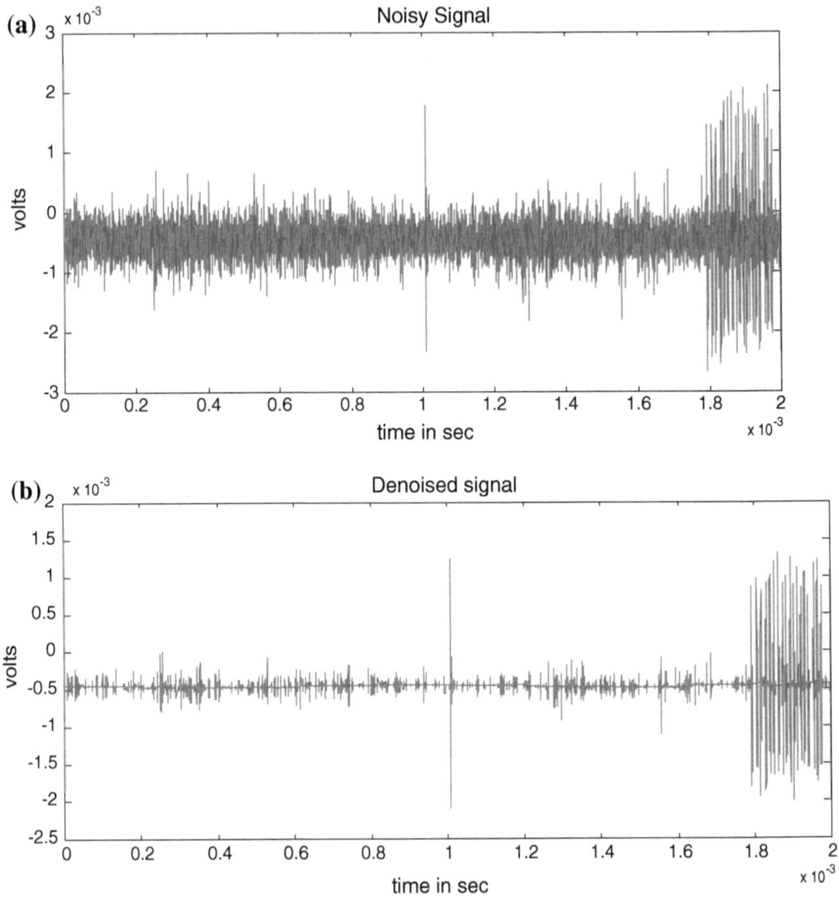

Fig. 34.5 **a** Corona discharges acquired at the applied voltage of 5.8 kV. **b** Denoised signal

34.4.3 Denoising Practical PD Signal

PD signals are obtained from experimental setup at CPRI, Bangalore. The applied voltage was 5.8 kV, and a sharp needle was used as the test sample to obtain the corona discharges. The corona discharges along with the noise generated in the experimental set-up were fed to the PD detector circuit. The noisy and denoised signals are as shown in the Fig. 34.5. When denoised by modified wavelet technique, the SNR obtained was 11.283 dB and RPA was obtained to be 33 %. Using GA for denoised signal, the positive SNR was obtained as 13.5 dB and RPA was 4.66 %. The same PD signal, when denoised using method mentioned in [7], the RPA obtained was 27.36 %.

34.5 Conclusion

PD signals were acquired by considering interferences which occur during onsite and online measurements. The signals were acquired in laboratory in a realistic manner. When denoising the practical PD signals obtained from corona discharges by the proposed method, positive signal-to-noise ratio (SNR) was obtained and also the reduction in pulse amplitude improved. So, it can be summarized that by applying the proposed method for denoising the PD signals, the denoising process improves as compared to the method proposed in [7].

References

1. Ma X, Zhou C, Kemp IJ (2002) Interpretation of wavelet analysis and its application in partial discharge detection. IEEE Trans Dielectr Electr Insul 9(3):446–457
2. Kreuger FH (1989) Partial discharge detection in high voltage equipment. Temple Press, London, 1964, Butterworths and Co, 1989
3. Zhang H, Blackburn TR, Phung BT, Sen D (2007) A novel wavelet transform technique for online partial discharge measurements Part 1: WT denoising algorithm. IEEE Trans Dielectr Electr Insul 14(1):3–14
4. Zhang H, Blackburn TR, Phung BT, Sen D (2007) A novel wavelet transform technique for online partial discharge measurements part 2: on-site noise rejection application. IEEE Trans Dielectr Electr Insul 14(1):15–22
5. Mathew TV Genetic algorithm. IIT Bombay
6. Srinivas M, Patnaik LM (1994) Adaptive probabilities of crossover and mutation in genetic algorithm. IEEE Trans Syst Man Cybern 24(4):656–666
7. Vidya HA (2009) Some investigations on denoising of partial discharge signals using wavelet transform. Thesis for degree of Doctor of Philosophy, Vishveshvaraya Technological University (2009)
8. Satish L, Nazneen B (2003) Wavelet based denoising of partial discharge signals buried in excessive noise and interferences. IEEE Trans Dielectr Electr Insul 10(2):354–367
9. Nagesh V, Gururaj BI (1993) Evaluation of digital filters for rejecting discrete spectral interference in onsite PD measurements. IEEE Trans Electr Insul 28(1):73–85

Chapter 35
Automatic Image Mosaicing Using Discrete Cosine Transform

Ananda Chakrasali and P. Manjunatha

Abstract The image mosaicing can be used to combine two or more pictures extracted at different time from different sensors or different views. The general problem of mosaicing is to create a single seamless image by aligning a series of spatially overlapped images; the result is an image with a field of view greater than that of a single image. This paper proposes a framework for creating particularly convenient way to generate mosaics is by stitching together many ordinary photographs(capturing static scenes), proposed algorithm uses creating visually pleasing mosaics using a discrete cosine transform (DCT), it is a separable linear transformation and phase-correlation method to estimate the displacement between two adjacent images, DCT performs along a single dimension and the other dimension for an input images and also correlation-based scheme is used which operates in the discrete domain for finding the transformed coordinates (translational and rotational parameters) and use them for image mosaicing.

Keywords Registration · Translation · Rotation · Mosaicing · Discrete cosine transform

35.1 Introduction

Image stitching is the process of recovering the existing camera motions between images and then composting them together. This technique has been successfully applied to many different applications like photogrammetry, satellite imagery

A. Chakrasali (✉) · P. Manjunatha
Jawaharlal Nehru National College of Engineering, Shimoga, Karanataka, India
e-mail: anandsc9@gmail.com

P. Manjunatha
e-mail: manjup.jnnce@gmail.com

V. Chakravarthi et al. (eds.), *Proceedings of International Conference on VLSI, Communication, Advanced Devices, Signals & Systems and Networking (VCASAN-2013)*, Lecture Notes in Electrical Engineering 258, DOI: 10.1007/978-81-322-1524-0_35,

video compression [1], video indexing [2, 3], object tracking [3], or creation of virtual environments [1–11]. For example, Shum and Szeliski [6] proposed methods to stitch a set of images together to form a panorama, and also which are used in biological and medical application.

Different types of transformations (Table 35.1) are used for finding the transformed coordinates (translational and rotational parameters),an affine camera model is used to approximate the possible motions between two consecutive frames, parameters of this model can be recovered from pair of images by two common methods, i.e. the correlation based approach and the optimization-based one. For the first approach, the measure correlation can be calculated in frequency domain or spatial domain. For example, Kuglin and Hines [5] presented a phase-correlation method to estimate the displacement between two adjacent images in frequency domain, and also large displacement between two images can be calculated. For example Hsieh [3] presented the edge alignment problem is solved by global alignment involves calculation of transform, which align two images. If the image is rotated by some angles that problem can be avoided using image registration technique, that technique is presented by Fonesca and Costa [1].

In this paper, we present an image mosaicing script that will stitch two images together to create one larger image. This stitching will be performed using point or transformed co-ordinates (translational and rotational parameters) between the two images. First use of discrete cosine transforms (DCT) to the overlapped images, it enables an image into a discrete Fourier domain. Then the function fftshift is used to shift quadrants of DCT around to see the lowest frequencies in the centre of the plot, second step is used for the search for the unknown motion parameters (Image calibration), therefore large motions can be recovered with no priority information. Third the approach phase correlation is a method and then obtains the normalized cross-correlation, this is used to determine and select the best correlation point between any second image and the first image. Uses a discrete frequency-domain approach to estimate the relative translate offset between two similar images. Final step is to complementary image blending algorithm is used, based on a

Table 35.1 Hierarchy of 2D coordinate transformations [4]

Name	Matrix	#D.O.F	Preserves:	Icon
Translation	$[\ I \mid t\ ]_{2x3}$	2	Orientation +···	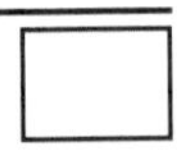
Rigid (Euclidean)	$[\ R \mid t\ ]_{2x3}$	3	Lengths +···	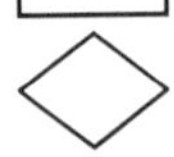
Similarity	$[\ sR \mid t\ ]_{2x3}$	4	Angle +···	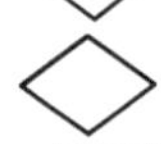
Affine	$[\ A\ ]_{2x3}$	6	Parallelism +···	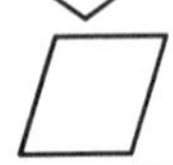
Projective	$[\tilde{H}]_{2x3}$	8	Straight line	

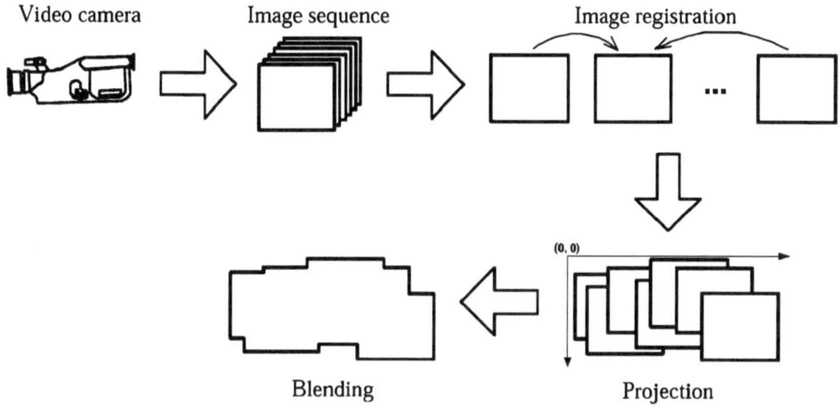

Fig. 35.1 Three basic steps of image mosaicing [4]

translational method to join multiple images. Thus it involves three different steps in the image mosaicing [4] as shown in Fig. 35.1.

35.2 Steps of Image Mosaicing (Stitching)

These are steps involved in the image mosaicing as shown in Fig. 35.1.

1. Image calibration (perspective correction, vignette correction, chromatic aberration correction). Images are processed in this stage to improve results.
2. Image registration (analysis for translation, rotation, and focal length). Direct or feature-based image alignment methods may be used. Direct aliment methods search for image orientations that minimize the sum of absolute differences between overlapping pixels. Feature-based methods determine proper image orientations by identifying features that appear in multiple images.
3. Image blending; combining the sections.

35.2.1 Image Calibration

35.2.1.1 Finding Projective Transform for Two Images

Assume that input images are captured by a video camera from a static scene. Then, the relationship between two adjacent images can be described by a [4] planar perspective motion model as follows as shown in (x′, y′) the coordinate of its corresponding point in the next frame; and $M = (m_0, m_1, m_2, \ldots m_8)$the parameters associated

$$\begin{pmatrix} x' \\ y' \\ d' \end{pmatrix} = \begin{pmatrix} m_0 & m_1 & m_2 \\ m_3 & m_4 & m_5 \\ m_6 & m_7 & m_8 \end{pmatrix} x \begin{pmatrix} x \\ y \\ d \end{pmatrix} \tag{35.1}$$

With the focal length, rotation angle, and scaling of the camera. Clearly, Eq. (35.2) is a nonlinear transformation In the past, the parameters of this model were obtained by minimizing the [1] error function E (M), as shown in the Eq. (35.3). Different types of transformations [4] are shown in Table 35.1.

$$x' = \frac{m_0 x + m_1 y + m_2}{m_6 x + m_7 y + 1} \text{ And } y' = \frac{m_3 x + m_4 y + m_5}{m_6 x + m_7 y + 1} \tag{35.2}$$

$$E(M) = \sum_i [I_1(x', y') - I_0(x, y)]^2 = \sum_i e_i^2 \tag{35.3}$$

35.2.2 Image Registration

Image registration is the process of transforming different sets of data into one coordinate system. Data may be multiple photographs, data from different sensors, from different times, or from different viewpoints [2]. It is used in computer vision, medical imaging, military automatic target recognition, and compiling and analyzing images and data from satellites. Registration is necessary in order to be able to compare or integrate the data obtained from these different measurements.

Frequency domain approaches for finding displacement and rotation/scale are computationally efficient but can be sensitive to noise. These methods also require the overlap extent to occupy a significant portion of the images (e.g. at least 50 %).

35.2.3 Blending the Images

The next and last step in mosaicing is image blending, which modifies the image gray levels in the vicinity of common boundary to obtain a smooth transition between images by removing the seams. Creating a blended image requires determining how pixels in an overlapping area should be presented. Performing blending in the entire overlapped region between the images is not only time consuming but also leads to poor image quality. Further, when the overlap is very large, it could lead to false seams (presented in the "Results" section). Our aim is to blend only the regions near transition.

35.3 Algorithm for DCT Based Image Mosaicing

1. Load the images.
2. Convert the color image into grayscale image.
3. Compute discrete cosine transform each image and take the FFTSHIFT (fft-shift-zero frequency component to centre of spectrum for vectors) F1 and F2 of images A and B respectively to that image. Given an image, S, in the spatial domain, the pixel at coordinates (x, y) is denoted Syx, to transform S into an image in the frequency domain F, we can use Eq. (35.5).

$$C_u = \begin{cases} 1/\sqrt{2} & \text{if } u = 0 \\ 1 \rightarrow \text{else} \end{cases} \text{ And } C_v = \begin{cases} 1/\sqrt{2} & \text{if } v = 0 \\ 1 \rightarrow \text{else} \end{cases} \tag{35.4}$$

$$F(v,u) = \frac{1}{4} C_v C_U \sum_{y=0}^{N-1} \sum_{x=0}^{N-1} S_{yx} \cos\left(v\pi \frac{2y+1}{2N} \right) \cos\left(u\pi \frac{2x+1}{2N} \right) \tag{35.5}$$

4. Let P (u, v) be the phase correlation value of F1 and F2 as shown in Eq. (35.6).

$$P = \frac{F1. * F2}{|F1. * F2|} \tag{35.6}$$

5. Compute the inverse discrete cosine transform T (u, v) of P (u, v): to rebuild an image in the spatial domain from the frequencies obtained above, we use the IDCT.

$$S(y,x) = \sum_{y=0}^{N-1} \sum_{x=0}^{N-1} C_v C_U F_{vu} \cos\left(v\pi \frac{2y+1}{2N} \right) \cos\left(u\pi \frac{2x+1}{2N} \right) \tag{35.7}$$

6. The total offset or translation between images depends on the location of the peak in the cross-correlation matrix and on the size and position of the images. That maximum value gives the translation (Xtrans, Ytrans). Where Xtrans = x-axis translation and Ytrans = y-axis translation the image.

7. Find the index of maximum peak from the value stored in the vector. It gives the angle of rotation. Let it be angle (theta). Using the theta value we can find the rotation of the image using matrix R.

$$R(\theta) = \begin{pmatrix} \cos\theta & -\sin\theta & 0 \\ \sin\theta & \cos\theta & 0 \\ 0 & 0 & 1 \end{pmatrix} \tag{35.8}$$

8. Transformations [4]

Suppose the two images, image1 and image2 to be registered having both translation and rotation with angle of rotation being'theta' between them. When image2 is rotated by theta, there will be only translation left between the images. So by rotating image2 by one degree each time and computing the correlation peak for that angle, we reach a stage where there is only translation left between the images. That angle becomes the angle of rotation [4]. Different types of transformations are shown in Table 35.1.

(a) Translation Transform

$$x' = x + b \tag{35.9}$$

(b) Affine Transform

$$x' = ax + b \tag{35.10}$$

(c) Bilinear Transform

$$\begin{aligned} x' &= m_1xy + m_2x + m_3y + m_4 \\ y' &= m_5xy + m_6x + m_7y + m_8 \end{aligned} \tag{35.11}$$

(d) Projective Transform

$$x' = ax + b/cx + 1 \tag{35.12}$$

9. Blending the images: The two images are combined using translation parameter and combination of pixels in the areas of overlap, which would produce a satisfactory result using Matlab7.9 software.

35.4 Results

The algorithm have been tested on different sets of images,are shown in the Figs. 35.2, 35.3, 35.4 and 35.5 and also different types of quality measurement values are computed for DCT and FFT based image mosaicing shown in the Tables 35.2 and 35.3. The algorithm for image mosaicing has been implemented using MATLAB 7.9 software.

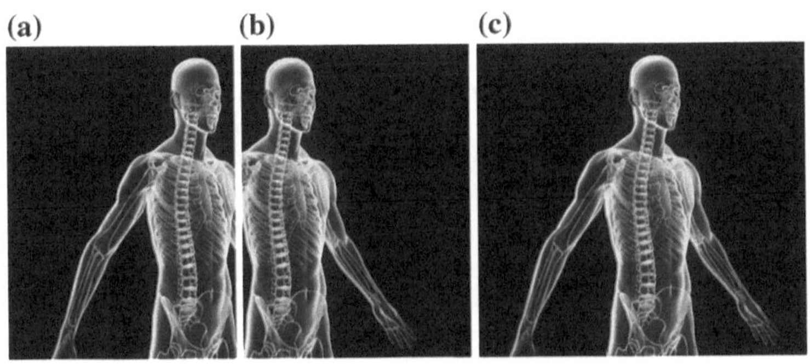

Fig. 35.2 X-ray image (**a**) First image (**b**) Second image (**c**) Mosaiced image

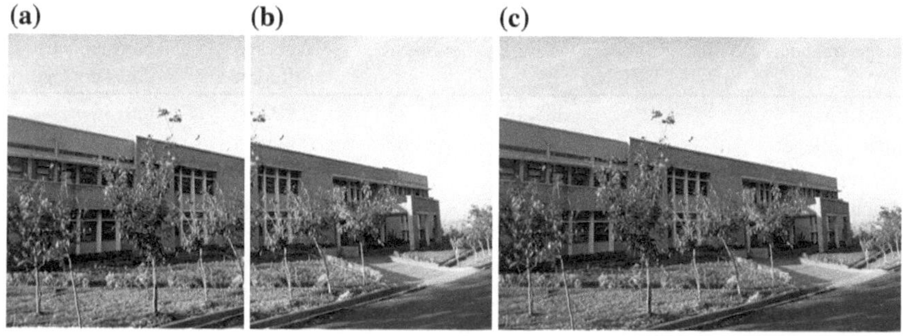

Fig. 35.3 Junce college image (**a**) First image (**b**) Second image (**c**) Mosaiced image

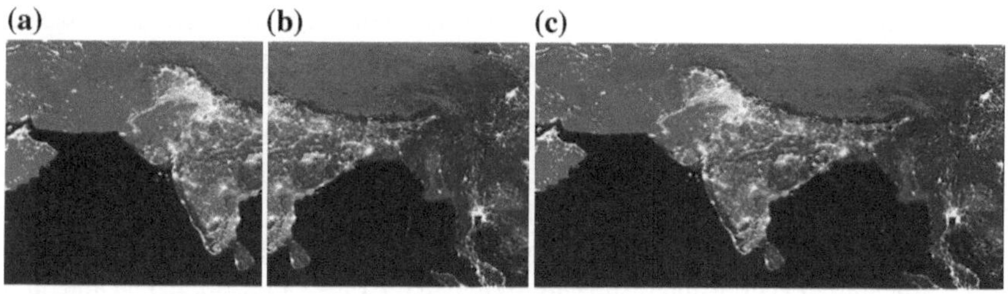

Fig. 35.4 Satellite map of India. (**a**) First image (**b**) Second image (**c**) Mosaiced image

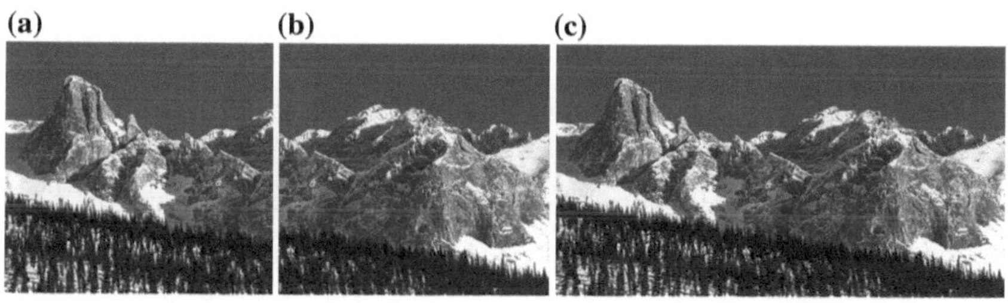

Fig. 35.5 Mountain Image (**a**) First image (**b**) Second image (**c**) Mosaiced image

Table 35.2 Quality measurement of DCT based image mosaicing

Image mosaicing using DCT	MSE	PSNR	Average difference	Normalized cross-correlation
X-ray image	388.09	22.45	0.9924	0.0120
Junce college	40.13	32.14	0.9993	0.0983
Satellite map of India	353.21	22.67	0.9546	0.5484
Mountain image	848.49	18.86	0.9946	1.1811

Table 35.3 Quality measurement of FFT based image mosaicing

Image mosaicing using DCT	MSE	PSNR	Average difference	Normalized cross-correlation
X-ray image	388.46	22.23	0.9927	0.0134
Junce college	40.51	32.05	0.9995	0.0994
Satellite map of India	353.41	22.64	0.9568	0.5497
Mountain image	848.59	18.84	0.9954	1.1825

35.5 Conclusion

In this paper, we have presented three algorithms for still images taken with a low cost digital camera. The first is a simple and reliable algorithm for finding rotation and transformations of planar transformations based on the phase correlation. we got the better result compared to image mosaic using FFT, comparison values are shown in the Tables 35.2 and 35.3, mosaiced image quality measured between the algorithms DCT and FFT computed using MSE, PSNR, average difference and normalized cross-correlation. In that analysis DCT based method assessed the best results. Ultimately we can conclude that we have implemented algorithms for construction of panoramic image mosaics which provide wide views, even exceeding human vision, and results are shown in the Figs. 35.2, 35.3, 35.4 and 35.5.

35.6 Drawbacks

- The input images given by the user should have at least 10 overlap for good results.
- The brightness of the images given by the user should not differ too much.

References

1. Fonesca LM, Costa MHM (1997) Automatic regestration of satellite images. In: Proceedings on IEEE transaction of computer society, pp 219–226
2. Irani M, Anandan P (1998) Video indexing based on mosaic representation. In proceedings of IEEE 86 (May), pp 905–921
3. Hsieh J-W (2004) Fast stitching algorithm for moving object detection and mosaic construction. Image Vis Comput 22:291–306
4. Szeliski R (2006) Image Alignment and Stitching: A Tutoriall, Microsoft Re-search, Microsoft Corporation, One Microsoft Way Redmond, WA 98052. http://www.research.microsoft.com
5. Kuglin C, Hines D (1975) The phase correlation image alignment method. In: Proceedings of IEEE international conference on cybernetics and society, 163165

6. Szeliski R, Yeung H (1997) Creating full view panoramic image mosaic and environment. In: Computer graphic proceeding, annual conference series, pp 251–258
7. Patidar MD, Jain A (2011) Automatic image mosaicing: an approach based on FFT. Int J Sci Eng Technol (ISSN: Applied) 1(1):01–04
8. Bonnet M (1999) Mosaic representation for video shot description. In: Proceedings of MPEG-7 evaluation. Ad hoc meeting 636 (Feb)
9. Sawhney H, Ayer S (1997) Compact representation of video through dominant and multiple motion estimation. IEEE Trans Pattern Anal Machine Intell 18(Aug):814–830
10. Chen SE (1995) Quicktime VR, an image based approch to virtual environment navigation. In: Proceedings of SIGGRAPH Conference, pp 29–38
11. Davis J (1998) Mosaics of Scenes with Moving Objetcs. In: Proceedings of IEEE international conference on PAMI, vol 20. pp 354–376

Chapter 36
Visual Cryptography for Color Images with Meaningful Shares Using Image Fusion Technique

S. B. Prashanth and S. V. Sathyanarayana

Abstract In this paper, we consider a novel cryptographic scheme, called visual cryptography, which can decode cascaded images without any cryptographic computations. We extend this scheme into an application, in which the two shares of a secret image are embedded with a text like names of the participants to whom these shares belong, using image fusion technique. This makes the shares more meaningful. The paper discusses the results obtained. The scheme is perfectly secure and easy to implement.

Keywords Visual cryptography · Image fusion · Meaningful shares

36.1 Introduction

Usually, multimedia data is transferred via Internet, as e-commerce is getting popular day by day. So, there is a need of secure e-transfer in today's increasing open network environment. Traditional cryptosystems are normally used to protect information on the network, which uses a secret key [1]. Naor and Shamir [2] proposed visual cryptography in 1994, which has a notable feature, where it can recover a secret image without any computation. Visual cryptography is a cryptographic technique which allows visual information (picture, text) to be encrypted in such a way that the decryption can be performed by the human visual system (HVS) without the aid of computers. Thus, it overcomes the complex computation required in traditional cryptography techniques. The threshold scheme makes the application of visual cryptography more flexible. With the t out of n threshold

S. B. Prashanth (✉) · S. V. Sathyanarayana
Jawaharlal Nehru National College of Engineering, Shimoga, Karnataka 577204, India
e-mail: prashanth.sb.69@gmail.com

S. V. Sathyanarayana
e-mail: svs.tce@gmail.com

V. Chakravarthi et al. (eds.), *Proceedings of International Conference on VLSI, Communication, Advanced Devices, Signals & Systems and Networking (VCASAN-2013)*, Lecture Notes in Electrical Engineering 258, DOI: 10.1007/978-81-322-1524-0_36,

scheme ($t \leq n$), the manager can first produces n copies of transparency drawn from the secret image, one for each of his members. If any t of them stacks their transparencies together, the content of the secret image will show up. If the number of transparencies is less than t, the content of the secret image will remain hidden [2]. Lin and Tsai [3] proposed a new visual cryptography scheme suitable for gray-level images. Here, instead of using gray sub-pixels directly to construct shares, a dithering technique is used first to convert a gray-level image into approximate binary image, and then existing visual cryptography technique for binary images is applied for creating shares. The overall effect of this method is the achievement of visual encryption and decryption functions for gray-level images. Young-Chang Hou [4] proposed three methods for visual cryptography for color images. These methods have the backward compatibility with the previous results in the black-and-white visual cryptography such as t out of n threshold scheme and can be applied to gray-level and color images easily. Chetan Hegde et al. [5] proposed a technique of processing the signature of a customer and then dividing it into shares for banking applications. Kumari and Bhatia [6] presented multi-pixel visual cryptography for color images to generate two meaningful shares. Here, some filters are proposed for better visual quality of recovered image and a simple watermarking algorithm is proposed to generate meaningful shares. In the proposed scheme, the visual cryptography concept is used to embed a text in the color image shares using a simple image fusion technique which is computationally efficient. The results obtained are very impressive which is suitable for real-time applications, and they are discussed in detail in this paper.

36.2 Visual Cryptography Model

The simplest version of the visual secret sharing problem assumes that the message consists of a collection of black-and-white pixels and each pixel is handled separately.

Each original pixel appears in n modified versions (called shares), one for each transparency. Each share is a collection of m black-and-white sub-pixels, which are printed in close proximity to each other so that the human visual system (HVS) averages their individual black-and-white contributions [2]. Typically, the black-and-white visual cryptography decomposes every pixel in a secret image into a 2×2 block in the two transparencies as shown in Fig. 36.1 [4]. When a pixel is white, the method chooses one of the six combinations for white pixels to form the content of the block in the two transparencies; when a pixel is black, it chooses one of the other two combinations. Then, the characteristics of two stacked pixels are as follows: black and black is black, white and black is black, and white and white is white. Therefore, when stacking two transparencies, the blocks corresponding to black pixels in the secret image are full black, and those corresponding to white pixels are half-black and half-white, which can be seen as 50 % gray pixels. As for information security, there are six possible patterns from which every block in a

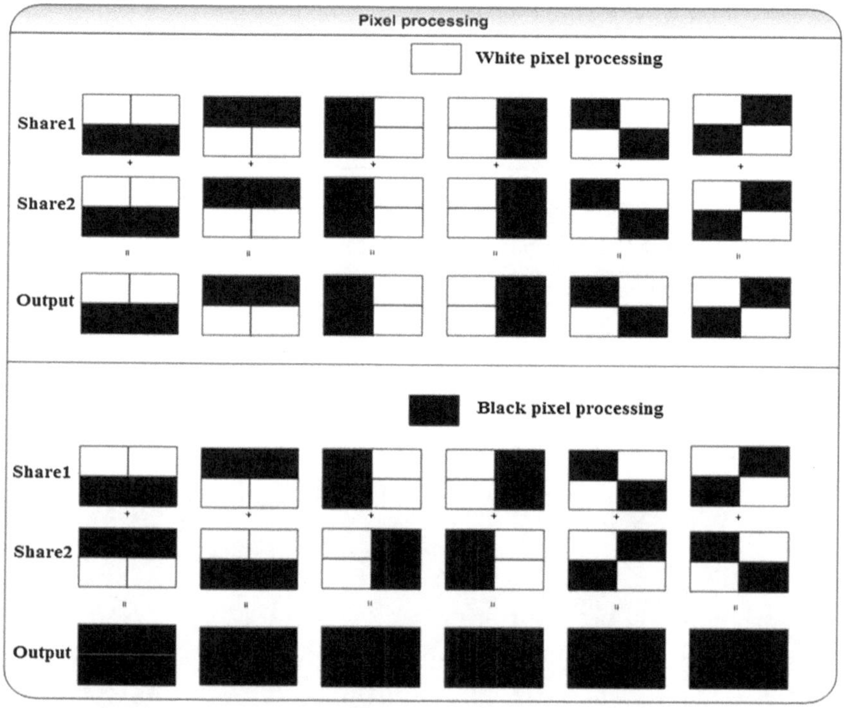

Fig. 36.1 Pixel processing and stacking [2]

transparency can randomly choose, so the secret image cannot be identified from a single transparency [4].

36.3 Visual Cryptography for Gray-level Images

Since most printers have to transform gray-level images into halftone ones before printing, and the transformed halftone images are black and white only, such an image format is very suitable for the traditional method to generate the shares of visual cryptography [4].

Halftone Technology: The method that uses the density of the net dots to simulate the gray level is called "halftone" and transforms an image with gray level into a binary image before processing (Fig. 36.2).

Algorithm:

1. Transform the gray-level image into a black-and-white halftone image.
2. For each black or white pixel in the halftone image, decompose it into a 2 × 2 block of the two transparencies according to the rules in Fig. 36.1 If the pixel is

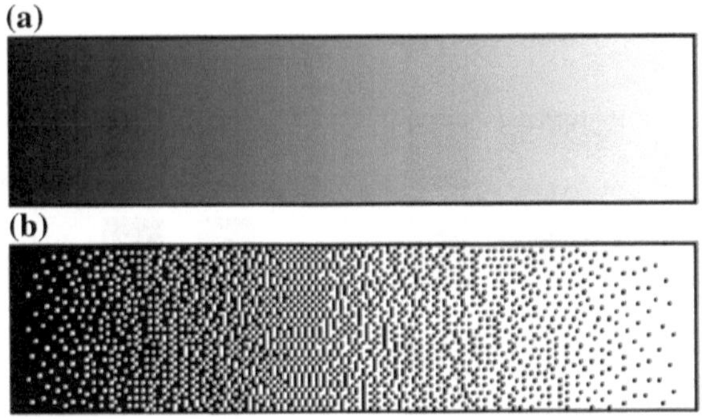

Fig. 36.2 **a** Continuous tone and, **b** halftone [4]

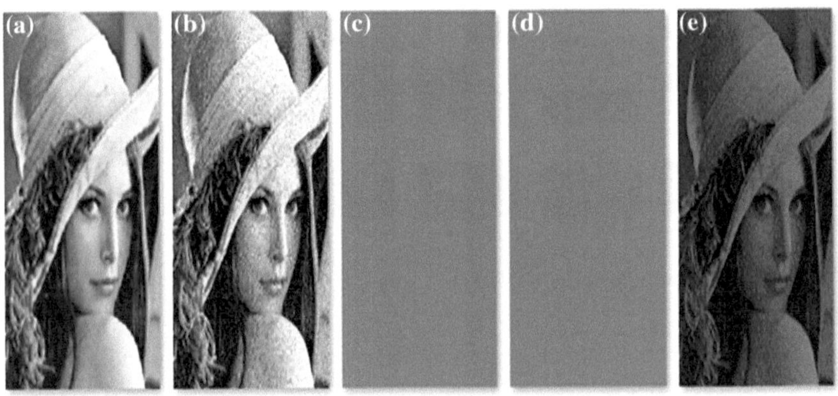

Fig. 36.3 **a** Input image, **b** Halftone image, **c** Share 1, **d** Share 2, **e** Decrypted image

white, randomly select one combination from the former two rows in Fig. 36.1 as the content of blocks in Shares 1 and 2. If the pixel is black, randomly select one combination from the latter two rows as the content of the blocks in the two transparencies.

3. Repeat the Step 2 until every pixel in the halftone image is decomposed, which results in two transparencies of visual cryptography to share the secret image.
4. After stacking the two image shares, the secret image can be decrypted.

A sample result is obtained by using the above-said algorithm as shown in Fig. 36.3.

36.4 Visual Cryptography for Color Images

For describing the constitution of colors, additive and subtractive are commonly used [7]. In the additive system, the primaries are red, green, and blue (RGB), with desired colors being obtained by mixing different RGB components. By controlling the intensity of red (green or blue) component, we can modulate the amount of red (green or blue) in the compound light. The more the mixed colored lights, the more is the brightness of the light. When mixing all red, green, and blue components with equal intensity, white color will result. The computer monitor is a good example of the additive model. In the subtractive model, color is represented by applying the combinations of colored lights reflected from the surface of an object (because most objects do not radiate by themselves). By mixing cyan (C) with magenta (M) and yellow (Y) pigments, we can produce a wide range of colors. C, M, and Y are the three primitive colors of pigment, which cannot be composed from other colors. The color printer is a typical application of the subtractive model. In visual cryptography, we use sharing images as the decryption tool; that is, the final outputs are transparencies. Because the subtractive model is more suitable for printing colors on transparencies, we will use the CMY model to represent colors in what follows [4].

Algorithm:

1. Transform the color image into three halftone images: C, M, and Y.
2. For each pixel Pij of the composed image, do the following:
   (a) According to the traditional method of black-and-white visual cryptography, expand Cij, Mij, and Yij into six 2×2 blocks, C1ij, C2ij, M1ij, M2ij, and Y1ij, Y2ij.
   (b) Combine the blocks C1ij, M1ij, and Y1ij and fill the combined block corresponding to Pij in Share 1.
   (c) Combine the blocks C2ij, M2ij, and Y2ij and fill the combined block corresponding to Pij in Share 2.
3. Repeat the Step 2 until every pixel of the composed image is decomposed, hence, obtaining two visual cryptography transparencies to share the secret image.
4. After stacking the two sharing images, the secret image can be decrypted.

What follows are the proposed work and the results obtained, which is an application of visual cryptography.

36.5 Proposed Scheme

An application of visual cryptography for color images, where text is embedded as a watermark in the colored shares, is proposed in [6]. In this paper, we propose a

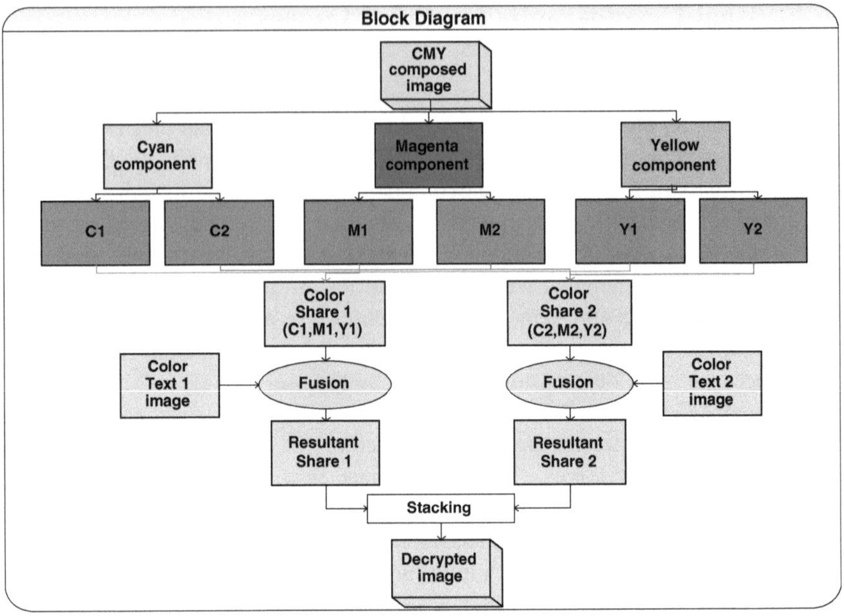

Fig. 36.4 Block diagram of the scheme proposed

scheme in which color text is embedded in a color share using image fusion technique to make it meaningful. We first decompose C, M, and Y color planes from the input color image as explained in previous section. These color planes look like gray-level images. Then, we apply visual cryptography technique for each color planes as explained in Sect. 36.3, so that we get two shares for each color components. This scheme is useful in scenario where the names of the participants to whom the shares belong can be embedded. That is, Share 1 contains the name of the participant to whom the Share 2 belongs and vice versa. This helps the participants to identify their co-participants so that secret message is revealed easily. The complete scheme is shown in Fig. 36.4. This is done using image fusion technique. That is, text to be embedded is also represented as an image, which is fused with the colored share image. It should be noted that image fusion technique is computationally efficient compared to actual watermark embedding [6], as it is a simple image adding technique.

36.6 Results and Discussion

The complete scheme is programmed using MATLAB 7.0 on Windows-based computer environment with core i3, 2.2 GHz processor. The results are shown in Fig. 36.5. The input image of size 215 $\times$ 263 is applied to the proposed algorithm,

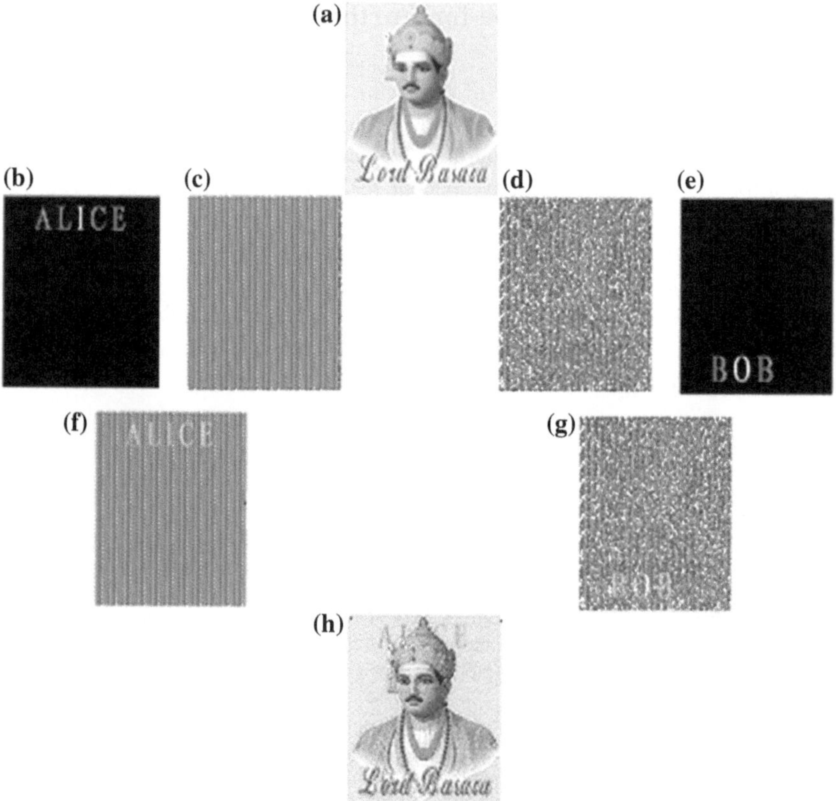

Fig. 36.5 **a** Input image, **b** Colored text 1 image, **c** Share 1, **d** Share 2, **e** Colored text 2 image, **f** Resultant image 1, **g** Resultant image 2, and **h** Decrypted image

and various levels of the outputs are obtained. Figure 36.5c and d is two image shares of the input image Fig. 36.5a. Figure 36.5b and e is the images of the text to be embedded in the shares. Figure 36.5f and g is the shares embedded with text, which has been obtained by adding images Fig. 36.5b with Fig. 36.5c, e with Fig. 36.5d, respectively, uses image fusion technique. Figure 36.5h shows the decrypted image which is obtained by stacking images in Fig. 36.5f, g. We can see that it contains residual information of embedded text that does not harm the original secret image; hence, it almost resembles the input image. The proposed algorithm executes by taking negligible execution time, which is very encouraging for real-time applications.

36.7 Conclusion and Scope for Further Work

Visual cryptography provides one of the secure ways to communicate images over open network environment. Here, computational complexity involved in decryption process of traditional cryptography is removed. This paper deals with the application of visual cryptography concept for color image processing. The concept is further extended for embedding a colored text in color image shares so that the shares become meaningful shares. The embedding process is computationally efficient where a simple image fusion technique is used. The proposed algorithm executes by taking negligible execution time, which is suitable for real-time applications. The authors are intended to continue the work in the area of biometric applications and key distribution of traditional cryptography.

References

1. Menezes AJ, Van Oorschot PC, Vanstone SA (2010) Handbook of applied cryptography, CRC press, US
2. Naor M, Shamir A (1995) In: De Santis A (ed) Visual cryptography, advances in cryptology: Eurpocrypt94, lecture notes in computer science, vol 950. Springer, Berlin, pp 1–12
3. Lin CC, Tsai WH (2003) Visual cryptography for gray-level images by dithering techniques. Pattern Recogn Lett 24:349–358
4. Hou YC (2003) Visual cryptography for color images, pattern recognition, 36
5. Hegde C, Manu S, Shenoy PD, Venugopal KR, Patnaik LM (2008) Secure authentication using image processing and visual cryptography for banking application. Sixteenth international conference, pp 65–72
6. Kumari K, Bhatia S (2010) Multi-pixel visual cryptography for color images with meaningful shares. Int J Eng Sci Technol 2(6)
7. Poynton CA (2011) Frequently Asked Questions about Color. Available at http://www.inforamp.net/poynton

Chapter 37
Automatic Detection of Microaneurysms from Fundus Images Using Morphological Operations

G. G. Rajput, Preethi N. Patil and Ramesh Chavan

Abstract Diabetic retinopathy is an ocular disorder resulting in patients with long history of diabetes. It is a progressive disease characterized by numerous features like, microaneurysms (MA), hard exudates, soft exudates, veins bleeding, and hemorrhages. Presence of microaneurysms is the early signs of Diabetic retinopathy. In this paper, automatic detection of microaneurysms, that alternates the tedious and time-consuming manual process, is presented. Thresholding and morphological operations are used for microaneurysms detection from fundus images. In the first step, optic disk and blood vessels are eliminated to facilitate the detection of MA. Secondly, the candidate features are extracted based on their size. Experiments are performed on a set of 100 fundus images and have yielded encouraging results.

Keywords Diabetic retinopathy · Diabetes · Microaneurysms · Optic disk · Morphological operation

37.1 Introduction

Diabetic retinopathy (DR) is the most common diabetic eye disease and a leading cause of blindness. DR drastically changes the texture and appearance of the retina. DR is characterized by numerous features like, microaneurysms (MA),

G. G. Rajput · P. N. Patil (✉) · R. Chavan
Department of Computer Science, Gulbarga University, Gulbarga, Karnataka 585102, India
e-mail: Preethip15@gmail.com

G. G. Rajput
e-mail: ggrajput@yahoo.co.in

R. Chavan
e-mail: ramesh.chauhan1050@gmail.com

V. Chakravarthi et al. (eds.), *Proceedings of International Conference on VLSI, Communication, Advanced Devices, Signals & Systems and Networking (VCASAN-2013)*,
Lecture Notes in Electrical Engineering 258, DOI: 10.1007/978-81-322-1524-0_37,

hard exudates, soft exudates, veins bleeding, hemorrhages. Appearance of microaneurysms indicates the early sign of DR in diabetic patients. Microaneurysms are small areas of balloon-like swelling in the retina's tiny blood vessels. Detection of MA not only helps ophthalmologist in treating DR and there by prevent blindness but also alternates the time-consuming manual process which is prone to observer's errors. Several methods and procedures for detecting MA have been reported in the literature. A brief review of the methods available in the literature is given below.

In [1, 2], initial segmentation of microaneurysms is achieved by using a bilinear top-hat transformation and matched filtering followed by thresholding. Region-growing algorithm is then used for the analysis of size, shape, and energy characteristics of each candidate results in the final segmentation of microaneurysms. Morphological operations are used to extract microaneurysms in [3]. Morphological opening with linear structuring elements in different directions is applied to the input image and top-hat transformation is then applied to extract details corresponding to microaneurysms candidates. Use of morphological-based operations to identify microaneurysms is also presented in [4, 5]. In [6], mathematical morphology is used to segment microaneurysms within fluorescein angiograms. Gaussian matched filters are used to retain candidate MA from fundus images in [7]. In [8], back propagation neural network is applied to sub-images to extract microaneurysms. An automated system for detection of diabetic retinopathy using recursive region–growing segmentation (RRGS) is reported in [9]. Classification of DR stages based on diameter closing and kernel density estimation is presented in [10]. In [11], a diameter-closing approach to segment microaneurysms using k-nearest neighbors (kNN) is reported. A detection system based on pixel classification is proposed in [12]. In his paper, thresholding and morphological operations have been used for detecting microaneurysms from fundus images. The following sections provide the details of the proposed method.

37.2 Proposed Method

The block diagram of the proposed method for extracting microaneurysms is shown in Fig. 37.1. In the preprocessing stage, the input image is resized and the green channel is extracted. Contrast enhancement is achieved using adaptive histogram equalization method. Segmentation stage involves edge detection of microaneurysms and blood vessels. Thresholding is applied for removal of exudates and noise. Blood vessels are eliminated by using morphological operation. Finally, optic disk is detected and eliminated, leaving microaneurysms in the resulting image. The details for microaneurysms detection is presented below.

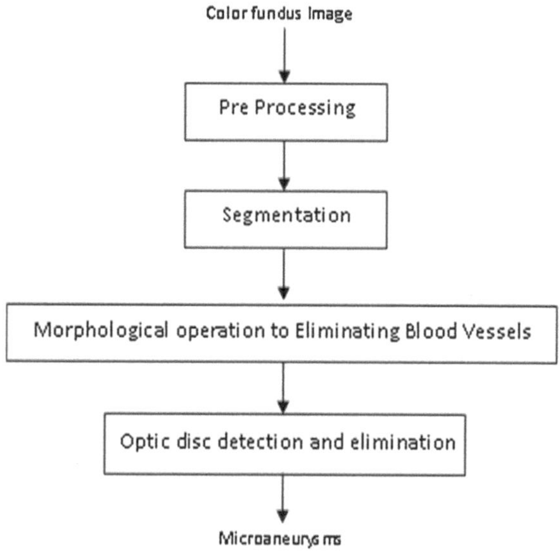

Fig. 37.1 Block diagram for microaneurysms extraction

37.2.1 Material

A total of 100 digital color fundus photographs, each of dimension 4288 × 2848 pixels, captured by NIKON D300 camera are used for experimentation. The images are provided by Karnataka Institute of Diabetology, Bangalore [13].

37.2.2 Preprocessing

In order to bring the uniformity in the size of input images, the input image is resized to a standard size. From the literature, it is known that green channel of an RGB image is best suited for feature extraction of DR. Hence, we extract green channel of RGB for further processing (Fig. 37.2b). After extracting the green

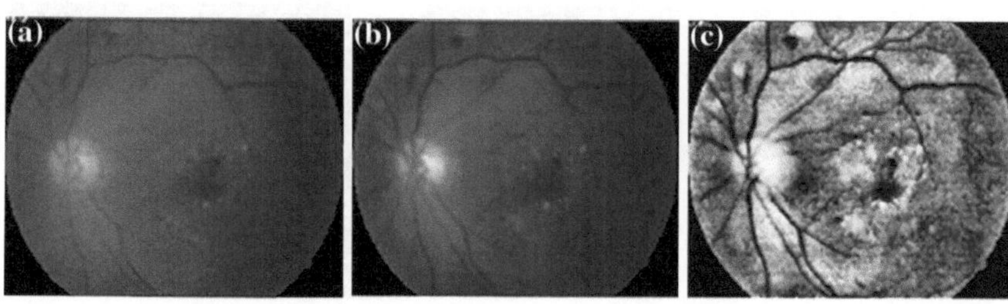

Fig. 37.2 **a** Original image. **b** Green channel image. **c** Adaptive histogram applied image

channel, the contrast of the resulting gray-scale image is adjusted using adaptive histogram equalization method (Fig. 37.2c).

37.2.3 Segmentation

The Canny edge detector is one of the most commonly used image processing tools for detecting edges in a very robust manner [14]. Its parameters allow it to be tailored to recognition of edges of differing characteristics. Blood vessels and microaneurysms have been detected using canny edge detector (Fig. 37.3a). Next, the candidate microaneurysms are selected by filling only the circular region. Exudates are removed by performing logical AND operation. Thresholding is used to remove the noise (Fig. 37.3b).

37.2.4 Blood Vessel Elimination

The preprocessed image is binarized and noise is eliminated. The resulting image is ANDed with binarized image obtained in segmentation (Figs. 37.3a, 37.4).

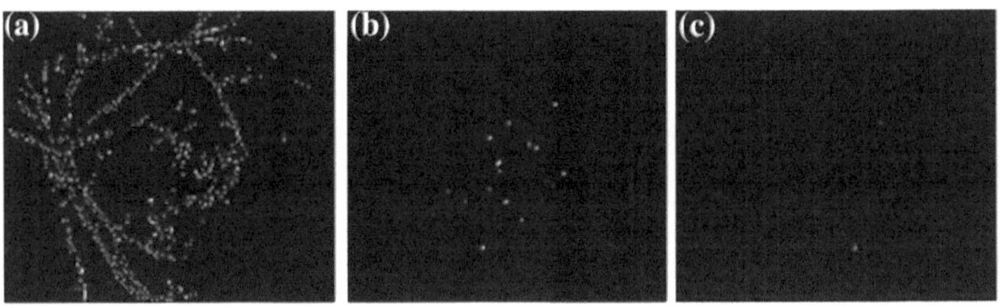

Fig. 37.3 **a** Edge-detected image. **b** Image with MA, exudates, and noise. **c** Image after removing exudates and noise

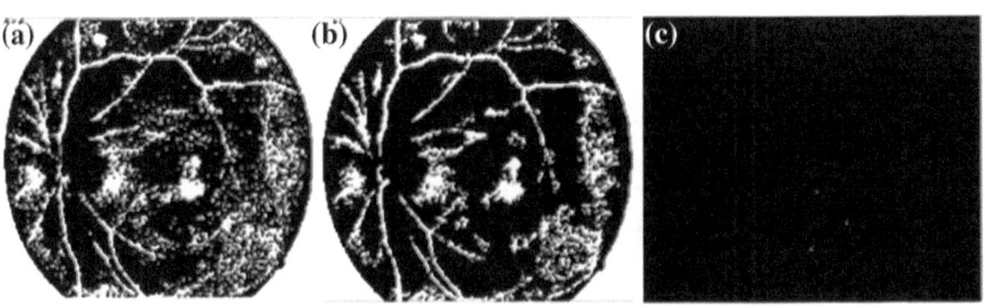

Fig. 37.4 **a** Binarized image. **b** Small noise removed. **c** Image after removing blood vessels

37.2.5 Optic Disk Detection and Elimination

Optic disk appears in color fundus images as a bright yellowish or white and is more or less circular in shape interrupted by the outgoing vessels (Fig. 37.5a). This feature can be used to trace the optic disk in the preprocessed image. From the preprocessed image, the brightest pixel value is determined. A mask is created with an appropriate radius (90 pixels) (Fig. 37.5). The mask is applied to the ROI to remove the optic disk. We have also used multistage active contour method to

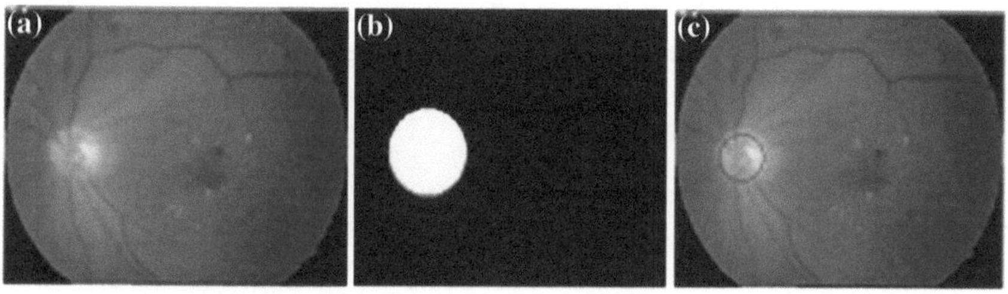

Fig. 37.5 **a** Original image. **b** Mask. **c** Optic disk detected

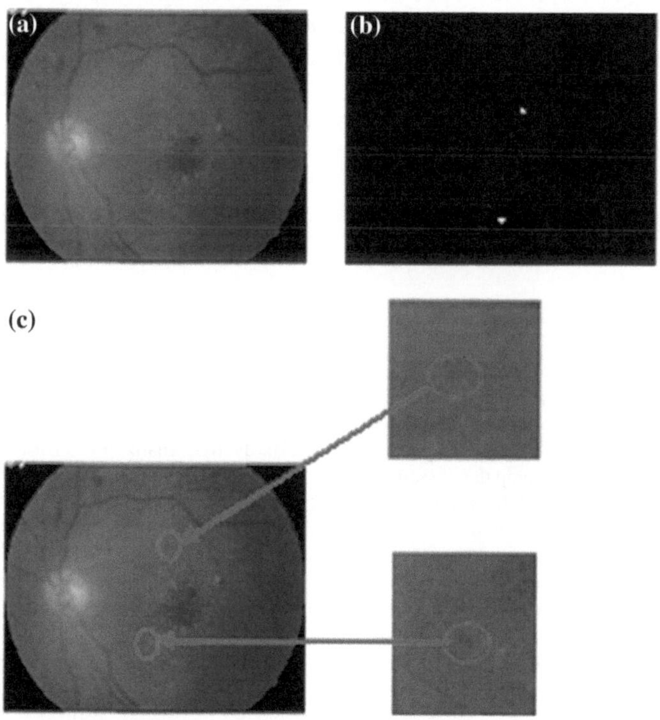

Fig. 37.6 **a** Original RGB image, **b** MA extracted image, **c** MA position-marked image

detect the optic disk which has yielded encouraging results. It provides a better segmentation for images with weak boundaries when compared to the mask-based methods and other parametric methods.

37.3 Experimental Results

Experiments are performed on 100 color fundus images collected for Karnataka Institute of Diabetology, Bangalore. The MA in the original image and the MA in the resulting images can be visibly seen (Fig. 37.6). The proposed method detects and eliminates the blood vessels and optic disk successfully without affecting the microaneurysms present in the original image.

37.4 Conclusion

Morphological operations-based method is proposed to identify microaneurysms from the fundus images. The proposed method has successfully detected and removed blood vessels and optic disk leaving behind microaneurysms. The proposed method is an initial step toward automatic detection of DR stages. Detection of exudates from fundus images is reported in [15]. We are working on combining both these methods and grading the DR disease as proliferative and non-proliferative using artificial neural network. The performance of the proposed method can also be tested on the images from the standard databases available in the literature.

References

1. Spencer T, Olson JA, McHardy KC, Sharp PF, Forrester JV (1996) An image processing strategy for the segmentation and quantification of microaneurysms in fluorescein angiograms of the ocular fundus. Comput Biomed Res 29:284–302
2. Frame AJ, Undrill PE, Cree MJ, Olson JA, McHardy KC, Sharp PF, Forrester JV (1998) A comparison of computer based classification methods applied to the detection of microaneurysms in ophthalmic fluorescein angiograms. Comput Biol Med 28:225–238
3. Baudoin CE, Lay BJ, Klein JC (1984) Automatic detection of microaneurysms in diabetic fluorescein angiographies. Revue D' Épidémiologie et de Sante Publique 32:254–261
4. Spencer T, Olson JA, McHardy KC et al (1996) An image-processing strategy for the segmentation and quantification of microaneurysms in fluorescein angiograms of the ocular fundus. Comp Biomed Res 29:284–302
5. Cree MJ, Olson JA, McHard KC et al (1997) A fully automated comparative microaneurysms digital detection system. Eye 11:622–628

6. Frame A, Undrill P, Cree M et al (1998) A comparison of computer based classification methods applied to the detection of microaneurysms in ophthalmic fluorescein angiograms. Comput Biol Med 28:225–238
7. Hipwell JH, Strachan F, Olson JA et al (2000) Automated detection of microaneurysms in digital red-free photographs: a diabetic retinopathy screening tool. Diabet Med 17:588–594
8. Gardner G, Keating D, Williamson TH et al (1996) Automatic detection of diabetic retinopathy using an artificial neural network: a screening tool. Br J Ophthalmol 80:940–944
9. Sinthanayothin C, Boyce JF, Williamson TH et al (2002) Automated detection of diabetic retinopathy on digital fundus image. Diabet Med 19(2):105–112
10. Walter T, Massin P, Erginay A et al (2007) Automatic detection of microaneurysms in color fundus images. Med Image Anal 11(6):555–566
11. Dupas B, Walter T, Erginay A et al (2010) Evaluation of automated fundus photograph analysis algorithms for detecting microaneurysms, haemorrhages and exudates, and of a computer-assisted diagnostic system for grading diabetic retinopathy. Diabet Metab 36(3):213–220
12. Niemeijer M, van Ginneken B, Staal J et al (2005) Automatic detection of red lesions in digital color fundus photographs. IEEE Trans Med Imaging 24(5):584–592
13. http://kidbangalore.in/
14. Gonzalez RC, Woods RE, Eddins SL (2009) Digital image processing using matlab, Pearson Education India
15. Dr Rajput GG, Patil PN, Chavan R (2013) Automatic detection of hard and soft exudates from fundus images using morphological operations, 2nd national conference on innovative paradigms in engineering and technology (NCIPET), pp 1–4

Chapter 38
AERB SG D-25 and IEC 60880 for Certification of Software in Safety Systems of Indian NPP

Gopinath Karmakar and Yogesh Nirgude

Abstract In the nuclear domain, strict adherence to standards and guides is mandatory for safety–critical software. IEC 60880 standard provides requirements for the safety–critical (Class IA) software of the computer-based I&C systems. But, each country has its own guides that need to be followed for licensing/certification of safety–critical software. This work aims to bring out the essential regulatory requirements for certification of software for Class IA systems in Indian nuclear domain. Also, this work attempts to determine whether there are any additional regulatory requirements for certification of safety–critical software vis-à-vis adhering to the IEC 60880. Finally, this work attempts to identify the objectives, fulfillment of which can form the basis for certification of Class IA software.

Keywords Class IA · Software certification · Nuclear safety · Indian NPP

38.1 Introduction

Certification of software that goes into computer-based system (CBS) of nuclear power plants (NPP) requires independent verification and validation (IV&V). Independent evaluation of software is practiced since early 1990s, especially in Europe and in the United States [1]. In Indian nuclear domain, Atomic Energy Regulatory Board (AERB) is the agency, which certifies safety–critical software for NPP.

G. Karmakar (✉) · Y. Nirgude
Reactor Control Division, Bhabha Atomic Research Centre, Mumbai 400085, India
e-mail: gkarma@barc.gov.in

Y. Nirgude
e-mail: ynirgude@barc.gov.in

V. Chakravarthi et al. (eds.), *Proceedings of International Conference on VLSI, Communication, Advanced Devices, Signals & Systems and Networking (VCASAN-2013)*, Lecture Notes in Electrical Engineering 258, DOI: 10.1007/978-81-322-1524-0_38,

The most important regulatory requirement for CBS in NPP is adherence to standards and guides. IEC 60880 [2] is the most relevant standard for safety–critical software in the nuclear domain. AREB safety guide AERB SG D-25 [3] deals with the computer-based systems of pressurized heavy water reactors (PHWR). It provides guideline for development of computer-based system and explains the Standard Regulatory Review Process (SRRP) for computer-based I&C systems covering both hardware and software. Appendix-6 and appendix-2 of SG D-25 discuss the applicable regulatory requirements for the software development phases and verification and validation (V&V), respectively.

As on date, the computer-based systems (hardware and software) for safety–critical application (safety class IA [4, 5]) pertaining to I&C of nuclear power plants are developed by the various organizations within the Department of Atomic Energy (DAE) for reasons of availability of expertise and domain knowledge.

It is likely that in near future, Indian industry will participate in development of software related to systems involving nuclear safety. It motivated us to provide a clear understanding of what is required in terms of objectives and deliverables for certification of class IA software.

This paper makes the following contributions.

- Brings out the essential regulatory requirements for certification of software for safety class IA systems of Indian NPP.
- Clearly identifies the relationship between AERB SG D-25 and IEC 60880 from a software developer's perspective.
- Identifies the specific objectives, fulfillment of which can form the basis for certification by AREB.

38.2 Regulatory Requirements and Related Standards

In this section, we discuss i) the categorization of software according to their importance for safety and ii) the essential regulatory requirements.

The software development process and the review of the evidences that the required process has been followed in its development plays the most important role for certification. The rigorousness of review of software varies according to the importance of the computer-based system in which the software is deployed. IEC 61226 [5] categorizes the I&C systems of nuclear plants as shown in Table 38.1.

Table 38.1 Software criticality levels

Safety category	Failure condition	Implication of failure in nuclear context
Class IA	Safety–critical	Can directly lead to accident conditions, which may cause unacceptable consequences
Class IB	Safety-related	Can lead to actuation of system(s) performing class IA functions
Class IC	Non-safety	No nuclear safety hazard

38.2.1 AERB SG D-25

This guide describes approach to design and review of computer-based systems (CBS) when they are to be deployed for performing functions important to safety in the nuclear power plant. The elements presented in this guide are set of goals and good practices that form the basis of acceptance of the computer-based system. It utilizes international practices as described in IAEA NUSS series and IEC standards.

Like many other safety standards, a safety case is a requirement of SG D-25 for certification of computer-based system. SG D-25 defines safety case and states the implementation of a safety case as follows:

The Safety Case, i.e., the arguments and evidence in support of system and in particular the software safety and integrity shall be based on designs created and design documents produced during the system development and results of analysis of specifications, algorithms, designs and implementation.

Recognizing the difficulties associated with quantitative estimation of software reliability, SG D-25 places high level of importance on demonstration of qualitative attributes of software.

38.2.2 IEC60880

This standard provides requirements for the software of computer-based I&C systems of nuclear power plants performing functions of safety category A as defined by IEC 61226 [5].

This standard provides requirements for the purpose of achieving highly reliable software. It addresses each stage of software generation and documentation, including requirements specification, design, implementation, verification, validation, and operation.

It may happen that a category A system performs additional functions of lower category. In such cases, IEC 60880 refers to IEC 61513 [6] and makes it mandatory to upgrade the qualification requirements of those functions also to Category A level.

38.2.3 Relationship Between AERB SG D-25 and IEC 60880

In order to facilitate certification of safety–critical software, the relationship between the regulatory guide SG D-25 and the IEC standard 60880 needs to be defined. This is necessary to provide answer to the following questions that are likely to arise from the point of view of the software developers of class IA I&C systems.

- Are there any additional regulatory requirements vis-à-vis IEC 60880?
- Is there any difference in the strictness of the requirements between the two documents?
- What are the necessary objectives to be satisfied for certification?

38.3 Objectives to be Satisfied for Certification

In this section, we propose a set of objectives, which we find, are essential for certification of class IA software. We compared the requirements of IEC 60880 and AERB SG D-25 against each identified objectives. We also identified the applicable standards, which provide guidelines/framework for realizing the objectives identified. We refer to these standards as working standards.

The essential objectives necessary to build a safety case along with the working standards are captured in Table 38.2, 38.3, 38.4, 38.5, and 38.6.

It can be observed from Table 38.2 that SG D-25 does not provide any strict guideline on SQAP and leaves it the QA policy of the development organization. But, it provides detailed requirement for SCMP as against IEC 60880.

From Table 38.3, it can be observed that there is no significant difference between AERB SG D-25 and IEC 60880 so far as strictness in software requirement specification is concerned. But, it may be noted that IEC 60880 provides more detailed requirement for SRS.

Table 38.2 Objectives and comparison (*Software development plan*)

Objectives	IEC 60880	SG D-25	Working standards
Software quality assurance plan (SQAP)	Requires SQAP at an early stage of the software life cycle and details out its attributes	Requires SQAP within the framework of organizational level QA plan	IEEE STD 730
Software configuration management plan (SCMP)	Describes general guidelines for SCMP	Explicit details of SCMP are provided	IEEE STD 828
Software verification and validation plan (SVVP)	Specific guidelines for verification of software and software aspects of system validation	No significant difference	IEEE STD 1012
Programming guidelines (PG)	Detailed guidelines for software coding are provided	Refers to IEC 60880	Depends on the language and the guideline adopted. (For example, MISRA-C [7] is acceptable in Indian nuclear domain)

Table 38.3 Objectives and comparison (*Software requirements*)

Objectives	IEC 60880	SG D-25	Working standards
Software requirements specifications (SRS)	Specific details of software requirements, including self-supervision and periodic testing is provided	No significant difference. Brief description under system architectural design is provided	IEEE STD 830
Requirement analysis	Recommends use of formal or application-oriented language for SRS	No specific recommendation	Depends on selected specification language. [For example, UML [8] (semi-formal)]
Traceability to user/system requirements	Traceability to previous SDLC document required	No difference	Part of SRS.

Table 38.4 Objectives and comparison (*Software design*)

Objectives	IEC 60880	SG D-25	Working Standards
Software architectural design (SAD)	Required	Guidelines provided	IEEE STD 1016 can be referred for architectural viewpoints
Software detailed design (SDD)	Detailed mandatory requirements provided	Brief guideline only	IEEE STD 1016
–Design of dynamic behavior	–Required	–Required	
–Traceability to SRS	–Required	–Required	

Table 38.4 compares AERB SG D-25 and IEC 60880 against the objectives related to software design. It can be observed in Table 38.4 that IEC 60880 has stricter guidelines on SDD in the sense it specifies mandatory requirements for compliance.

Table 38.5 shows that use of static analysis tool is only recommended by both the documents. But, the authors suggest that use of static analysis tools should be mandatory for reasons discussed in Sect. 38.3.2.

It can be observed from Table 38.6 that SG D-25 refers to IEC 60880 for its regulatory requirements related to software testing.

38.3.1 IEC 60880 Compliance

It is important to note that SG D-25 has a mandatory requirement of generating an IEC 60880 compliance matrix for class IA software. Therefore, we conclude that

Table 38.5 Objectives and comparison (*Software implementation*)

Objectives	IEC 60880	SG D-25	Working standards
Compliance to PG	Required	Required	As per PG (example MISRA-C)
Source code	Deliverable	Deliverable	As per the language selected
Static analysis report for compliance with quality metrics specified in SQAP (e.g., acceptable nesting depth, complexity, and comments per lines of code)	Use of static analysis tool is recommended	No difference	As specified in PG and/or SQAP

Table 38.6 Objectives and comparison (*Software testing*)

Objectives	IEC 60880	SG D-25	Working standards
Software unit test plan and report (SUTP/R) –Traceability to SDD –Statement coverage, branch coverage	Required	Required	IEEE STD 1008 and IEEE STD 829
Software integration test plan and report (SITP/R) –Functional testing –Traceability to SDD –Code coverage and branch coverage	Software integration is considered to be part of system integration	Refers to IEC 60880	IEEE STD 829

IEC 60880 shall supersede wherever the corresponding regulatory requirements is not specified or not elaborated in SG D-25.

38.3.2 Use of CASE Tools

Both AERB SG D-25 and IEC 60880 only recommend the use of computer-aided software engineering (CASE) tools for improved reviewability.

But, it may be noted that except for applications, which are very simple and small in terms of code size, the use of CASE tools becomes essential. This is because, use of CASE tools not only strengthens the case for certification with well-documented specifications and design, but also offers the inherent benefits that the tools provide in developing dependable software as discussed below.

Compliance to Programming Guidelines (PG): It becomes a huge task to verify a large code for its adherence to PG, unless a standard guideline (e.g.,

MISRA-C [7]) and a corresponding compliance checking tool is used. Let us take a few examples of commonly violated required rules of programming from a real-world application.

- Use of bit operator on signed type.
- Pointer arithmetic done other than on an array.
- Multiple declaration of an identifier with external linkage (same identifier used for both global and local variable).
- Expression needs brackets.

There are more than 100 such *required* programming rules specified in MISRA-C (2004) in addition to the *advisory* rules. Manual checking of compliance to PG of any real-world software can lead to unacceptable delay in verification, and it is prone to human error. Therefore, we conclude that checking compliance of programming guidelines is not feasible without a CASE tool.

Requirement Analysis and Design: A tool is necessary for requirement modeling and design using a modeling language like Unified Modeling Language (UML) [8]. Also, a tool is required for verification of a design carried out using formal/semi-formal modeling language.

Quality Metrics: A static analysis tool is necessary to generate metrics related to testability and maintainability of code.

38.4 Conclusion

In this paper, we brought out the essential objectives, fulfillment of which can form the basis for certification of safety–critical software in the Indian nuclear domain. It has been established that adherence to IEC 60880 alone can form the basis for certification of class IA software. It was expected and can be justified that AERB SG D-25 is a guide for computer-based system as a whole and not a document dedicated to development of software.

Additionally, we reasoned that the use of CASE tools is essential in the development of class IA software, even though it has not been made mandatory by AREB SG D-25 and IEC 60880.

Further work related to certification of pre-developed software (PDS) and class IB software is part of our ongoing activity.

Acknowledgments The authors would like to thank Shri Y. S. Mayya, Head RCnD, BARC and Dr. A. P. Tiwari, Head CSES, RCnD, BARC for their continuous support and encouragement. We also thank Shri Ashutosh Kabra, SO/D, RCnD, BARC for many a useful discussions.

References

1. Lahtinen J, Johansson M, Ranta J, Harju H, Nevalainen R (2010) Comparison between IEC 60880 and IEC 61508 for certification purposes in the nuclear domain. In: Proceedings of 29th international conference on computer safety, reliability, and security, SAFECOMP 2010, LNCS 6351
2. IEC 60880 (Ed. 2.0) (2006) Nuclear power plants—I&C systems important to safety—software aspects for computer based systems performing category A functions. In: Proceedings of international electrotechnical commission
3. AERB Safety Guide No. AERB/NPP-PHWR/SG/D-25 (2010) Computer based systems of pressurised heavy water reactors. Atomic Energy Regulatory Board, Mumbai, India
4. AERB Safety Guide No. AERB/NPP-PHWR/SG/D-1 (2003) Safety classification and seismic categorisation for structures, systems and components of pressurised heavy water reactors. Atomic Energy Regulatory Board, Mumbai, India
5. IEC 61226 (Ed. 3.0) (2009–07) Nuclear power plants—instrumentation and control important to safety—classification of instrumentation and control functions. In: Proceedings of international electrotechnical commission
6. IEC 61513 (Ed. 2.0) (2011) Nuclear power plants—instrumentation and control important to safety—general requirements for systems. In: Proceedings of international electrotechnical commission
7. MISRA-C (2004) Guidelines for the use of the C language in critical systems. The Motor Industry Software Reliability Association, UK
8. UML 2.0 (2005) Infrastructure and superstructure. Object Management Group

Chapter 39
Selective Rotation-Based CORDIC Architecture for High-Speed Applications

K. S. Geetha, Gayathri Jeyaram and A. Bhagyashree

Abstract The CORDIC algorithm is an efficient method for computing trigonometric, logarithmic, hyperbolic, and exponential functions. It is used when hardware multipliers are not available, mainly in FPGAs. It finds innumerable applications in digital communication, image processing, artificial neural networks, and robotics. In this paper, the conventional CORDIC algorithm is optimized using selective rotation (SR) techniques which employ rotation selection algorithm (RSA) for faster convergence. A coarse LUT (Look up Table) is incorporated in the design to obtain coarse values which are fine-tuned using SR CORDIC stages. This architecture is simulated and verified as a part of a numerically controlled oscillator (NCO) using MATLAB and SIMULINK. This novel architecture is area-efficient and reduces the number of iterations required, to converge to a value, by 50 %.

Keywords CORDIC · Selective rotation · Rotation selection algorithm · NCO

39.1 Introduction

CORDIC algorithm was first proposed by Volder in 1959 [1]. Some of the prominent early applications of CORDIC were in the design of computers and calculators. Modern applications include calculation of activation functions, for

K. S. Geetha · G. Jeyaram (✉) · A. Bhagyashree
Electronics and Communication Department, R.V. College of Engineering, Mysore road, R.V Vidyaniketan Post, Bangalore 560059, India
e-mail: gayathri.jeyaram.k@gmail.com

K. S. Geetha
e-mail: geethaks@rvce.edu.in

A. Bhagyashree
e-mail: bhagya1410@gmail.com

V. Chakravarthi et al. (eds.), *Proceedings of International Conference on VLSI, Communication, Advanced Devices, Signals & Systems and Networking (VCASAN-2013)*, Lecture Notes in Electrical Engineering 258, DOI: 10.1007/978-81-322-1524-0_39,

example tan, sigmoidal and exponential functions in Neural networks, computation of various transforms in image processing, to build NCOs that are used in the field of digital communication and several other digital signal processing applications.

A numerically controlled oscillator (NCO) is used to synthesize digital sinusoidal signals of desired frequencies. The Frequency Control Word (FCW) of the NCO is accumulated every clock cycle to traverse angles over a period. For each angle, the CORDIC algorithm is employed to generate sine and cosine values. Hence, the number of iterations taken by CORDIC, to compute the amplitude corresponding to each phase, decides the maximum sampling rate for a given clock frequency. Lower the number of iterations, higher the achievable sampling frequency. Hence, the reduction in number of iterations is necessary for high-speed applications.

This paper introduces a novel method to reduce the number of iterations required. A coarse LUT is used initially to obtain coarse values, followed by SR stages to fine-tune it. Rotation selection algorithm used within each SR CORDIC stage dynamically chooses the angle of rotation for that iteration, thereby converging to the required value in drastic steps. This method produces more accurate results at lower computation times when compared to the existing CORDIC algorithm. The rest of the paper is organized as follows. In Sect. 39.2, the conventional CORDIC is briefly explained. The proposed method for implementing SR-based CORDIC is discussed in detail under Sect. 39.3. Design and simulation for the same are explained in Sect. 39.4. Under Sect. 39.5, the results of the new architecture are displayed. The paper is concluded in the Sect. 39.6.

39.2 Conventional CORDIC

Trigonometric and exponential functions are some of the most frequently evaluated functions in digital computations. Hence, they must be computed accurately using minimum time and resources. CORDIC algorithm is well suited for this purpose as it uses only addition, subtraction, and bit-shift operations. This algorithm functions in two modes, rotation mode, and vectoring mode.

In rotation mode, the CORDIC machine takes in an angle and computes the sine and cosine values. Here, an initial vector $[1\ 0]^T$ is rotated by $\tan^{-1}(2^{-i})$ in the ith iteration. The direction of rotation is chosen to diminish the residual angle. This operation converts polar coordinates to Cartesian. The equations that govern the CORDIC machine working in rotation mode are as follows [2]:

$$x_{i+1} = x_i - y_i.\sigma_i.2^{-i} \tag{39.1}$$

$$y_{i+1} = y_i + x_i.\sigma_i.2^{-i} \tag{39.2}$$

$$z_{i+1} = z_i - \tan^{-1}(2^{-i}) \tag{39.3}$$

where

$$\sigma_i = \begin{cases} +1, & z_i > 0 \\ -1, & otherwise \end{cases} \tag{39.4}$$

In vectoring mode, the inputs to this algorithm are sine and cosine values of a particular angle. The algorithm performs certain vector rotations and accumulates the angle by which it rotates per iteration. This operation converts Cartesian coordinates to polar.

This algorithm requires considerable number of rotations to achieve a specified accuracy. Therefore, its major drawback is its low processing speed. This paper proposes a new method to accelerate its speed.

39.3 SR-based CORDIC

In conventional CORDIC algorithm, regardless of the angle, the initial vector is rotated by certain predetermined angles, leading to increase in the number of iterations required to converge at the desired angle. One of the methods to increase the speed would be the use of pipelined architecture. This introduces pipeline latency and area-overhead (Fig. 39.1).

To overcome these drawbacks, SR-based CORDIC method is proposed. Instead of rotating the initial vector by all the predetermined angles (elementary angles), SR (selective rotation) dynamically selects the angle of rotation that is closest to the desired angle, from the set of predetermined angles. The vector is then rotated by that angle. The residual angle is computed by subtracting the desired angle from the rotation angle [3]. In the next iteration, the SR stage selects the angle that is closest to the residual angle and the process continues till convergence is achieved, with the desired accuracy. This is called rotation selection algorithm (RSA).

When a vector is rotated by an angle $'\emptyset'$, it is scaled by a factor 'K' whose magnitude is equal to $'\cos\emptyset'$. Hence, skipping the initial angles will result in considerable amount of error and SR stages cannot be used independently. After certain iterations, this scaling factor becomes nearly equal to 1 and the scaling effects can be ignored. A small LUT is incorporated into the design to overcome this problem. It stores the coarse values of sine and cosine values, which compensates for initial iterations and the scaling effects associated with it. The size of the LUT is dependent on the number of stages after which the SR stages can be used, which is determined by the word-length of the output [1]. It is given by

$$i > \left\lceil \frac{N-1}{2} \right\rceil \tag{39.5}$$

where N is the size of the output word-length and $2^i X2$ is the size of the LUT required to store coarse sine and cosine values.

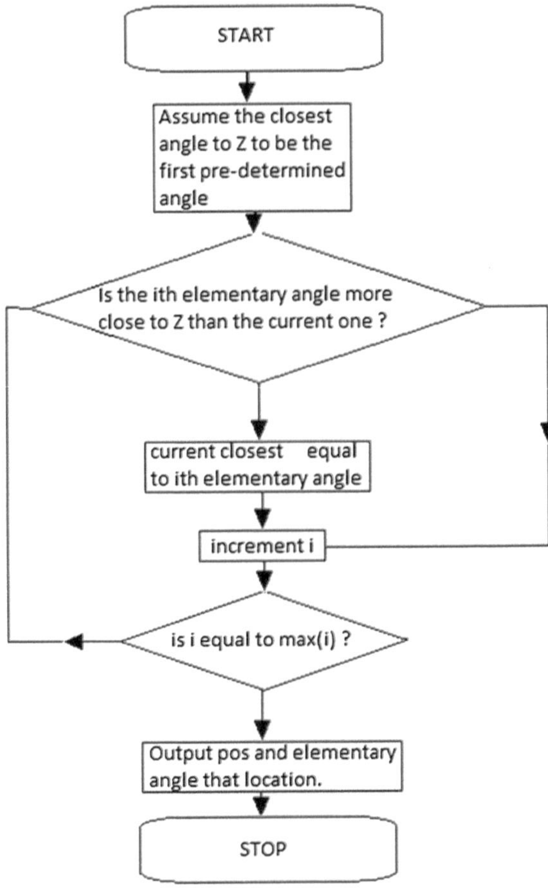

Fig. 39.1 Flowchart of rotation selection algorithm (RSA)

39.4 Design and Simulation

The proposed idea was designed and simulated using MATLAB [4] and SIMULINK. Fixed point arithmetic was used. The input phase was 32 bits wide and output values were 16 bits wide. A functional block diagram of the idea is shown below (Fig. 39.2).

This CORDIC machine computes the sine and cosine values for angles in the first quadrant and the same values are manipulated to get the values for the angles in other quadrants by the quad block.

The FCW is the phase increment that should be given to the CORDIC machine per iteration. It decides the frequency of the output sinusoidal signals. FCW can be calculated as shown [1],

$$FCW = \frac{f_0}{f_{clk}} X\, 2^{32} \tag{39.6}$$

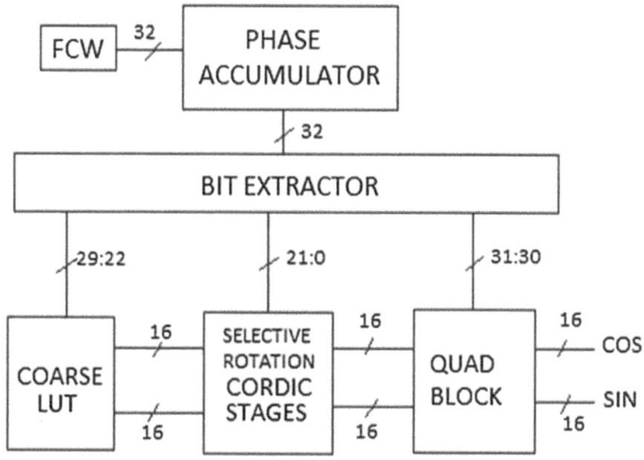

Fig. 39.2 Top-level block diagram of the selective rotation-based CORDIC machine

where f_0 is the desired output frequency and f_{clk} is the clock frequency.

The FCW is added to the previous value stored in the Phase accumulator, in every cycle. The bits of this accumulated 32 bit value are used as inputs for the other blocks.

The two MSB bits [31:30] are used to determine the quadrant to which the phase value belongs. The quad block takes these two bits as input and operates on the sine and cosine value computed, to give the corresponding values in other quadrants [5]. The next 8 bits [29:22] are fed into the LUT to get coarse values of sine and cosine. The remaining bits [21:0] are given to 3 elective rotation CORDIC stages, which drive the vector to the closest possible value.

To contain the remaining phase error within $\tan^{-1}(2^{-12})$, the error after 12 iterations in conventional CORDIC, a design with three stages of SR CORDIC is sufficient. More stages can be added to reduce the remaining phase further. If many SR CORDIC stages are used, parallelism in Z path should be used [5, 6].

The rotation selection algorithm (RSA) used in the SR CORDIC stages finds the angle nearest to the residual angle using a linear search operation. The angle is given as the phase to the next stage and the position of the angle chosen is given to the barrel shifter to perform that many shifting operations. Thus, vector rotations are converted to simple addition and shifting operations using this CORDIC algorithm.

39.5 Results

Conventional CORDIC is compared to selective rotation CORDIC for an input phase of 67°. The results obtained are tabulated in Tables 39.1 and 39.2. In SR CORDIC architecture simulated, Bit extractor and LUT use a single iteration each. It is observed that for a remaining phase in the order of 10^ − 6, the number of iterations required in conventional CORDIC and SR CORDIC are 12 and 5,

Table 39.1 Angles of rotation and the corresponding remaining phase after each iteration using conventional CORDIC algorithm for a phase of 67°

Number of iteration	Rotation angle (rad sfix32_31)	Remaining phase (radsfix32_31)
1	0.250000000000000	0.372222222387791
2	0.147583617828786	0.122222222387791
3	0.077979130204767	−0.025361395440996
4	0.039583424106240	0.052617734763771
5	0.019868524279445	0.013034310657531
6	0.009943947661668	−0.006834213621914
7	0.004973187111318	0.003109734039754
8	0.002486745361239	−0.001863453071564
9	0.001243391539901	6.232922896750150e-004
10	0.000621698331088	−6.200992502259849e-004
11	0.000310849398375	1.599080860614777e-006
12	0.000155424699187	−3.092503175139427e-004
13	0.000077712349594	−1.538256183269415e-004
		−7.611326873294152e-005

Table 39.2 Angles of rotation and the corresponding remaining phase after each iteration using SR CORDIC algorithm for a phase of 67°

Number of iteration	Rotation angle (rad sfix32_31)	Remaining phase (rad sfix32_31)
1	0.001243391539901	0.001128
2	7.771234959363937e-005	−0.0001149
3	3.885617479681969e-005	−3.721e-5
		1.649e-6

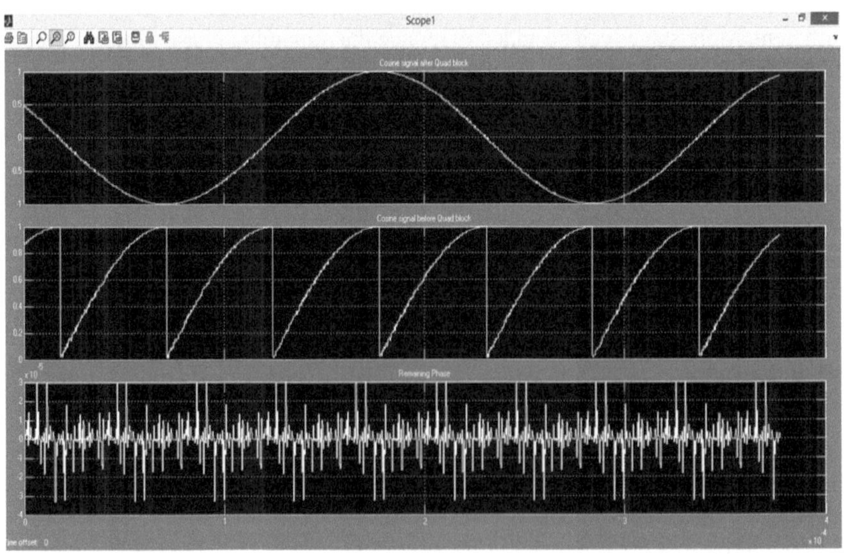

Fig. 39.3 Cosine wave generated by a NCO using SR-based CORDIC simulated in SIMULINK

respectively. Therefore, SR CORDIC algorithm takes lesser iterations to produce a more accurate result.

Figure 39.3 is a screenshot of simulation conducted using SIMULINK. Cosine signal is generated using a NCO set to a FCW of 0.003 and sampling frequency of 3.13 MHz. The first wave represents cosine signal after the quadrature operation has been performed and the second wave represents the output of the CORDIC machine which always lies in the first quadrant. The last signal is the phase error encountered while calculating the amplitude.

39.6 Conclusion

Thus, SR-based CORDIC technique of generating sinusoidal signals in a NCO provides more accuracy and faster computations when compared to conventional CORDIC. This is essential in the area of digital signal processing where calculating elementary functions is often required. In the field of image processing, it is used in many transforms, like Hough transform, to generate sinusoidal signals. The activation functions of a neuron in neural network use this algorithm too.

This paper briefly introduces the concept of CORDIC algorithm and NCO and proposes an optimized version of CORDIC and SR-based CORDIC algorithm. It also provides the simulation in MATLAB and SIMULINK to validate the concept of SR-based CORDIC algorithm. The simulation results suggest a good amount of accuracy with respect to remaining phase angle when compared to conventional CORDIC in NCO.

References

1. CORDIC-based Numerically Controlled Oscillator (NCO), Jie Qin, Charles Stroud, Foster Dai, Auburn University
2. A Survey of CORDIC algorithm for FPGA based Computers, Ray Andraka, Andraka Consulting Group, Inc
3. CORDIC designs for fixed angle of rotation. Pramod Kumar Meher, Senior Member, and Sang Yoon Park, Member, IEEE
4. Application of MATLAB for NCO design-based on FPGA, Shufang Zhang, Jun Han school of mechanical and electronic engineering, Suzhou University, Suzhou, Anhui province, China
5. An angle recoding method for CORDIC algorithm implementation, Hu Yu Hen, department of electrical and computer engineering, Wisconsin University, Madison, WI, USA
6. Implementation of Para-CORDIC algorithm and its applications in satellite communication. In: Sharma S, Ravichandran PN, Kulkarni S, Vanitha M, Lakshminarasimhan P (eds) Digital systems group, ISRO Satellite Centre, Bangalore, India

Chapter 40
Bio-Inspired Image Processing for Contour Enhancement and Fourier Spectrum Model for Orientation and Motion Detection

D. R. Vishnuvardhan and Shanthi Prince

Abstract A modeling of the image processing occurring at the level of the retina is sought to be developed. The aim is to show the advantages of using such a modeling in order to develop efficient and fast bio-inspired modules for low-level image processing in computer vision. The retinal model produces a contour-enhanced profile which is used to detect motion and orientation of contours. For this, a discrete Fourier transform–based approach has been developed.

Keywords Bio-inspired · Retinal model · DFT · Phase delay · Spatial delay · Orientation angle

40.1 Introduction

Computer vision (CV) is a field that includes methods for acquiring, processing and analyzing and understanding images, and in general, high-dimensional data from real world, in order to produce numerical and symbolic information.

To develop image processing modules for CV, various bio-inspired models are being explored. The motivation being that such models are deemed to be effective and efficient computationally. Of the various bio-inspired models is the human visual system (HVS) [1]. The image acquisition and processing capabilities of the human eye have been extensively studied and modeled into different layers of processing [1–4]. The image processing occurring at the retinal level is used to

D. R. Vishnuvardhan (✉) · S. Prince
Department of Electronics and Communication, SRM University, Kattankulathur, Kancheepuram, Tamil Nadu 603203, India
e-mail: ranjith.vishnuvardhan@gmail.com

S. Prince
e-mail: shanthi.p@ktr.srmuniv.ac.in

V. Chakravarthi et al. (eds.), *Proceedings of International Conference on VLSI, Communication, Advanced Devices, Signals & Systems and Networking (VCASAN-2013)*, Lecture Notes in Electrical Engineering 258, DOI: 10.1007/978-81-322-1524-0_40,

extract the contours in an image. The resulting contour-enhanced profile is used for orientation analysis, so as to find out the orientation angle of different pixel sizes. A discrete Fourier transform model analysis is used to identify motion and orientation.

40.2 Biological Perspectives and Retinal Model

The three layers of processing in the retina have been classified as follows [1]:

1. Photoreceptor layer
2. Outer plexiform layer (OPL)
3. Inner plexiform layer (IPL)

40.2.1 The Photoreceptor Layer

The photoreceptor layer is made up of two types of neurons capable of phototransduction, namely the rods and cones. The rods are extremely sensitive and, thus, are responsible for scotopic vision. Cones on the other hand function best under relatively bright light. They are responsible for color vision. Accordingly, there are three types of cones each for red, green, and blue.

The photoreceptors have the ability to adjust their sensitivity with respect to the luminance of their neighborhood. The image acquisition is modeled by an enzyme relation called the Michaelis–Menten relationship [1].

40.2.2 Outer Plexiform Layer

In the OPL, the photoreceptors pass on their information to horizontal cells and bipolar cells through a synaptic interaction called the synaptic triad. The gap junctions occurring in the horizontal cell layer result in a low-pass spatiotemporal filter [2]. The cellular interactions of the OPL are modeled as a band-limited high-pass filter whose lower and upper cutoff frequencies are determined by the horizontal cell layer model and the photoreceptor cell layer model, respectively.

The OPL transfer function in the spatial frequency domain is given by

$$F_{OPL}(f_s) = F_{ph}(f_s)[1 - F_h(f_s)] \tag{40.1}$$

where

f_s spatial frequency,

$F_{OPL}(f_s)$ response of the OPL,

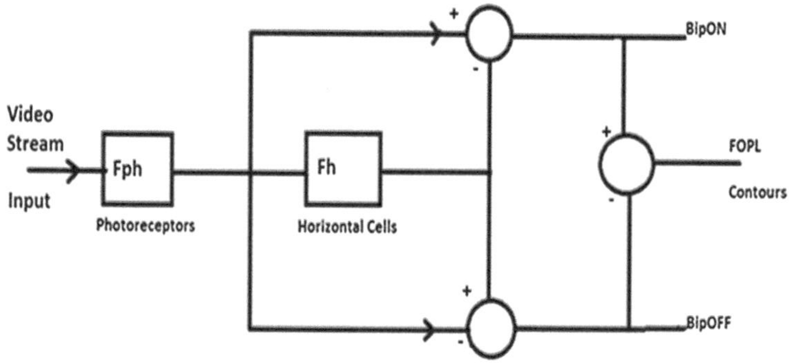

Fig. 40.1 OPL response model

$F_{ph}(f_s)$ response of the photoreceptor cell layer,
$F_h(f_s)$ response of the horizontal cell layer.

The OPL filter model can be considered as a difference between two low-pass spatial filters that model the photoreceptor network and the horizontal cell layer. The linked bipolar cells in the OPL perform the final subtraction giving the positive and negative parts of the difference between the photoreceptor cell layer output and the horizontal cell layer output. The overall model of the OPL is given by the block diagram in Fig. 40.1 [1]:

The output from the horizontal cells is excitatory for a group of bipolar cells and inhibitory for another group of bipolar cells [3]. This is represented by the BipON BipOFF output. The subtracted output from these two groups forms the final OPL output.

40.3 Mathematical Model

40.3.1 Photoreceptor Layer Model

The photoreceptor layer, which performs the first level of filtering, is modeled as a low-pass filter with a cutoff frequency that is higher corresponding to the upper cutoff frequency of the OPL filter model. In this work, it is modeled as a five-point truncated sinc filter in a single dimension in spatial domain as shown in Fig. 40.2a. The corresponding frequency domain response is a low-pass filter with a cutoff frequency at 0.25 f_s.

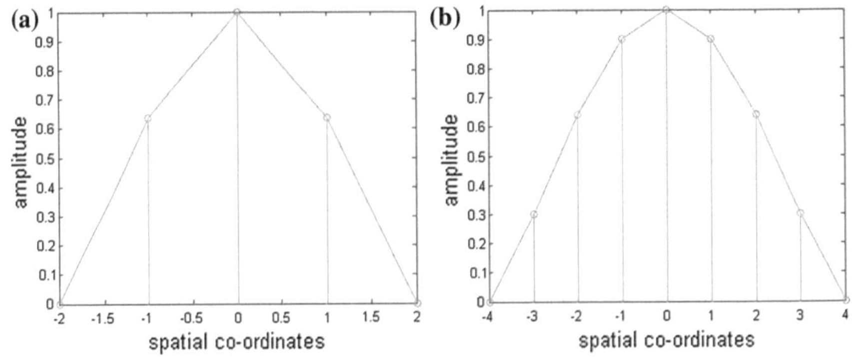

Fig. 40.2 **a** Photoreceptor layer model. **b** Horizontal cell layer model

40.3.2 Horizontal Cell Layer Model

The horizontal layer is modeled as a low-pass filter with a cut-off frequency that is lower, corresponding to the lower cutoff frequency of the OPL filter model. Here, it is modeled as a nine-point truncated sinc filter in a single dimension in spatial domain as shown in Fig. 40.2b. The corresponding frequency domain response is a low-pass filter with a cutoff frequency at $0.125\,f_s$.

Truncated sinc filter is used to convolve with the original data to realize a low-pass filtered data as it effectively models the response of the cells with a definite rise time and fall time to input impulses. Moreover, for circuit-level implementations of the above models, RLC circuits can be used to get similar response of the above models by tuning the time constant of the circuits.

40.3.3 Implementation of the OPL Model

The block diagram of Fig. 40.1 is implemented using two-dimensional convolution in spatial domain. The BipON and BipOFF equations are given by (40.4)

$$BipON = F_{ph}(f_s) - F_{ph}(f_s)F_h(f_s) \tag{40.2}$$

$$BipOFF = F_{ph}(f_s)F_h(f_s) - F_{ph}(f_s) \tag{40.3}$$

The BipOFF output is subtracted from the BipON output to simulate the final OPL output. Thus, the response in the desired frequency range is enhanced. The OPL thus gives spatial band-pass effect, retaining high spatial frequencies, which produce a contour-enhanced output that is shown in Fig. 40.3.

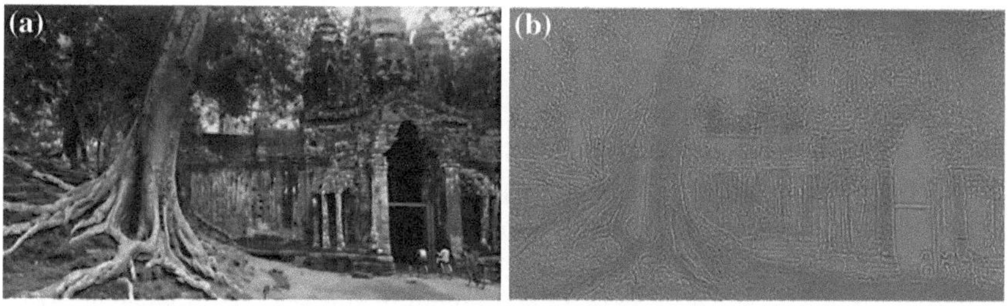

Fig. 40.3 **a** Original image. **b** Contour-enhanced OPL output

40.4 Motion Detection from Contour-Enhanced Data Using DFT

Fourier transform [5] is an ideal tool to identify various aspects of contours [6]. The amplitude changes and phase angle change related to each frequency lead to identifying the contour orientation changes as well as energy changes.

Motion changes can best be identified by observing the changes in two orthogonal orientations. The basic data are contour-enhanced image of M*N pixels. Two profiles are created, an X-oriented profile of M elements (in which the value of each element is the sum all pixels of the corresponding column) and a Y-oriented profile of N elements (in which the value of each element is the sum of all the elements of the corresponding row). DFT is carried out on the two profiles in the spatial frequency range of 0.1–0.25.

The orientation angle $\emptyset$ of the wave surface (contour) for each frequency can be shown to be related to phase angles θ_x and θ_y obtained through DFT for X-project and Y-project profiles as

$$\tan(\emptyset) = \frac{\text{spatial delay along X}}{\text{spatial delayalong Y}} = \frac{\theta_x/2\pi f_s}{\theta_y/2\pi f_s} = \frac{\theta_x}{\theta_y} \tag{40.4}$$

40.5 Analysis

40.5.1 Rotation of Frame

The orientation program is tested to identify the rotation of object about the camera axis perpendicular to the picture frame. The amplitude and orientation angle for each frequency is obtained for the vertical position of a frame. The frame is then rotated by 90° clockwise, and DFT analysis is carried out for this frame. The results show that the orientation of all frequencies has shifted by 90°. The precision of the results validates the model.

40.5.2 Head Motion Analysis

A head motion analysis is done on video frames taken on the rotation of head. The video is taken as the head rotates from left to right (from the observer's point of view). Two frames are considered for analysis. In frame1, the head is looking toward +180° direction (left) and in frame2, after rotation, is looking toward −180° K direction (right) in the Cartesian XY coordinates. From the DFT analysis of the two frames, it is observed that the orientation angles in the frequencies 0.12, 0.13, and 0.14 show sign changes from frame1 to frame2. Also in the high frequencies, i.e., 0.20, 0.21, 0.23, 0.24, orientation angles show left to right sign change.

Orientation lines are drawn in both frames for frequencies 0.12, 0.13, 0.14, and the wave fronts perpendicular to these orientation lines in the respective frames are identified. Thus, in frequency 0.12, orientation angle is changed from −170° to 164° from frame1 to frame2, and the perpendicular lines at the rear portion of the head (behind fore head) correspond to the respective orientation lines in both frames. In left frame, the rear portion of head is looking toward right side (−170°), and in the right frame, the rear side of the head is looking toward left side (+164°), matching with the signs of the angle as well. Also, the portion of the head on frame1 is more close to horizontally oriented contour (contour is more vertical) than the contour on frame2 which is inclined a little.

The contour lines joining the lips, nose, and forehead correspond to the respective orientation lines of +171° in frame1 and −172° in frame2 for the spatial frequency 0.13, and the frame features are on the opposite directions and inclined at a small angle of 8° from the vertical.

Similarly, orientation lines of frequency 0.14 which are nearly horizontal are at +178° (looking toward left) in frame2 and −176° (looking toward right) in frame1 correspond to contour lines along ear portion.

Results also show, as mentioned already, oppositely oriented features in high frequencies 0.20, 0.21, 0.23, and 0.24.

It can be observed, there is an increase in amplitude values in frame2 compared to frame1 in all the frequency which have exhibited orientation changes. In real time, the video was taken when head motion was from left to right (from observer's point of view). The change in energy in frame2 corresponds to the real situation. As the rotation was going in slant direction, energy changes are observed for both X-project and Y-project amplitude values. The energy change in 0.12, 0.13, and 0.14 spatial frequencies is 0.43, 0.28, and 0.55 dB, respectively. Small energy changes are observed in 0.20, 0.21, 0.23, and 0.24 frequencies as well. The energy change in frequency 0.14, identified to correspond to ear region, has the maximum value. As the head rotates about the vertical axis, which runs over the head, the lateral velocity would be more for object away from the axis ($v = r\omega$) for an angular velocity ω. Maybe this explains the more energy changes corresponding to ear portion, as the rate of change of contours are faster along this region (Fig. 40.4, Table 40.1).

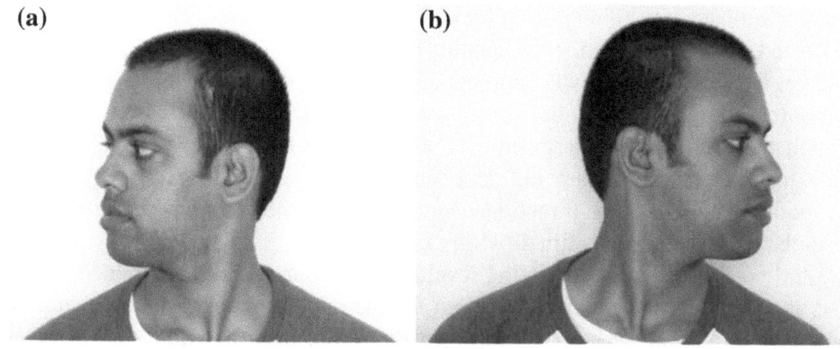

Fig. 40.4 **a** Frame1. **b** Frame2 used in head motion analysis

Table 40.1 Results of head motion analysis

Spatial frequency	Frame 1		Frame 2		Orientation angle	
	X-project amplitude (ax1)	Y-project amplitude (ay1)	X-project amplitude (ax2)	Y-project amplitude (ay2)	Frame1	Frame2
0.12	2.15	3.47	2.27	3.64	−170.22	164.44°
0.13	2.01	3.47	2.13	3.56	171.22°	−172.21°
0.14	1.91	3.23	1.98	3.37	−176.63°	178.65°
0.20	1.27	2.22	1.24	2.31	−161.78°	173.71°
0.21	1.13	2.06	1.12	2.14	−167.01°	173.64°
0.23	0.97	1.88	0.97	1.89	173.09°	−167.71°
0.24	0.91	1.73	0.92	1.74	171°	−176.69°

$$\text{Change in energy (dB)} = 10\ \log_{10}\frac{P_2}{P_1} \tag{40.5}$$

where

$$P_2 = ax2^2 + ay2^2;$$
$$P_1 = ax1^2 + ay1^2.$$

40.6 Conclusion

Contour-enhanced profiles obtained using IPL model come out exceedingly well in the model which incorporates sinc filters for low-pass filtering. The DFT orientation analysis brings out features of the contour in different types of motion. It gives out results close to actual feature changes. Compared to earlier models described in literature [1, 4] for optical motions, which compute projected optical power of all frequency in certain directions and assuming the direction of motion in a generalized way (direction of maximum projected power), the present model described gives exactly the orientation angle with sign change for direction change. This is achieved with comparatively less computations, and this model computes fine feature changes along with energy changes. It is seen that even though basic change of energy in 0.12, 0.13, and 0.14 spatial frequencies is not wide, orientation angles are different and precisely obtained. Going by energy computations to identify direction cannot identify the different features of an ensemble as has been identified by this model.

While communicating the initial results of the Fourier spectrum model, further tests of the model have been carried out for the orientation and motion change in the case of blinking of human eye, from the initial open condition, blink down, remain closed, and again open up obtained from 25 video frames in sequence. The analysis shows reasonable change of energy whenever the eye blinks up or down. The orientation and energy changes have been observed to take place in the spatial frequency range of 0.12–0.17. The results agree closely with the actual feature changes. These results give further validation to the Fourier model, which can be very effectively used to identify even small orientation and energy changes occurring in biological events.

Identifying the orientation angle as $\tan\emptyset = \frac{\theta_x}{\theta_y}$, i.e., relating the contours orientation from phase changes (spatial delays) in two orthogonal directions, is a much simpler and straight forward approach than implementing multiple numbers of energy/orientation filters done in earlier works. Our model is more suitable to identify small orientation changes occurring in biological events.

System parameters like distance from the object, zoom, number of pixels per frame, etc., are required to identify exact physical feature to its corresponding frequency. Enhancement of the model to include object identification parameters is the next step to be done.

References

1. Benoit A, Caplier A, Durette B, Herault J (2010) Using human visual system modeling for bio-inspired low level image processing. Comput Vis Image Underst 114(7):758–773
2. Herault J, Durette B (2007) Modeling visual perception for image processing. Comput Ambient Intell 4507:662–775
3. Herault J (1996) A model of color processing in the retina of vertebrates: From photoreceptors to color opposition and color constancy phenomena. Neurocomputing 12:113–129
4. Benoit A, Caplier A (2010) Fusing bio-inspired vision data for simplified high level scene interpretation: application to face motion analysis. Comput Vis Image Underst 114:774–789
5. Oppenheim AV, Shafer RW, Buck JR (1999) Discrete time signal processing. Prentice Hall, New Jersey
6. Gonzalez RC, Woods RE (2002) Digital image processing, 2nd edn. Prentice Hall, New Jersey

Chapter 41
Efficient Color Image Retrieval with Selective Relevance Feedback

Jayashree Khanapuri and Linganagouda Kulkarni

Abstract Image retrieval has become an important aspect in today's world as there is rapid growth of digital data. It is required to have efficient search system which delivers fast retrieval to cater to the need of end user with low computational cost and more accuracy. A new content-based search system is required to address the needs. In this paper, a new retrieval algorithm based on the statistical parameters like energy, standard deviation, and entropy of complex wavelets is presented. The performance is further enhanced by selective relevance feedback. The retrieval is carried out by decomposing the image using complex wavelet transform and computing the energy, standard deviation, moments, and the entropy of the subbands as feature vectors. The average retrieval accuracy of each of the class is improved by relevance feedback by training only the selected query images with poor retrieval accuracy.

Keywords Complex wavelet transform · Energy · Standard deviation · Entropy · Moments · Relevance feedback · Retrieval accuracy

41.1 Introduction

With the vast growth in the use of Internet and World Wide Web, the large number of users is able to access the digital image data. The image database is growing in a rapid manner creating the need to develop effective and efficient image retrieval

J. Khanapuri (✉)
K J Somaiya Institute of Engineering and Information Technology, Ayurvihar Complex, Eastern Express Highway, Sion, Mumbai, India
e-mail: jaya2312@yahoo.co.in

L. Kulkarni
Hirasugar Institute of Technology, Nidasoshi, India
e-mail: linganagouda@yahoo.co.uk

V. Chakravarthi et al. (eds.), *Proceedings of International Conference on VLSI, Communication, Advanced Devices, Signals & Systems and Networking (VCASAN-2013)*, Lecture Notes in Electrical Engineering 258, DOI: 10.1007/978-81-322-1524-0_41,

system. The color, texture, and shape of the image play the most powerful role as features in the content-based image retrieval. In a typical system, visual contents of the image are retrieved in the form of feature vector. The feature vector of each of the images is extracted and stored in the database and is compared with features of the query image. In order to have efficient retrieval, it is necessary to capture the important characteristics of an image which makes it different from the other similar images.

Much of the research work has been done in this direction in terms of texture analysis, classification, and segmentation. The texture characteristics are extracted taking into account the color bands giving the information about color texture [1, 2]. The texture features and color features extracted from the moments of the histogram are used in joint color texture feature extraction [3, 4].

Many algorithms are developed using different techniques for the efficient extraction of color and texture features together. The wavelet approach has emerged as efficient over other methods in terms of computational advantages. The different approaches can be conventional discrete wavelet transform, Gabor wavelet, M-band wavelets, etc. The standard DWT becomes inefficient in analyzing the high-frequency signal with narrow bandwidth. It also lacks the sensitivity and directionality. The Gabor wavelet introduces redundancy and memory requirement as the basis functions are not orthogonal. The M-band wavelet decomposition improves the retrieval accuracy, but at the cost of computational complexity. Retrieval accuracy can be further improved by relevance feedback [5]. Relevance feedback based on nearest neighbor approach ranks each image with reference to relevance score based on nearest neighbor distance [6]. Another method uses Markov model mediators' frameworks to cluster the database in order to improve the performance of query processing and matching [7]. Region-based image retrieval with relevance feedback system learns the importance of region based on feedback from the user to improve the performance [8].

The objective of this paper is to present an efficient method which provides better average retrieval accuracy with less computational time and cost. The selective relevance feedback is used to improve the performance of the system. An investigation is presented in the paper using color and texture feature retrieval by complex wavelet transform and selective relevance feedback to further enhance the performance.

The paper is organized as follows. The theory related to dual-tree complex wavelets and moments is discussed in Sect. 41.2. The design and implementation of proposed algorithm are discussed in Sect. 41.3. The experiment and its results are discussed in Sect. 41.4 followed by conclusion in Sect. 41.5.

41.2 Wavelet Decomposition

41.2.1 Dual-Tree Complex Wavelet Transform

In multi-resolution analysis of wavelets, the low-pass information consists of approximated version of high-resolution image. The high-pass information gives sharper variation details. In dual-tree complex wavelets, two trees are used in parallel to generate the output interpreting them as real and imaginary part of complex coefficient. The transform decomposes the image into subbands providing two smoothed versions of image and the information in six directions at each stage. All the filters used in the analysis are real and orthogonal [8, 9]. The filter outputs are given as

$$\Phi_1(x,y) = \Phi_h(x).\Phi_h(y) \quad \Phi_2(x,y) = \Phi_g(x).\Phi_g(y) \tag{41.1}$$

$$\Psi_{1,1}(x,y) = \Phi_h(x).\Psi_h(y) \quad \Psi_{2,1}(x,y) = \Phi_g(x).\Psi_g(y) \tag{41.2}$$

$$\Psi_{1,2}(x,y) = \Psi_h(x).\Phi_h(y) \quad \Psi_{2,2}(x,y) = \Psi_g(x).\Phi_g(y) \tag{41.3}$$

$$\Psi_{1,3}(x,y) = \Psi_h(x).\Psi_h(y) \quad \Psi_{2,3}(x,y) = \Psi_g(x).\Psi_g(y) \tag{41.4}$$

The resultant six wavelets generated using Eqs. (41.2), (41.3), and (41.4) are as follows:

$$\Psi_i(x,y) = \Psi_{h,i}(x) + \Psi_{g,i}(y) \tag{41.5}$$

$$\Psi_{i+3}(x,y) = \Psi_{h,i}(x) + \Psi_{g,i}(y) \tag{41.6}$$

for $1 \leq i \leq 3$. These six wavelets are strongly oriented in {+ 15°, +45°, +75°, −15°, −45°, −75°} to capture information in respective direction and are shift invariant.

41.2.2 Moments

The moments are one of the important factors that are successfully used in many color-based image retrieval systems [11]. The moments up to third order play important role in the image retrieval. The first-, second-, and third-order moments, i.e., mean, variance, and skewness are proved to be efficient and effective in representing color distributions of images [12]. The color moments are defined on the histogram of quantized energy submatrix of image. Given the histogram as $H = \{h_1, h_2, \ldots h_i, \ldots h_n\}$. The moments are defined as

$$M1 = \frac{1}{n}\sum_{1}^{n} h_i \tag{41.7}$$

$$M2 = \left[\frac{1}{n}\sum\nolimits_1^n (h_i - M1)^2\right]^{1/2} \quad (41.8)$$

$$M3 = \left[\frac{1}{n}\sum\nolimits_1^n (h_i - M1)^3\right]^{1/3} \quad (41.9)$$

41.2.3 Selective Relevance Feedback

The relevance feedback plays an important role in improving the retrieval accuracy of the system. But it increases the computational time and cost. The proposed selective relevance feedback method will train the query images with poor retrieval accuracy. The top 100 retrievals of each query image with poor retrieval accuracy are classified as relevant and non-relevant image groups [13]. The feature vectors corresponding to the relevant and non-relevant group of query image are calculated as

$$F_{relevant} = \frac{\sum \textit{Respective feature vector components of all relevantimages}}{\textit{Number of relevant images}} \quad (41.10)$$

$$F_{non\_relevant} = \frac{\sum \textit{Respective feature vector components of all non relevant images}}{\textit{Number of non relevant images}} \quad (41.11)$$

41.3 Feature Extraction

41.3.1 Texture Feature Extraction

The DT-CWT algorithm is used for the texture feature extraction by decomposing the image up to third level using Selsenick's first- and higher-stage low-pass and high-pass filters. As DT-CWT is oriented in six directions, the filter coefficients are obtained by passing the image through real and imaginary trees. The energy and standard deviation are calculated for all the subbands and the LL component at each stage. The calculations for the energy and standard deviation of kth subband are given as:

$$E_k = \frac{1}{MXN}\sum\nolimits_{i=1}^{M}\sum\nolimits_{j=1}^{N} |W_k(i,j)| \quad (41.12)$$

$$\sigma_k = \frac{1}{MXN}\left[\sum_{i=1}^{M}\sum_{j=1}^{N}\left(W_k(i,j) - \mu_k\right)^2\right]^{1/2} \tag{41.13}$$

where $W_k(i, j)$ is the kth wavelet decomposed subband, M x N is the size of the kth subband, μ_K is the mean of kth subband. The feature vector constructed using the energy and standard deviation of the subbands is given by

Feature vector at each stage consists of {E_1, E_2...E_6, σ_1, σ_2... σ_6, E(LL(tree 1), LL(tree 2)), σ(LL(tree 1), LL(tree 2))}.

The final texture feature vector with L decomposition levels will be given as

$$\textbf{\textit{Feature Vector}}1 = [\textbf{\textit{L}} \times \textbf{\textit{size of Feature Vector at each stage}}] \tag{41.14}$$

41.3.2 Color Feature Extraction

The moments are extracted as features from LL component by decomposing the image up to third level using DT-CWT. The energy matrix is constructed by calculating the energy associated with the LL subband. The quantized energy matrix of each subband is used for the generation of moment features. The subbands are named as {A1, A2,...A9, A10} (Fig. 41.1).

The energy of the subbands is given by

$$E_k = \frac{1}{MXN}\sum_{i=1}^{M}\sum_{j=1}^{N}|W_k(i,j)| \tag{41.15}$$

where k = 1, 2...10. Let P (i, j) is the coefficient of a particular subband, then it is set to either '0' or '1' based on the equation

$$P(i,j) = \begin{cases} 0 \\ 1 \end{cases} |W_k(i,j) \leq T| \tag{41.16}$$

The moments are calculated for each subband by quantizing the coefficients of the subband to 0/1 to provide the quantized submatrix. The quantization is carried out taking the mean of the subband as threshold (T) as given in (41.16). The threshold is self-adaptive with respect to subbands. The subbands are converted to same size by up sampling before quantization. The histogram is constructed using subbands. The 10-D vector is generated by combining the all the elements from the

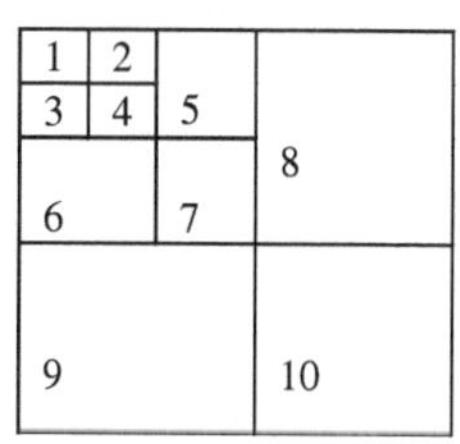

Fig. 41.1 Three-level decomposition with 10 subbands

same location from all the quantized submatrices. This 10-D vector is converted from binary to decimal. The moments are calculated based on the histogram of the matrix using Eqs. (41.7), (41.8), and (41.9). The entropy is calculated using the histogram as

$$E_t = -\sum_{i=1}^{n} P_i log_2 P_i \tag{41.17}$$

The color feature vector is given as

$$Feature\ Vector2 = \{M1, M2, M3, E_t\} \tag{41.18}$$

The final feature vector obtained by combining texture and color features as

$$Final\ Feature\ Vector = \{Feature\ Vector1, Feature\ vector2\} \tag{41.19}$$

Thus, each image in the database consists of 52 features in the feature vector.

The texture and color features are extracted to form the final feature vector for all the images in the database. The similarity measure between the query and the database images is carried out using Canberra distance. If D and Q are the feature vectors of the database and query images, respectively, then the Canberra distance is given by

$$Canb(D, Q) = \sum_{i=1}^{n} \frac{|D_i - Q_i|}{|D_i| + |Q_i|} \tag{41.20}$$

The retrieval accuracy is further improved by relevance feedback. In the proposed algorithm, only query images having retrieval accuracy less than the average retrieval accuracy of that particular class without relevance feedback are considered for training to improve retrieval accuracy and reduce the computation time. The vectors $F_{relevant}$ and $F_{non\_rrelevant}$ are calculated for query images under training to construct the modified feature vector. The modified feature vector of the query image is given as

$$F_{new} = F_{original} + F_{relevant} - F_{non\_relevant} \tag{41.21}$$

where $F_{original}$ is the original feature vector associated with the query image before modification. The image retrieval is carried out with modified feature vector using Canberra distance to obtain improved retrieval accuracy.

41.4 Experimental Results

The experiment was conducted on Wang database consisting of 1000 images belonging to 10 different classes. The preprocessing involves resizing to [256, 256] and converting to Y Cb Cr domain. The feature vector is constructed for all images in the database. The similarity measure is executed using Canberra distance to calculate the retrieval accuracy of query image and the average retrieval accuracy of all the classes. The retrieval accuracy and the average retrieval accuracy are calculated as

$$\textbf{\textit{Retrieval Accuracy}} = \frac{\textbf{\textit{No.of Relevant images Retrieved}}}{\textbf{\textit{Total No.of Relevant Images}}} \tag{41.22}$$

$$\textbf{\textit{Average Retrieval Accuracy}} = \frac{1}{n}\sum\nolimits_{i=1}^{n} \textbf{\textit{Retrieval Accuracy}} \tag{41.23}$$

where n = No. of images under consideration

The performance of the system is further improved by training the query images with poor retrieval accuracy with relevance feedback. This decreases the computational cost and time, and also improves the retrieval accuracy. However, in the experiment, only five iterations are used to improve the retrieval accuracy of selected query images. The retrieval accuracy of trained query image is used to replace its existing value without feedback to improve the average retrieval accuracy.

The number of images considered for training with relevance feedback and the average retrieval accuracy achieved for different iterations is shown in Table 41.1. The average retrieval accuracy with and without selective relevance feedback using Canberra distance as similarity measure is tabulated in Table 41.2. Table 41.3 indicates the percentage of improvement achieved by the proposed method. Table 41.4 tabulates the average and total time required for training the selected images and the entire class.

Table 41.1 Average retrieval accuracy of selected images for selective relevance feedback

Sr. No.	Class	I	Average retrieval accuracy with selective relevance feedback				
			Iter 1	Iter 2	Iter 3	Iter 4	Iter 5
1	People	52	16.00	18.94	20.56	21.70	22.56
2	Beach	36	6.26	8.30	9.48	10.18	10.82
3	Building	38	7.77	9.47	10.60	11.84	12.76
4	Bus	42	27.64	33.16	35.64	36.25	36.46
5	Dinosaur	29	27.70	27.95	28.02	28.05	28.05
6	Elephant	47	16.81	20.21	22.49	24.16	24.97
7	Rose	21	16.18	18.69	18.91	18.91	18.91
8	Horse	51	22.51	27.81	30.09	31.17	31.61
9	Mountain	43	9.10	11.29	12.84	13.95	14.85
10	Food	39	9.96	11.91	13.30	14.75	15.91

I= No. of images considered for training under selective relevance feedback

Table 41.2 Average retrieval accuracy of classes without and with selective relevance feedback

Sr. No	Class	Average retrieval accuracy					
		Without feedback	With selective relevance feedback				
		Iter 0	Iter 1	Iter 2	Iter 3	Iter 4	Iter 5
1	People	21.79	29.87	32.81	34.43	35.57	36.43
2	Beach	29.93	30.96	32.73	33.91	34.61	35.24
3	Building	24.69	26.75	28.43	29.56	30.80	31.72
4	Bus	58.56	67.00	72.52	75.00	75.61	75.82
5	Dinosaur	86.88	92.75	93.00	93.07	93.10	93.10
6	Elephant	29.54	36.35	39.75	42.03	43.70	44.51
7	Rose	75.28	82.52	85.03	85.25	85.25	85.25
8	Horse	36.99	48.68	53.68	55.96	57.04	57.48
9	Mountain	21.49	24.20	26.39	27.94	29.05	29.95
10	Food	24.65	28.08	30.02	31.42	32.87	34.03

Table 41.3 Comparison of outputs without and with selective relevance feedback (SRF)

Sr. No.	Class	Average retrieval accuracy without SRF	Average retrieval accuracy with SRF	Improvement in average retrieval accuracy SRF
1	People	21.79	36.43	14.64
2	Beach	29.93	35.24	5.31
3	Building	24.69	31.72	7.03
4	Bus	58.56.	75.82	17.26
5	Dinosaur	86.88	93.10	6.22
6	Elephant	29.54	44.51	14.97
7	Rose	75.28	85.25	9.97
8	Horse	36.99	57.48	20.49
9	Mountain	21.49	29.95	8.46
10	Food	24.65	34.03	9.38

Table 41.4 Computation time with and without selective relevance feedback

Sr. No	Class	Computational time for selected query images (s)	Computational time for entire class (s)
1	People	245.769	465.278
2	Beach	172.553	461.214
3	Building	180.263	461.321
4	Bus	197.298	461.429
5	Dinosaur	134.199	454.986
6	Elephant	221.956	475.905
7	Rose	98.198	469.046
8	Horse	240.463	474.479
9	Mountain	197.448	468.207
10	Food	184.813	461.112
Total time (s)		1675.662	4652.997
Average time (s)		167.566	465.299

41.5 Conclusion

In this paper, the implementation of proposed method and its comparison with retrieval without relevance feedback is carried out. The results indicate that the average retrieval accuracy is improved with selective relevance feedback. It is observed that horse and beach classes have maximum and minimum retrieval accuracy. The retrieval accuracy increases with each iteration. Since relevance feedback is carried out on the selected images, the computational time and cost are less with improved average retrieval accuracy as compared to training the entire class.

References

1. Paschos G (2000) Fast color texture recognition using chromaticity moments. Pattern Recogn Lett 21:837–841
2. Van de Wouver G, Scheunders P, Livens S, Dyck DV (1999) Wavelet correlation signatures for color texture characterization. Pattern Recogn Lett 32:443–451
3. Dubuisson-Jolly MP, Gupta A (2000) Color and texture fusion: application to aerial image segmentation and GIS updating. Image Vis Comput 18:823–832
4. Drimbean A, Whelan PF (2001) Experiments in color texture analysis. Pattern Recogn Lett 22:1161–1170
5. Karthik S, Saha S, Chaitra G (2013) A new relevance feedback based approach for efficient image retrieval. Int J Comput Appl 61(14):1–6
6. Giacinto G (2007) A nearest neighbor approach to relevance feedback in content based image retrieval. In: CIVR-2007, pp 456–463
7. Shyu M-L, Chen S-C, Zhang C (2004) A unified framework for image database clustering and content based retrieval, In: MMDB-04, pp 19–27
8. Jing F, Zhang B, Lin F, Ma W-Y (2001) A novel region based image retrieval method using relevance feedback. In: international conference on multimedia information retrieval, pp 28–31
9. Kingsbury NG (1998) The dual tree complex wavelet transform: a new technique for shift invariance and directional filters, In: Proceeding 8th IEEE DSP workshop, Utah, 9–12 Aug 1998
10. Prakash KN, Satyaprasad K, Band M (2012) Dual tree complex wavelet transform for texture image indexing and retrieval. Int J Adv Eng Technol 4(1):661–671
11. Miller JT, Li C-C (1998) Adaptive multi wavelets initialization. IEEE Trans Signal Process 46(12):3282–3291
12. Zhiyong A, Feng Z (2009) Image retrieval based on energy and entropy of multi wavelet transform. In: international conference on information technology and computer science, pp 544–577
13. Patil P, Kokare M (2011) Semantic image retrieval using relevance feedback. Int J Web Semant Technol 2(4):139–148

Chapter 42
Nonlinear Dynamical Analysis of Speech Signals

N. Chaitra, D. Murali Mohan and D. Narayana Dutt

Abstract The major difficulty in handling time series measurements is in identifying the system, whether it is purely deterministic, chaotic, or random. Proper identification of system characteristics and application of appropriate signal processing tools can lead to superior performance in signal analysis; especially when one handles real speech signals, these issues become even more important and crucial. Evidence for chaotic behavior with speech signals has been claimed and disputed. Over the past two decades, researchers have come out with efficient nonlinear dynamical tools applicable to time series. In this paper, different nonlinear dynamical tools like phase-space plot, running correlation dimension, and running Lyapunov exponent applied to speech signals are discussed which provide a convenient framework for speech signal analysis.

Keywords Nonlinear dynamical tools · Lyapunov exponent · Correlation dimension · Phase-space plot

42.1 Introduction

Chaos can be defined as effectively unpredictable long-time behavior arising in a deterministic dynamical system because of sensitivity to initial conditions. It must be emphasized that a deterministic dynamical system is perfectly predictable given

N. Chaitra (✉)
ECE Department, BNM Institute of Technology, Bangalore, India
e-mail: chaitranagraj@gmail.com

D. Murali Mohan
NXP Software, Bangalore, India

D. Narayana Dutt
Indian Institute of Science, Bangalore, India

V. Chakravarthi et al. (eds.), *Proceedings of International Conference on VLSI, Communication, Advanced Devices, Signals & Systems and Networking (VCASAN-2013)*, Lecture Notes in Electrical Engineering 258, DOI: 10.1007/978-81-322-1524-0_42,

perfect knowledge of the initial condition and is in practice always predictable in the short term. The key to long-term unpredictability is a property known as sensitivity to (or sensitive dependence on) initial conditions.

For a dynamical system to be chaotic, it must have a large set of initial conditions which are highly unstable. No matter how precisely the initial conditions in these systems are measured, prediction of its subsequent motion goes radically wrong after a short time. The order and degree of chaos can be studied through correlation dimensions and Lyapunov exponents.

When the number of state variables defining a system is more (greater than three) or when each and every state variable is not known, it becomes very difficult for visualizing system's trajectory and hence to analyze such systems. If we analyze a complex system characterized by the nonlinear interaction of a set of variables, it is usual in speech processing systems that the complete description of such a set is unknown. However, Takens' theorem [1, 2] states that we can recreate a topologically equivalent picture of the original multidimensional system behavior, using the time series of a single observable variable, by means of the method of time delays. Phase-space plot and Poincare plot are the methods for obtaining the trajectories of the system.

In this paper, different nonlinear dynamical tools like phase-space plot, running correlation dimension, and running Lyapunov exponent are applied to speech signals. The next section discusses about these nonlinear tools in detail, and finally, the tools applied to speech are discussed in the results section.

42.2 Nonlinear Dynamical Tools

42.2.1 Phase Space

Phase space is the collection of possible states of a dynamical system. A phase space can be finite (e.g., for the ideal coin toss, there are two states heads and tails), countably infinite (e.g., state variables are integers), or unaccountably infinite (e.g., state variables are real numbers). Implicit in the notion is that a particular state in phase space specifies the system completely; it is all that is needed to know about the system to have complete knowledge of the immediate future. Thus, the phase space of the planar pendulum is two-dimensional, consisting of the position (angle) and velocity. According to Newton, specification of these two variables uniquely determines the subsequent motion of the pendulum.

The path in phase space traced out by a solution of an initial value problem is called an orbit or trajectory of the dynamical system. If the state variables take real values in a continuum, the orbit of a continuous-time system is a curve, while the orbit of a discrete-time system is a sequence of points.

42.2.1.1 State-Space Reconstruction

Let the signal be represented by the time series X (k). Then, the dynamical behavior of the signal is reconstructed by succession of these points X (k) in the phase space. Phase-space reconstructions are based on the analysis of dynamic systems by delay maps. The vectors X (k) in the multidimensional phase space are constructed by time-delayed values of the time series, which determine the coordinates of the phase-space plot.

$$\mathrm{X}\ (\mathrm{k}) = \{\mathrm{x}\ (\mathrm{k}),\ \mathrm{x}\ (\mathrm{k}+\tau),\ \mathrm{x}\ (\mathrm{k}+2\tau), \ldots\ \mathrm{x}\ (\mathrm{k}+(\mathrm{E}-1)\tau)\} \tag{42.1}$$

where X (k) is one point of the trajectory in the phase space at time k, x (k + iτ) are the coordinates in the phase space corresponding to the time-delayed values of the time series, τ is the time delay between the points of the time series considered, and E is the embedding dimension, which is the number of coordinates of the phase-space plot. The attributes of the reconstructed phase-space plot depend on the choice of value of τ. If the value is too small, x (t + τ) is close to x (t) and the phase portrait is too decreased around the diagonal. On the contrary, if the value of τ is too large, deformation of the phase portrait occurs. Therefore, choice of τ must be optimal. Phase-space plots with different values of τ are given in results section.

42.2.2 Lyapunov Exponents and Correlation Dimension

Detecting the presence of chaos in a dynamical system is an important problem that is solved by measuring the largest Lyapunov exponent. Lyapunov exponents quantify the exponential divergence of initially close state-space trajectories and estimate the amount of chaos in a system. This method of calculation follows directly from the definition of the largest Lyapunov exponent and is accurate because it takes advantage of all the available data [3–5]. This method of calculation is fast, easy to implement, and robust to change in the following quantities: embedding dimension, size of data set, reconstruction delay, and noise level. This method is used to calculate the correlation dimension as well. Thus, one sequence of computations will yield an estimate of both the level of chaos and the system complexity.

This section mainly deals with the calculation of Lyapunov exponents and the correlation dimension for large data sets. The Lyapunov exponents give an estimate of the level of chaos in the dynamical system and dimension gives an estimate of system complexity. In this approach, intermediate calculations are used to estimate both dimension and entropy. For time series produced by dynamical systems, the presence of a positive Lyapunov exponent indicates chaos. In many applications, it is sufficient to calculate only the largest exponent.

42.2.2.1 Calculation of Lyapunov Exponents

The first step involves reconstructing the attractor dynamics from a reconstructed trajectory, X, it can be expressed as a matrix where each row is a phase-space vector.

That is,

$$X = [X_1 X_2 \ldots X_M]T \tag{42.2}$$

where X_i is the state of the system at discrete time *i*. For an *N*-point time series, $x_1, x_2, \ldots x_N$, each X_i *is* given by

$$X_i = [x_i x_{i+J} \ldots x_{i+(m-1)J}] \tag{42.3}$$

where *J* is the lag or reconstruction delay, and m is the embedding dimension. Thus, *X* is an *M x m* matrix, and the constants *m*, *M*, *J*, and *N* are related as

$$M = N - (m-1)J \tag{42.4}$$

It has been found that a good approximation of *J* to be equal to the lag where the autocorrelation function drops to $(1 - 1/e)$ of its initial value. After reconstructing the dynamics, the algorithm locates the nearest neighbor of each point on the trajectory. The nearest neighbor $\hat{X}_j$ is found by searching for the point that minimizes the distance to the particular reference point X_j. This is expressed as

$$d_j(0) = \min_{X_{\hat{j}}} \left\| X_j - X_{\hat{j}} \right\| \tag{42.5}$$

where $d_j(0)$ is the initial distance from the *j*th point to its nearest neighbor, and || || denotes the Euclidean norm. We impose the additional constraint that nearest neighbors has a temporal separation greater than the mean period of the time series:

$$|j - \hat{j} > mean - period| \tag{42.6}$$

This allows us to consider each pair of neighbors as nearby initial conditions for different trajectories. The largest Lyapunov exponent is then estimated as the mean rate of separation of the nearest neighbors.

The estimation of λ_1 is done as follows:

$$\lambda_1(i) = \frac{1}{i \cdot \Delta t} \cdot \frac{1}{(M-i)} \sum_{j=1}^{M-i} \ln \frac{d_j(i)}{d_j(0)} \tag{42.7}$$

where Δt is the sampling period of the time series, and $d_j(i)$ is the distance between the *j*th pair of nearest neighbors after *i* discrete-time steps, i.e., $i \cdot \Delta t$ s. In order to improve convergence, alternate form of estimating λ_1 is

$$\lambda_1(i,k) = \frac{1}{k \cdot \Delta t} \cdot \frac{1}{(M-k)} \sum_{j=1}^{M-k} \ln \frac{d_j(i+k)}{d_j(i)} \tag{42.8}$$

k is held constant, and λ_1 is extracted by locating the plateau of $\lambda_1(i,k)$with respect to i.

42.2.2.2 Calculation of Correlation Dimension

The algorithm normally employed in this analysis [6] aims at creating an artificial (or pseudo-)space of dimension M with delay vectors constructed by splitting a scalar time series s(t) with delay time as J.

The correlation function is the average number of data points within a distance R from a data point

$$C_M(R) \equiv \lim_{N\to\infty} \frac{1}{N(N-1)} \sum_{i}^{N} \sum_{j,j\neq i}^{N} H\left(R - \left|x_i - x_j\right|\right) \tag{42.9}$$

where x_j is the position vector of a point belonging to the attractor in the M-dimensional space, N is the number of reconstructed vectors, and H is the Heaviside step function.

The fractional dimension $D_2(M)$ is defined as

$$D_2 \equiv \lim_{R\to 0} \frac{d\log C_M(R)}{d\log(R)} \tag{42.10}$$

and is essentially the scaling index of $C_M(R)$ variation with R. The fractal dimension $D_2(M)$ can be used to differentiate between different temporal behavior, since for an uncorrelated stochastic system, $D_2 = M$, while for a chaotic system, $D_2(M) =$ constant for M greater than a certain dimension $M_{\max}$.

42.3 Application of Nonlinear Dynamic Tools to Speech Signals

Speech signal corresponding to sound "a" sampled at 8-kHz sampling rate was used in obtaining the phase-space plot. Voiced and unvoiced sounds corresponding to sound "a" and "f" sampled at 8 kHz were generated using cool edit and a microphone. These speech signals were used in obtaining the running correlation dimension and running Lyapunov exponents. Speech frame size of 300 samples was used in this analysis. In order to obtain running correlation dimension and Lyapunov exponents of longer speech data where there are more fluctuations in the speech signal, two sentences from ITU speech database sampled at 16 kHz was used. Two sentences corresponding to two different languages were used in this analysis.

42.3.1 Phase-Space Plot

Phase-space plots with different values of τ were obtained for sound "a". The attributes of the reconstructed phase-space plot depend on the choice of value of τ. Hence, phase-space plot was obtained for different values of τ for sound "a". A small value of τ was chosen ($\tau = 1$). Figure 42.1 corresponds to the phase-space plot for the small value of $\tau = 1$. It can be observed from the figure that x $(t + \tau)$ is close to x(t) and the phase portrait is too decreased around the diagonal.

Phase-space plots were also obtained for large value of τ ($\tau = 15$). Figure 42.1 depicts the phase-space plot for this case. It can be observed in Fig. 42.1 that the phase portrait appears to be deformed.

Therefore, choice of τ must be optimal. The optimal value of τ was chosen using the autocorrelation function. The phase-space plot was obtained for $\tau = 4$. Figure 42.2 gives the phase-space plot for the optimal τ value

42.3.2 Lyapunov Exponents

The running maximum Lyapunov exponents were obtained. The routine to calculate the maximum Lyapunov exponent was implemented in Matlab. Frame size of 300 samples at sampling rate 8 kHz was chosen for the evaluation of Lyapunov exponents.

The running Lyapunov exponents were calculated for the following cases:

Voiced sound (sound //a//), unvoiced sound (sound //f//), a small English sentence with background noise before and after the sentence (sentence //I never//).

The maximum Lyapunov exponent was obtained for each frame of data using GP algorithm. In order to visualize the correspondence between the frame and the

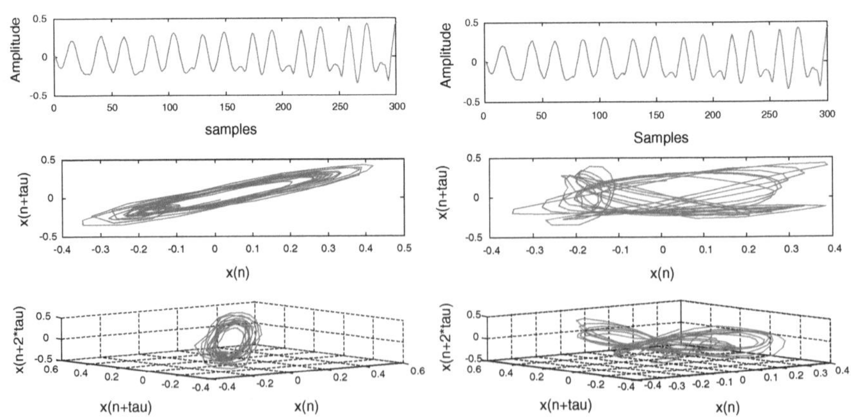

Fig. 42.1 Effect of *small* τ on phase plot ($\tau = 1$) and *large* τ on phase plot ($\tau = 15$)

Fig. 42.2 Phase plot for optimal τ ($\tau = 4$)

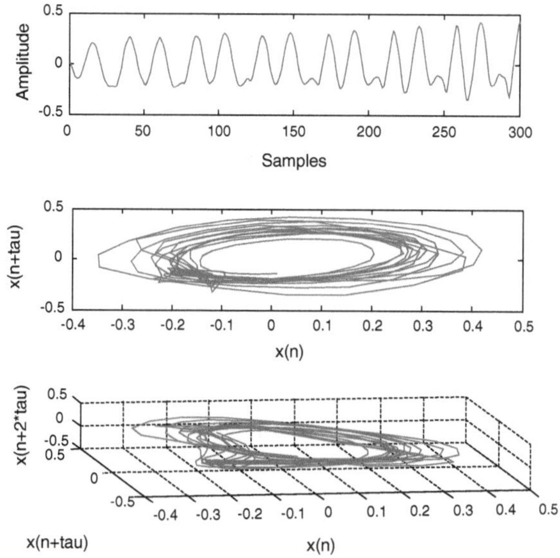

maximum Lyapunov exponents, each sample in a frame was mapped to the calculated maximum Lyapunov exponent of the corresponding frame.

The plot in the top row of Fig. 42.3 corresponds to the speech waveform corresponding to sound //a// and //f//. Clearly, the waveform corresponding to sound //a// has definite structure arising because of periodicity due to pitch in case of voiced sounds. The waveform corresponding to sound //f// appears more like noise as the noise excitation generates unvoiced sounds. The second row in Fig. 42.3 shows the plot of running maximum Lyapunov exponents corresponding to sounds //a// and //f//. It can be observed that as the sound //a// builds up and

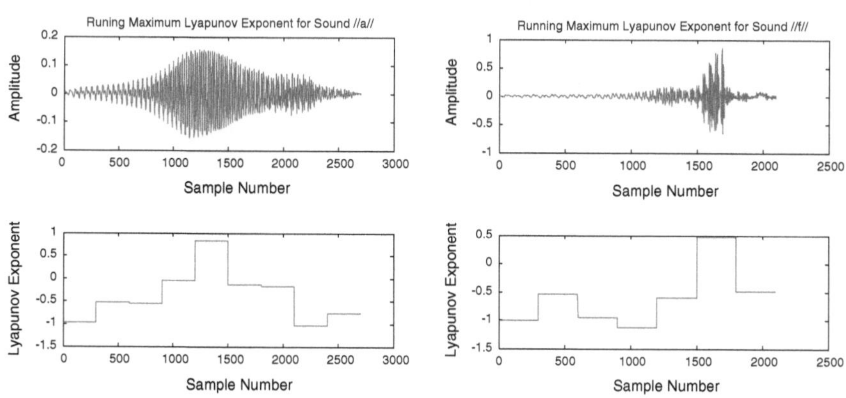

Fig. 42.3 Running maximum Lyapunov exponent for vowel sound "a" and running maximum Lyapunov exponent for sound "f"

attains maximum energy for frame around 1,200th sample, Lyapunov exponent also increases during this buildup and attains maximum value of around 0.85. Then, during the gradual fall in the energy of the signal beyond 5th frame, the Lyapunov exponent also falls off and attains minimum value of around −1. Similar observations can be seen even for unvoiced sounds. However, the maximum Lyapunov exponent value attained by unvoiced sound is around 0.5. The maximum Lyapunov exponent value for unvoiced sound //f// is lower than that of voiced sound //a//: The reason for this could be attributed to random noisy structure present in unvoiced speech. Both sounds (//a// and //f//) can be considered as generated by chaotic systems due to the presence of positive values of Lyapunov exponents observed in the stable regions. However, the figure also suggests that there are negative values of Lyapunov exponents; this suggests random nature of speech during the buildup and tape out of both voiced and unvoiced sounds. The maximum Lyapunov exponent calculated for sound //a// (0.85) and //f// (0.5) also suggests that divergence rate of the chaotic system that generates sound //a// is more divergent or more chaotic as compared to the process that generates sound //f//.

Clearly, the positive Lyapunov exponent during speech utterance suggests that speech is generated from a chaotic process and the negative exponent during background noise suggests it can be modeled using stochastic process.

42.3.3 Correlation Dimension

Similar to calculation of maximum Lyapunov exponents, correlation dimension was also calculated for speech frames of size 300 samples sampled at 8 kHz. Matlab routine was written for the calculation of correlation dimension. It can be observed from the Fig. 42.4 that as the voiced sound builds up, the complexity of

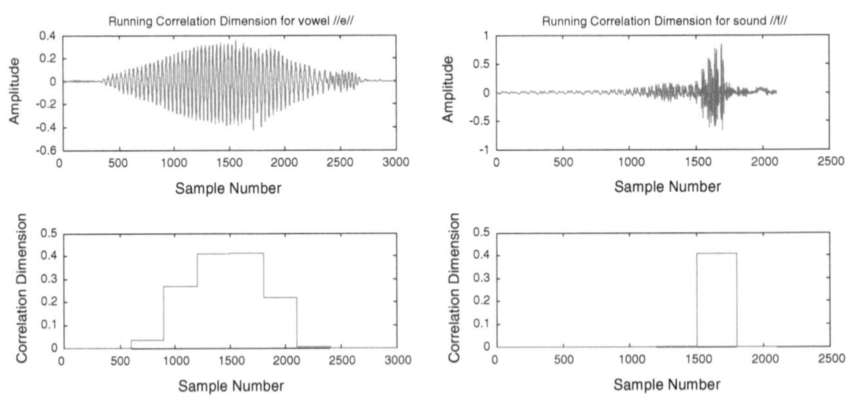

Fig. 42.4 Running correlation dimension for vowel "e" and running correlation dimension for unvoiced sound signal "f"

the underlying chaotic system which is quantified using correlation dimension increases and attains a maximum value of 0.4 during the stable part of sound //e//. Then, during the gradual fall in the energy of the signal after the stable region of sound //e//, the correlation dimension also falls off. The running correlation dimension was also calculated for unvoiced sound //f//. Figure 42.4 gives the running correlation dimension for sound //f//. It can be observed from this figure that unlike the Lyapunov exponents, the complexity of the chaotic system is same for both voiced and unvoiced sounds as the correlation dimension for stable region of both sound //e// and //f// is around 0.4.

42.4 Conclusion

In this paper, different nonlinear dynamical tools like phase-space plot, running correlation dimension, and running Lyapunov exponent applied to speech signals were discussed. These nonlinear dynamical tools could be used for providing a convenient framework for speech signal analysis. This paper provides evidence for chaotic behavior of speech generation process and suggests that chaotic model could be used for speech signal analysis.

References

1. Tsonis AA (1992) Chaos-from theory to applications. Plenum Press, New York
2. Takens F (1980) Detecting strange attractors in turbulence. In: Rand D, Young L (eds) Dynamical systems and turbulence, vol 898 of lecture Notes on Mathematics, Springer, Berlin, pp 366–381
3. Sano M, Sawada Y (1985) Measurement of the Lyapunov spectrum from a chaotic time series. Phys Rev Lett 55(10):1082–1085
4. Sato S, Sano M, Sawada Y (1987) Practical methods of measuring the generalized dimension and largest Lyapunov exponents in high- dimensional chaotic systems. Prog Theor Phys 77:1–5
5. Wolf A, Swift J, Swinney H, Vastano J (1985) Determining Lyapunov exponents from a time series. Phys D 16:285–317
6. Broomhead DS, King GP (1986) Extracting qualitative dynamics from experimental data. Phys D 20:217–236

Chapter 43
A Survey on Gaze Estimation Techniques

M. V. Sireesha, P. A. Vijaya and K. Chellamma

Abstract Biometric systems are becoming important, since they provide more efficient and reliable means to identity verification. In the recent studies, various biometric traits that include iris, face, gait, ear, palm, and knuckle joints have been investigated. Due to the intraclass variation, noisy sensor data, and susceptibility to spoofing attacks, the performance of these biometric systems is inaccurate. Gaze is one of the recent attractive topics in biometric research. In this paper, an attempt has been made to present an insight of different gaze estimation methods.

Keywords Biometrics · Gaze · Gaze estimation · Review paper · Human–computer interaction

43.1 Introduction

Biometrics is basically a technology that measures and analyzes human physiological and behavioral characteristics for personal identification. The significant reason for the acceptance of the biometrics as a tool for security is its universality, distinctiveness, permanence, and collectability. Main issues to be considered when implementing a biometric system are performance, acceptability, and circumvention [1]. Biometric characteristics are broadly grouped into two categories,

M. V. Sireesha (✉)
Department of TCE, BNMIT, Bangalore, India
e-mail: mv.sireesha@gmail.com

P. A. Vijaya · K. Chellamma
Department of ECE, BNMIT, Bangalore, India
e-mail: pavmkv@gmail.com

K. Chellamma
e-mail: selvi_chella@yahoo.co.in

V. Chakravarthi et al. (eds.), *Proceedings of International Conference on VLSI, Communication, Advanced Devices, Signals & Systems and Networking (VCASAN-2013)*, Lecture Notes in Electrical Engineering 258, DOI: 10.1007/978-81-322-1524-0_43,

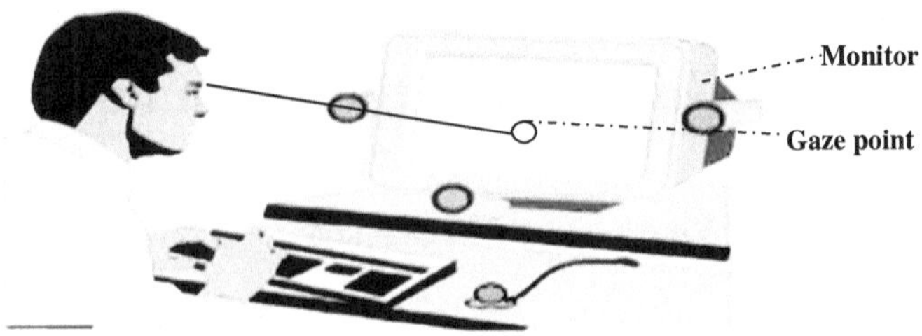

Fig. 43.1 Eye gaze tracker

namely physiological biometrics and behavioral. Physiological biometrics are based on the measure of anatomical biocharacteristics such as fingerprints, facial thermogram, vein structure, iris pattern, retina pattern, hand geometry, and facial recognition. Behavioral biometrics is based on the measure of trait during a period of time such as speech, signature, handwriting, keystrokes, and mouse dynamics [2]. Even though there are many biometric traits, these do not provide a perfect solution. This is likely due to number of reasons. Firstly, there are challenges with human interface supporting ICT systems and business process models that must be addressed with a practical implementation of any biometric system. Secondly, there are still several problems which are limiting the technology including problems such as noise in the sensed data (i.e., acquired voice may be corrupted by background noise in a voice-based biometric system), intraclass variation (e.g., testing lighting conditions may be different to enrollment lighting conditions in face biometrics), distinctiveness (e.g., two people may have very similar hand geometrics), and non-universality (e.g., certain people have very faint fingerprints which cannot be used to extract suitable features). Finally, most biometrics are not secret—they are often freely available and easy to acquire, and this makes them prone to spoof attacks (for example, a photo of a person's face, or a copy of their fingerprint can be used to foil a biometric system). While biometrics are harder to violate than pure knowledge-based or token-based systems, it is not impossible, ideally sager forms of biometrics would be based on non-visible and non-physiological information hidden deep within the person; such a biometric could be based on behavior or even thought process [3]. Gaze is a biometric which can avoid spoofing. The gaze tracking system captures intention of a person on the screen [4]. The gaze point determines the direction of where an individual is looking at as shown in Fig. 43.1.

43.2 Gaze as a Biometric Modality

Gaze is considered as a behavioral biometric. At the same time, it may also have physiological aspects determined by the tissues and muscles that determine its capabilities and limitations [5]. The main advantages of this biometric are firstly

Table 43.1 Different methods used for proving that gaze can be used as a biometric

Authors	Method used	Results
Farzin Dera et al.	Forward, backward, and branch and bound feature selection algorithms	Best performance obtained using BFS and B&B algorithms with 7 features
Zhang et al.	k-means clustering, linear and quadratic discriminant analysis, Naive Bayes rule, and k-nearest neighbors	High accuracy up to 90–100 % obtained
Bednarik et al.	Fourier spectrums, PCA, FFT + PCA	40–90 % of accuracy
Fooks et al.	Eigen gaze technique involved PCA or eigen decomposition approach.	Equal error rate of 8.9 % on small database of viewers
Maeder et al.	Spatial clustering algorithms	PIN was generated successfully
Bieder et al.	Considered processing time and relative conditional gaze entropy	Distinguished between informed and uninformed users for some stimuli and tasks
Fu et al.	Eye templates extracted using appearance-based technique in spatial domain. Gaze detection using center of iris or center of pupil	Reduced the total cost of the system
Weaver et al.	Eye dent system	Most successful authentication achieved using fixed minimum points per clusters of 9
Ali et al.	Facial landmark points were extracted using STASM. Centers of pupils were used	Genuine users showed smaller variances compared to fake attempts

such a biometric does not require physical contact between the viewer and a device. Secondly, such a biometric is difficult to detect by third person observation or surveillance due to the ambiguity of the visual attention process. Thirdly, such a biometric would be virtually impossible to spoof as the biometric measure itself. This process is heavily related to the higher cognitive, physiological, and neurological processes of a person and thus provides the opportunity for a close to ideal biometric [6]. Derav and Guness [5] examined gaze as a biometric. The drawback of this method is small database is used. Zhang et al. [7] suggested eye movements as biometric. The main drawbacks of this method are saccades may become slower with age. The second drawback is the eye moments might be affected by some disease. The third drawback is if subjects saccades vary too much at short intervals, this may cause difficulties in distinguishing his or her saccades from those of others. Bednarik et al. [8] suggested that there is discriminatory information in the eye movements. The main drawback is accuracy depends on the task and proper weight selection for fusion. Fooks and Sridharan [6] presented a technique to provide a simple and effective biometric for classifying individuals. The method is applied for small database. Maeder et al. [9] used gaze for personal computer applications. PIN choices are less due to 3 × 3 grid. Bieder et al. [10] used eye tracking to differentiate familiar user and unfamiliar with the system he operates. The drawback is processing time is less reliable. Fu and Yang [11] used gaze to control the display. The performance of this work is only in an acceptable level. Weaver et al. [12] utilized an eye tracker to enter passwords by nearly looking at the proper symbols on the computer monitor. The main drawback of this technique is to separate calibration step is needed before performing authentication. Ali et al. [13] used Gaze for liveness detection. This method requires user cooperation. The comparisons of different methods are shown in Table 43.1.

43.3 Gaze Estimation Methods

The main principle in the gaze estimation includes fixations and saccades. A fixation occurs when the gaze rests for some minimum amount of time on a small predefined area, within 2–5° of central vision approximately for at least 80–100 ms. The saccades are fast and jump like rotations of the eye between two fixated areas bringing objects of interests into the central few degrees of the visual field. During saccades, eye becomes blind. The existing gaze tracking techniques are broadly classified into feature based and appearance based as shown in Fig. 43.2.

43.3.1 Feature-Based Gaze Estimation

Gaze estimation methods using extracted local features such as eye corners, contours, and reflections from the eye image are called feature-based methods. The main reason for using feature-based methods is that the pupil and glints (under active light models) are relatively easy to find and that these features can, as

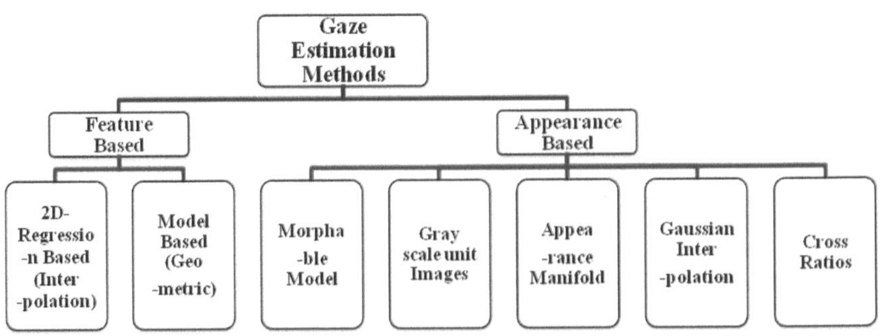

Fig. 43.2 Classification of gaze estimation methods

indicated above, be related to gaze. Two types of feature-based approaches exist, interpolation based (regression based) and model based (geometric).

43.3.1.1 2D-Regression-Based Gaze Estimation

The regression-based methods assume the mapping from image features to gaze coordinates (2D or 3D) have a particular parametric form such as a polynomial or a nonparametric form such as in neural networks. These methods avoid explicitly calculating the intersection between the gaze direction and gazed object [14]. Merchant et al. [15] proposed a real-time video-based eye tracker employing IR light using a single camera. Pupil-glint vector and a linear mapping were used to estimate the point of regard (POR). Nonlinearities with large pupil-glint angles were observed. Compensation was done for these by using polynomial regressions. Stampe et al. [16] also used same approach, without using glint information. Polynomial functions were used to model the correlation between pupil centers. White et al. [17] assumed a flat cornea surface and proposed a polynomial regression method for POR estimation. A first order linear regression was added. Higher order polynomial functions do not provide better calibration. Neural networks and their variants are popular tools for regression tasks. Ji and Zhu [18] suggested a generalized regression neural network–based method. The pupil-glint displacement, pupil parameters, orientation and ratio of the major and minor axes of the pupil ellipse, and glint coordinates were used to map the screen coordinates. The main advantage is only once calibration has to be done. Head movements were moderately allowed. Zhu et al. [19] suggested the use of Support Vector Machines. Mapping from the pupil and single glint to screen coordinates was done using SVM. The main drawback of two-dimensional methods is these do not handle head pose changes well.

43.3.1.2 Model-Based Gaze Estimation

Three-dimensional model-based approaches model the common physical structures of the human eye geometrically to calculate a 3D gaze direction vector. By using the gaze direction vector and integrating it with information about the objects in the scene, the point of regard is computed. Generally, these methods assume eye ball structure as spherical. Shih et al. [19], Guestrin and Eizenman [20] showed that by adding two or more light sources and known cornea curvature can find out the gaze estimation. Head movements were also allowed. The gaze tracking systems relying on this approach were consequently inflexible when attempting to change the geometry of light sources, camera (e.g., zoom), and screen to particular needs. In general, multiple light sources are faced with increased chance that one of the glints might disappear. In general, a large field of view is required to allow for free head motion, but a limited field of view is needed to capture sufficiently high resolution eye images to provide reliable gaze estimates. Multiple cameras are utilized to achieve these goals either through wide-angle lens cameras or movable narrow-angle lens cameras. Multiple cameras also allow for 3D eye modeling. The drawbacks of regression-based gaze estimation methods are these are sensitive to head pose changes. A direct solution to compensate for minor head movement is to use one camera for observing the head orientation and another camera for eye images and then combine the information [17, 22, 23]. The methods are more complex. Zhu and Ji [22] suggested that stereo makes 3D eye modeling directly applicable. The proposed method can estimate Gaze when the optical axis of the eye intersects or is close to the line connecting the nodal points of the two cameras. Combinations of stereo systems with pan-tilt have been suggested [23, 24]. Talmi and Liu [23] combined a stereo system for face modeling with a pan-tilt for detailed eye images. Ohno and Mukawa [24] utilized three cameras, two fixed stereo wide-angle cameras and a narrow-angled camera mounted on a pan-tilt unit. Two calibration points were necessary in order to estimate the visual axis. The use of multiple cameras seems to produce robust results but requires stereo calibration.

43.3.2 Appearance-Based Methods

Feature-based methods require detection of pupils and glints, but the extracted features may be prone to errors. Besides, there may be latent features conveying information about gaze that is not modeled by the chosen features. Appearance-based models for gaze estimation do not explicitly extract features, but rather use the image contents as input with the intention of mapping these directly to screen coordinates (PoR). Consequently, the hope is that the underlying function for estimating the point of regard, relevant features, and personal variation can be extracted implicitly, without requirements of scene geometry and camera calibration. Appearance-based methods typically do not require calibration of cameras and geometry data since the mapping is made directly on the image contents [14]. The major appearance-based methods are based on morphable model, appearance manifold, grayscale unit images, Gaussian interpolation, and cross-ratios [25]. A morphable model is developed by Rikert [26]. The texture for a set of prototype images is mapped to the reference image based on shape transformation. The parameters for shape and texture of the eye region are trained using neural networks. The high value of the parameter for a specific prototype during comparison of images gives the direction of gaze. The criterion for matching is the sum of squared differences error. Betke and Kawai [27] used grayscale to determine gaze direction. The extraction component consists of pupil center, ellipse parameters, unit images, and its centers along the outline of ellipse. The gaze mapping component involves learning process using self-organizing map based on correlation for different directions. The outcome of the learning phase gives the arrangement of the unit images for test image. The comparison of the pupil positions in model image with various regions in test image using correlation coefficient provides the pupil position. The difference in pupil position between model image and test image estimates the direction of gaze. Tan et al. [28] used appearance manifold per gaze estimation. The parameters of the test point are estimated by interpolation. Maintaining a high-dimensional space is computationally expensive in this method. Hansen et al. [29] used a specific method for eye typing applications. The mapping function is based on Gaussian process interpolation method specified by mean and variance functions. The mean value of the interpolated point corresponds to the estimated position of the eye gaze. The performance of mean shift tracker is low due to lack of handling multiple hypotheses. Yoo et al. [30] determined the GR based on glint position. The gaze mapping function is determined by dividing each side of the polygon into segments. Two intermediate points on each segment called as the crossing points are marked. In this method, improper threshold value during the extraction of pupil and glint results in a false estimation of GR. One of the key challenges in previous gaze estimators is the need for explicit personal calibration to adapt to individual users. In these existing systems, the users are always required to actively participate in calibration tasks by fixating their eyes on explicit reference points. The other problem that most estimation methods suffer from is calibration drift, and their calibration accuracy highly depends on the users and installation settings. Although the number of reference points for personal calibration can be reduced using specialized hardware such as multiple light sources and stereo cameras, it still requires a user to actively participate in the calibration task. More importantly, such hardware-based attempts add a strong constraint to the application settings, and this limits the user scenarios [31]. To overcome this, Sugano [31] used a gaze sensing method using visual saliency maps. This does not need explicit personal calibration. Gaze estimator was created using only the eye images captured from a person watching a video clip. This method took a set of images recorded in synchronization with any video clip as an input. From such an input, this method automatically determines the relationship between eye images and gaze direction. Accuracy about a 3.5° error with fixed head positions was obtained. The main drawback of this method is it cannot handle the head pose variations.

43.4 Conclusions

The present paper reviews the research work made in gaze tracking. The different methods used for proving gaze as a biometric are also discussed. The survey provides a platform to explore more sophisticated feature extraction techniques. Cost-effective, calibration free and to handle head pose changes eye tracking techniques need to be investigated.

References

1. Latifi S, Solayappan N (2006) A survey of unimodal biometric method. In: proceedings of the 2006 international conference on security and management SAM 2006, Las Vegas, Nevada, USA, 26–29 June 2006
2. Radhika KR, Sheela SV (2011) Fundamentals of biometrics and hand written signature and Iris. In: Wang PSP (ed) Pattern recognition, machine intelligence and biometrics. Springer, Berlin, pp 733–783
3. Sheela SV, Vijaya PA (2011) An appearance based method for eye gaze tracking. J Comput Sci 7(8):1194–1203
4. Fookes C, Sridharan S, Maeder A, Mamic G (2010) Gaze based personal identification. IGI global: book chapters
5. Deravi F, Guness SP (2011) Gaze trajectory as a biometric modality. In: proceeding of: BIOSIGNALS 2011—proceedings of the international conference on bio-inspired systems and signal processing, Rome, Italy, 26–29 Jan 2011
6. Fookes C, Sridharan S (2010) Eigengaze—covert behavioral biometric exploiting visual attention characteristics. In: proceedings of practical guide to testing wireless smart phone applications
7. Zhang Y, Rasku J, Juhola M (2012) Biometric verification of subjects using saccade eye movements. Int J Biometrics 4(4):317–337
8. Bednarik R, Tomi K, Mihaila A, Franti P (2005) Eye-movements as a biometric. SCIA 2005, LNCS 3540, pp 780–789
9. Maeder A, Fookes C, Sridharan S (2004) Gaze based user authentication for personal computer applications. International symposium on intelligent multimedia, video and speech processing, 20–22 Oct 2004
10. Biedert R, Frank M, Martinovic I, Song D (2012) Stimuli for gaze based intrusion detection. Future Inf Technol Appl Serv Lecture Notes Elect Eng 164:757–763
11. Fu B, Yang R (2011) Display control based on eye gaze estimation. In: 4th international congress on image and signal processing
12. Weaver J, Mock K, Hoanca B (2011) Gaze-based password authentication through automatic clustering of gaze points. In: IEEE international conference on SMC, pp 2749–2754
13. Ali A, Deravi F, Hoque S (2012) Liveness detection using gaze collinearity. In: proceedings of the 3rd international conference on emerging security technologies
14. Hansen DW, Ji Q (2010) In the eye of the beholder: a survey of models for eyes and gaze. IEEE Trans Pattern Anal Mach Intell 32(3):478
15. Merchant J, Morrissette R, Porterfield J (1974) Remote measurements of eye direction allowing subject motion over one cubic foot of space. IEEE Trans Biomed Eng 21(4):309–317
16. Stampe DM (1993) Heuristic filtering and reliable calibration methods for video-based pupil-tracking systems. Behav Res Methods Instrum Comput 25(2):137–142
17. White KP, Hutchinson TE, Carley JM (1993) Spatially dynamic calibration of an eye-tracking system. IEEE Trans Syst Man Cybern 23(4):1162–1168
18. Ji Q, Zhu Z (2002) Eye and gaze tracking for interactive graphic display. In: proceedings of 2nd international symposium on smart graphics, pp 79–85
19. Zhu Z, Ji Q, Bennett KP (2006) Nonlinear eye gaze mapping function estimation via support vector regression. In: proceedings of 18th international conference on pattern recognition, vol 1, pp 1132–1135
20. Shih SW, Wu Y-T, Liu J (2000) A calibration-free gaze tracking technique. In: Proceedings of the 15th international conference on pattern recognition, pp 201–204
21. Guestrin ED, Eizenman M (2006) General theory of remote gaze estimation using the pupil center and corneal reflections. IEEE Trans Biomedical Eng 53(6):1124–1133, June 2006
22. Zhu Z, Ji Q (2007) Novel eye gaze tracking techniques under natural head movement. IEEE Trans Biomed Eng 54(12):2246–2260

23. Talmi K, Liu J (1999) Eye and gaze tracking for visually controlled interactive stereoscopic displays. Signal Process Image Comm 14(10):799–810
24. Ohno T, Mukawa N (2004) A free-head, simple calibration, gaze tracking system that enables gaze-based interaction. In: proceedings of eye tracking research and applications symposium, pp 115–122
25. Sheela SV, Vijaya PA (2011) Mapping functions in gaze tracking. Int J Comput Appl 26(3):0975–8887
26. Rikert TD, Jones MJ (1998) Gaze estimation using morphable models. In: IEEE international conference on automatic face and gesture recognition, pp 436–441
27. Betke M, Kawai J (1999) International workshop on recognition, analysis, and tracking of faces and gestures in Real-time systems, pp 70–76
28. Tan K-H, Kriegman DJ, Ahuja N (2002) Appearance-based eye gaze estimation. In: IEEE workshop on applications of computer vision, pp 191–195
29. Hansen DW, Hansen JP, Nielsen M, Johansen AS, Stegmann MB (2002) Eye typing using markov and active appearance models. In: IEEE workshop on applications of computer vision, pp 132–136
30. Yoo DH, Kim JH, Lee BR, Chung MJ (2002) Non-contact eye gaze tracking system by mapping of corneal reflections. In: IEEE international conference on automatic face and gesture
31. Sugano Y, Matsushita Y (2013) Appearance-based gaze estimation using visual saliency

Chapter 44
Handwritten Script Recognition Using DCT, Gabor Filter, and Wavelet Features at Word Level

G. G. Rajput and H. B. Anita

Abstract In a country like India, many of the documents such as office letters, checks, envelopes, forms, and other types of manuscripts are multiscript in nature. A document consisting of English script and a regional script is quite common. Hence, automatic recognition of scripts present in a multiscript document has a variety of practical and commercial applications in banks, post offices, reservation counters, libraries, etc. In this paper, a multiple feature-based approach is presented to identify the script type from a multiscript document. Features are extracted using Gabor filters, discrete cosine Transform, and wavelets of Daubechies family. Nine popular Indian scripts are considered for recognition in this paper. Experiments are performed to test the recognition accuracy of the proposed system at word level for bilingual scripts. Using neural network classifier, the average success rate is found to be 97 %.

Keywords Handwritten script · Multiscript · Gabor filter · Discrete cosine transform · Neural network

44.1 Introduction

In present information technology era, document processing has become an inherent part of office automation process. Many of the documents in Indian environment are multiscript in nature. A document containing text information in

G. G. Rajput
Gulbarga University, Gulbarga, Karnataka 585106, India
e-mail: ggrajput@yahoo.co.in

H. B. Anita (✉)
Gulbarga University, #8, 5th Left from T.R. Hospital, Kothnur Dinne Main Road, S.K. Nataraja Layout, Bangalore, Karnataka 560076, India
e-mail: anitahb@yahoo.com

V. Chakravarthi et al. (eds.), *Proceedings of International Conference on VLSI, Communication, Advanced Devices, Signals & Systems and Networking (VCASAN-2013)*, Lecture Notes in Electrical Engineering 258, DOI: 10.1007/978-81-322-1524-0_44,

more than one script is called a multiscript document. Many of the Indian documents contain two scripts, namely the state's official script (regional script) and English. An automatic script identification technique is useful to sort document images, select appropriate script-specific OCRs, and search online archives of document images for those containing a particular script. Handwritten script identification is a complex task due to following reasons: complexity in preprocessing, complexity in feature extraction and classification, sensitivity of the scheme to the variation in handwritten text in document (font style, font size, and document skew), and performance of the scheme. Existing script identification techniques mainly depend on various features extracted from document images at block, line, or word level. Block-level script identification identifies the script of the given document in a mixture of various script documents. In line-based script identification, a document image can contain more than one script, but it requires the same script on a single line. In word-level script identification, the script of every word is identified. A brief description of the methods proposed in the literature at word level is given below.

To discriminate between printed text lines in Arabic and English, three techniques are presented in [1]. In the first technique, an approach based on detecting the peaks in the horizontal projection profile is considered. Secondly, moments of the profiles are computed and neural network is used for classification. Finally, a run length histogram–based approach using neural networks is described. An automatic scheme to identify text lines of different Indian scripts from a printed document is attempted in [2]. Features based on water reservoir principle, contour tracing, profile, etc. are employed to identify the scripts. Twelve Indian scripts have been explored to develop an automatic script recognizer at text line level in [3, 4]. Script recognizer classifies the input word using the characteristics and shape-based features of the script. Devanagari script was discriminated through the headline feature, and structural shapes were designed to discriminate English from the other Indian script. Using the combination of shape, statistical and water reservoirs, an automatic line-wise script identification scheme from printed documents containing five most popular scripts in the world, namely Roman, Chinese, Arabic, Devanagari, and Bangla, has been introduced in [5]. Identification of the script type from a trilingual printed document is proposed in [6]. The distinct characteristic features of Kannada, Hindi, and English scripts are studied from the nature of the top and bottom profiles and the model is trained to learn thoroughly the distinct features of each script. A survey on both global-based approach as well as local-based approach for script identification in document images is reported in [7]. Majority of the methods available in the literature for script recognition at line or word level are for printed documents. To the best of our knowledge, handwritten script identification at word level, for Indian scripts, has not been reported in the literature. This motivated us to design a system for Indian script identification from handwritten bi-script documents at word level. Further, the present work is extension to our work presented in [8, 9] for script identification from handwritten documents at block level. The method proposed in this paper employs analysis of word for script identification, cropped out manually from the scanned document images.

In many cases, the most distinguished information is hidden in the frequency content of the signal rather than in the time domain. Hence, in this paper, frequency-based features using Gabor, DCT/wavelets are presented for identification of script type from a multiscript handwritten document. Most of the documents are written in two scripts; three or higher number of scripts in a handwritten or printed document appear rarely. Therefore, we have considered bi-script documents. Nine Indian scripts including English are considered for experimentation. Neural network classifier is used for recognition of script type.

44.2 Method Description

44.2.1 Data Collection and Preprocessing

Handwritten documents were collected from persons belonging to different professions and different states. Restrictions were not imposed on the writers regarding the content of the text and use of pen. Handwritten documents were written in English, Devanagari, Kannada, Tamil, Bangla, Telugu, Punjabi, Odiya, and Malayalam scripts. The document pages were scanned at 300 dpi resolution and stored as gray-scale images. Noise is removed by applying median filter. The dataset at word level from the preprocessed document images is created as explained below.

The words were manually cropped out from different areas of the document image. The cropped handwritten words contained two or more characters with variable spaces between characters. These words are then binarized using well-known Ostu's global thresholding approach [10]. The resulting images are inverted so that text pixels represent value 1 and background pixels represent value 0. The salt and pepper noise around the boundary is removed using morphological opening. This operation also removes discontinuity at pixel level. The preprocessing steps for a sample word are shown in Fig. 44.1. A total of 9,000 handwritten word images, representing nine scripts, containing text with more than two characters are thus created. This includes 1,000 words per script. Similarly, dataset consisting of 900 word images containing two characters and 900 word images

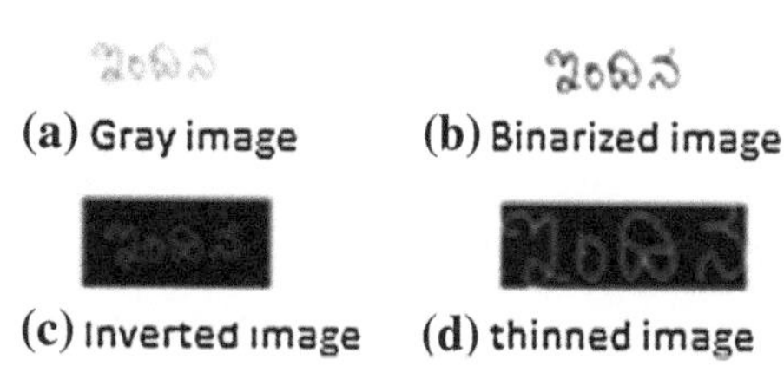

Fig. 44.1 Pipeline process for feature extraction

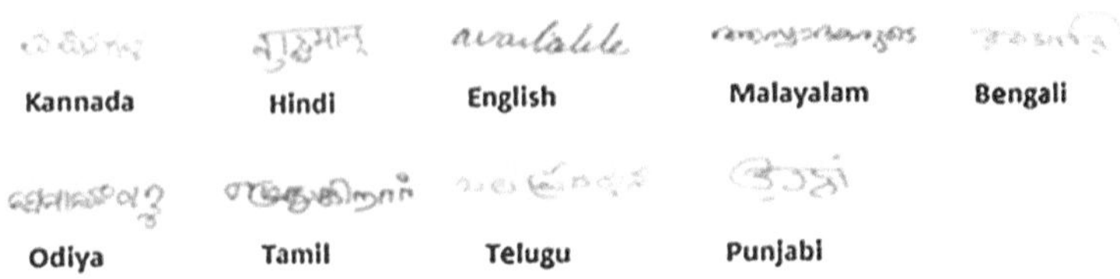

Fig. 44.2 Sample handwritten word images in different scripts

containing single character is created, with 100 images per script. A sample of word images of different scripts is shown in Fig. 44.2.

44.2.2 Feature Extraction

Features are the representative measures of a signal which distinguish it from other signals. The selected features should maximize the distinction between multi-scripts. In the proposed method, features are extracted by using two-dimensional Gabor functions by transforming the image in time domain to the image in frequency domain. Gabor filters are formed by modulating a complex sinusoid by a Gaussian function with different frequencies and orientations. The term frequency refers to variation in brightness or color across the image, i.e., it is a function of spatial coordinates, rather than time. The frequency information of image is needed to see information that is not obvious in time domain. A brief description of the transforms used for future extraction is given below.

1. Gabor Filter: Gabor filters are direction-specific band-pass filters. The Gabor wavelets have good directional selectivity for images that gives advantage of extracting features in any desired direction from the document image. In the proposed method, Gabor filters are used to encode the text images into multiple narrow frequency channels measured at different resolutions/orientation. Next, features are extracted from all these different resolutions and oriented images. A two-dimensional Gabor function consists of a sinusoidal plane wave of some frequency, orientation and modulated by a two-dimensional Gaussian and is given in Eq. (44.1).

$$g(x,y) = \left(\frac{1}{2\pi\sigma_x\sigma_y}\right)\exp\left(-\frac{1}{2}\left(\frac{x'^2}{\sigma_x^2}+\frac{y'^2}{\sigma_y^2}\right)\right)\exp(2\pi jWx') \tag{44.1}$$

$$x' = x\cos\theta + y\sin\theta$$
$$y' = -x\sin\theta + y\cos\theta$$

where $\sigma x2$ and $\sigma y2$ control the spatial extent of the filter, θ is the orientation of the filter, and w is the frequency of the sinusoid.

2. Cosine transforms: The discrete cosine transform (DCT) concentrates energy into lower-order coefficients. The DCT is purely real and expresses a sequence of finitely many data points in terms of a sum of cosine functions oscillating at different frequencies that are necessary to preserve the most important features [11]. With an input image, A_{mn}, the DCT coefficients for the transformed output image, Bpq, are computed according to Eq. (44.2). In this equation, A_{mn} is the input image having M-by-N pixels, A_{mn} is the intensity of the pixel in row m and column n of the image, and B_{pq} is the DCT coefficient in row p and column q of the DCT matrix.

$$B_{pq} = \alpha_p \alpha_q \sum_{m=0}^{M-1} \sum_{n=0}^{N-1} A_{mn} \cos\frac{\pi(2m+1)p}{2M} \cos\frac{\pi(2n+1)p}{2N}, \tag{44.2}$$

$$0 \le p \le M-1,\ 0 \le q \le N-1$$

$$\alpha_p = \begin{cases} 1/\sqrt{M}, & p=0 \\ \sqrt{2/M}, & 1 \le p \le M-1 \end{cases} \qquad \alpha_q = \begin{cases} 1/\sqrt{N}, & q=0 \\ \sqrt{2/N}, & 1 \le q \le M-1 \end{cases}$$

3. Wavelet transforms: The discrete wavelet transform (DWT), which is based on sub-band coding, is found to yield fast computation of wavelet transform [11]. The wavelet transforms are used to analyze the signal (image) at different frequencies with different resolutions. It represents the same signal, but corresponding to different frequency bands. Wavelets are used for multiresolution analysis, to analyze the signal at different frequencies with different resolutions, to split up the signal into a bunch of signals, representing the same signal, but all corresponding to different frequency bands, and to provide what frequency bands exist at what time intervals. Many wavelet families have been developed with different properties. For two-dimensional images, applying DWT corresponds to processing the image by two-dimensional filters in each dimension.

In this paper, we employ two-dimensional Gabor with DCT/wavelet filters to extract the features from input text word image for identification for script type. The preprocessed input binary image is convolved with Gabor filters considering six different orientations (0°, 30°, 60°, 90°, 120°, and 150°) and three different frequencies (a = 0.125, b = 0.25, c = 0.5) with $\sigma_x = 2$ and $\sigma_y = 4$ *Fig. 44.3. The values of these parameters are fixed empirically. From the 18 output images, we compute the standard deviation to obtain features of dimension 18. It was observed that these features extracted by using Gabor filters (Algorithm-1) are not sufficient in terms of recognition accuracy. Hence, to improve the accuracy, two separate algorithms were proposed using DCT (Algorithm-2) and wavelets (Algorithm-3), respectively. The first set of features is obtained by performing Algorithm-1 and Algorithm-2 on the input word image. The second set of features is obtained by performing Algorithm-1 and Algorithm-3 on the input word image.

Algorithm 1 Input: Image in gray scale at word level.
Output: Feature vector
Method:

1. Apply median filter to remove noise (Fig. 44.2a).
2. Binarize the image using Otsu's method and invert the image to yield text representing binary 1 and background binary 0 (Fig. 44.2b and c).
3. Remove small objects around the boundary using morphological opening (Fig. 44.2d).
4. Apply thinning operation (Fig. 44.2d).
5. Crop the image by placing bounding box over the portion of word.
6. Create Gabor filter bank by considering six different orientations and three different frequencies.
7. Convolve the input image with the created Gabor filter bank (Fig. 44.3).
8. For each output image obtained in step 7, perform following steps.

   a) Extract cosine part and compute the standard deviation (18 features).
   b) Extract sine part and compute the standard deviation (18 features).
   c) Compute the standard deviation of the entire output image (18 features).

   This forms feature vector of length 54

Algorithm 2 (*Gabor combined with DCT*)
Input: Image in gray scale at word level.
Output: Feature vector
Method:

1. Perform steps 1 through 7 of Algorithm-1 to obtain the preprocessed and convolved images (total 18).
2. Perform following steps.

   (a) Apply DCT to the preprocessed image and compute the standard deviation of the DCT.
   (b) Apply DCT for each convolved input images and compute the standard deviation. This gives us 18 features.

Fig. 44.3 Gabor-filtered images for zero degree orientation and frequencies a, b, and c

3. Concatenate features obtained in step 2 (a) and (b) to get the feature vector of length 19.

Algorithm 3 (*Gabor combined with wavelets*)
Input: Image in gray scale at word level.
Output: Feature vector
Method:

1. Perform steps 1 through 7 of Algorithm-1 to obtain the preprocessed convolved images (total 18).
2. Perform following steps.
   (a) Apply wavelet to the preprocessed image and compute the standard deviation for each frequency bands. This forms 4 features.
   (b) Apply DCT to the preprocessed image. Then, apply wavelet to the DCT image and compute the standard deviation of each frequency bands, namely approximation coefficients (cA), vertical coefficients (cV), horizontal coefficients (cH), and diagonal coefficients (cD). This forms 4 features.
   (c) Perform wavelet (Daubechies 9) decomposition for each convolved input images to obtain approximation coefficients (cA), vertical coefficients (cV), horizontal coefficients (cH), and diagonal coefficients (cD). Compute the standard deviation for each frequency band separately for all 18 images. This forms $4 \times 18 = 72$ features.
3. Concatenate features obtained in step 2(a), (b), and (c) to get the feature vector of length 80.

44.3 Script Recognition

Neural network classifiers are adopted for recognition purpose. The classifier appropriately learns how to weight multiple features and at the same time produces a generalized mapping which is not over-fitted. A 1-hidden-layer feed-forward neural network with 3-hidden-layer neurons is considered and trained. The input and target samples are divided into training, validation, and test sets. The training set is used to train the network. Training continues as long as the network continues improving on the validation set. The test set provides a completely independent measure of network accuracy. The details about the Artificial Neural Network can be found elsewhere [12].

44.4 Experimental Results

The performance of the proposed algorithms for script identification from bi-script document is evaluated by using the dataset of word obtained as described in Sect. 44.2. The dataset consists of images with one character and images with two or more characters. The complete dataset is manually processed to generate the ground truth for testing and evaluation of the algorithm. For bi-script documents, we have considered one Indian script and English script. Samples of one script are input to our system, and performance is noted in terms of recognition accuracy. For each dataset of word images of a particular script, 60 % images are used for training and remaining 40 % images are used for testing. Identification of the test script is done using neural network classifier. The proposed method is implemented using Matlab software. Two sets of features are extracted from the proposed algorithms. Features obtained by performing Algorithm-1 and Algorithm-2 are combined in first feature set. Features obtained by performing Algorithm-1 and Algorithm-3 are combined in second feature set. Experiments are performed on these two sets separately using neural network classifier. The results obtained by performing experiments on more than two characters, two characters, and one character are presented in Tables 44.1, 44.2, and 44.3, respectively. The second column of Tables 44.1, 44.2, and 44.3 presents the recognition accuracy for the first feature set. Likewise, the third column of Tables 44.1, 44.2, and 44.3 presents the recognition accuracy for the second feature set. The results clearly shows that features extracted by using Gabor function yield good results. Further, it is observed that both the feature sets yield good recognition rate for images with two or more than two characters in a word compared to recognition rate for one character word images.

Table 44.1 Recognition results of bi-script document images for more than two characters

Bi-scripts	DCT applied to Gabor convolved images (Algorithm-1–54 Features, Algorithm-2–19 Features) (%)	Wavelets applied to Gabor convolved images (Algorithm-1–54 Features, Algorithm-3–80 Features) (%)
Kannada, English	96	96
Hindi, English	99.7	99.5
Malayalam, English	95.5	97.9
Punjabi, English	97.2	98
Tamil, English	95.6	98.6
Telugu, English	97.2	95.5
Bengali, English	95.2	95
Odiya, English	97	97.5

Table 44.2 Recognition results of bi-script document images for two character word image

Bi-scripts	DCT applied to Gabor convolved images (Algorithm-1–54 Features, Algorithm-2–19 Features) (%)	Wavelets applied to Gabor convolved images (Algorithm-1–54 Features, Algorithm-3–80 Features) (%)
Kannada, English	90	87
Malayalam, English	83.5	87
Tamil, English	89	80
Telugu, English	90	87
Bengali, English	93	95.5
Punjabi, English	95.5	84
Hindi, English	91	100
Odiya, English	91	82

Table 44.3 Recognition results of bi-script document images for one character word image

Bi-scripts	DCT applied to Gabor convolved images (Algorithm-1–54 Features, Algorithm-2–19 Features) (%)	Wavelets applied to Gabor convolved images (Algorithm-1–54 Features, Algorithm-3–80 Features) (%)
Kannada, English	85	82.5
Malayalam, English	90.2	80.5
Tamil, English	92.5	80
Telugu, English	82.5	82.5
Bengali, English	72	82.5
Punjabi, English	90	85
Hindi, English	85	87.5
Odiya, English	87.5	82.5

44.5 Conclusion

In this paper, script identification from multiscript handwritten documents based on Gabor and DCT/wavelets is presented. Experiments are performed at word level for bi-script documents considering nine Indian scripts including English script. Neural network classifier is used in recognition phase. The average recognition rate of 97 % is obtained using features computed from Algorithm-1 and Algorithm-2. The proposed method is robust and independent of style of handwriting. In future, we extend our work for other Indian regional scripts and we also consider tri-script documents for experimentation.

References

1. Elgammmal AM, Ismail MA (2001) Techniques for language identification for hybrid arabic-english document images. In: Proceedings of sixth international conference on document analysis and recognition, pp 1100–1104
2. Pal U, Sinha S, Chaudhuri BB (2003) Multi-script line identification from indian documents, ICDAR, vol 2, pp 880, Seventh International Conference on Document Analysis and Recognition, vol 2
3. Pal U, Chaudhuri BB (1999) script line separation from indian multi-script documents, 5th ICDAR, pp 406–409
4. Pal U, Chaudhury BB (2002) Identification of different script lines from multi-script documents. Image Vis Comput 20(13–14):945–954
5. Pal U, Chaudhuri BB (2001) Automatic identification of English, Chinese, Arabic, Devanagari and Bangla script line. In: Proceedings of 6th international conference on document analysis and recognition (ICDAR'OI), pp 790–794
6. Padma MC, Vijaya PA (2010) Script identification from trilingual documents using profile based features. Int J Comput Sci Appl, Technomath Res Foundation 7(4):16–33
7. Abirami S, Manjula D (2009) A survey of script identification techniques for multi-script document images. Int J Recent Trends Eng 1(2)
8. Rajput GG, Anita HB (2010) Handwritten script recognition using DCT and wavelet features at block level, IJCA, Special issue on RTIPPR (3):158–163
9. Rajput GG, Anita HB (2010) Kannada, English, and Hindi handwritten script recognition using multiple features. In: Proceedings of national seminar on recent trends in image processing and pattern recognition, ISBN: 93-80043-74-0, pp 149–152
10. Rajput GG, Anita HB (2009) A two step approach for deskewing handwritten and machine printed document images using histograms and geometric features. In: Proceedings of Second International Conference on Signal and Image Processing, pp 414–417
11. Gonzalez RC, Woods RE (2008) Digital image processing, 3edn. Pearson Education, New Jersey
12. Yegnanarayana B (2004) Artificial neural network, PHI Publications, India

Chapter 45
Word-Based LID Using HMM and Bi-gram Modeling

J. M. Patil (Hatte J.S) and Priyadarshini K. Desai

Abstract Language identification is the task of automatically identifying the language of the speech signal uttered by an unknown speaker. An N language identification task is to classify an input speech utterance, spoken by unknown speaker and of unknown text, as belonging to one of the N languages. LID has applications as a front-end for machines of multi-lingual information retrieval system, multi-lingual speech recognition system and speech to speech translation system. In this paper, hidden Markov model is used for speech recognition and language modeling, i.e., bi-gram model which is the special case of N-gram model ($n = 2$ for bi-gram). The maximum-likelihood classifier is used to identify the language of given test speech.

Keywords Word recognition system · Language modeling and a classifier

45.1 Introduction

Automatic language identification (LID) has become an important research problem over the last decade with several promising solutions [1, 2]. An N language LID task is to classify an input speech utterance (typically spoken by an unknown speaker and of unknown text) as belonging to one of N languages L1…LN. Among the various approaches to LID, the phone recognition approach offers considerable promise, as it incorporates sufficient knowledge of the phonology of the languages to be identified. Three main frameworks can be identified

---

J. M. Patil (Hatte J.S) (✉) · P. K. Desai
E&CE Department, M. S. BidveEngg College, Latur, Maharashtra, India
e-mail: mayurrish29@rediffmail.com

P. K. Desai
e-mail: priyadarshini.p.r@gmail.com

V. Chakravarthi et al. (eds.), *Proceedings of International Conference on VLSI, Communication, Advanced Devices, Signals & Systems and Networking (VCASAN-2013)*, Lecture Notes in Electrical Engineering 258, DOI: 10.1007/978-81-322-1524-0_45,

within the phone recognition approaches, namely (1) phone recognition followed by language modeling (PRLM), (2) parallel PRLM (P-PRLM) and, (3) parallel phone recognition (PPR). Several phone recognition-based systems reported in the literature fall within these categories.

In this paper, we actually implement language identification using CSLU database. We are basically trying to implement speaker-independent but text-dependent system. PRLM, P-PRLM, and PPR systems have been studied using the OGI-TS database. The OGI-TS is a multi-speaker telephone speech corpus in 11 languages, specifically designed for LID research.

We are using the word-level approach. The system remains same as parallel phone recognition system, where the HMM models, language modeling, and maximum-likelihood classifier are used. The continuous speech signal is given as input to the system, and it is converted into isolated words. The corresponding text file is created, which consists of the sentences involving the words present in the continuous speech signal. The HMM model is created for each word.

The Bi-gram model is calculated for words of the corresponding languages from the text file. Both the probability of the HMM model and the probability of the language model are compared at the classifier during recognition for the given test data. The result is given in favor of the language which has the maximum log-likelihood probability.

45.2 Objective of the Paper

The main objective of this paper is to identify the languages using HMM and language modeling. It is word-based approach, because of which the system requires word labels. This system offers up to 80 % of result by word-level approach.

45.3 Methodology

Language identification is the process of identifying natural language. Language identification can be done using several approaches. It can be identified based on spectral features of the speech signal, lexicons, phoneme or statistical modeling. Initially, we started with isolated word recognition [2], where training of word model using HMM was done. Later, we have applied the same concept for language identification.

The language identification is done in four stages.

The Feature Extraction Stage: The mel-frequency cepstral coefficient of the speech signal is obtained during this stage

The word recognition stage: The HMM models are created for the words and are recognized.

The language modeling: The bi-gram probability for each text file containing words is calculated.

Classifier: Maximum-likelihood classifier is used for differentiating the three languages.

45.4 Feature Extraction Stage

In this stage, the pre-processing of speech signal and feature extraction of the speech signal are carried out. Voice samples behave as stationary variable within the duration of 10–25 ms. Thus, we take speech sample of 10–25 ms duration. In pre-processing, the following steps are carried out which are also part of voice activity detection:

Pre-emphasis: The speech signal x(n) is passed through a high-pass filter x2(n) = x(n) – a × x(n−1), where x2(n) is the output signal and the value of 'a' is usually between 0.9 and 1.0. The z-transform of the high-pass filter is H(z) = 1 – a × z − 1. The pre-emphasis is carried out to enhance the high-frequency content of the speech signal that was suppressed during speech recording.

Frame blocking: The input speech signal is segmented into frames. Each frame is of 20–30 ms duration, with overlap of approximate 1/3–1/2 of the frame size. The sample rate taken is 16 kHz, and the frame consists of 320 sample points, and then, the duration of each is 320/16000 = 0.02 s = 20 ms. Additional, if the overlap is 160 points, then the frame rate is 16000/(320 − 160) = 100 frames per second.

Hamming windowing: In order to keep the continuity of the first and the last points in the frame, each frame is multiplied with a hamming window. Hamming window is defined by $w(n, a) = (1 - a) - a\ \cos\left(\frac{2pn}{N-1}\right),\ 0 \leq n \leq N - 1$

MFCC: The very next step after the speech signal is given is the feature extraction. Here, we are extracting the MFCC feature vectors. The mel-frequency cepstrum (MFC) is a representation of the short-term power spectrum of a sound, based on a linear cosine transform of a log power spectrum on a nonlinear mel scale of frequency. Mel-frequency cepstral coefficients (MFCCs) are coefficients that collectively make up an MFC. They are derived from a type of cepstral representation of the audio clip (a nonlinear "spectrum of a spectrum"). The difference between the cepstrum and the mel-frequency cepstrum is that in the MFC, the frequency bands are equally spaced on the mel scale, which approximates the human auditory system's response more closely than the linearly spaced frequency bands used in the normal cepstrum. This frequency warping can allow for better representation of sound, for example, in audio compression.

MFCCs are commonly derived as follows:

1. The Fourier transform of a signal is taken.
2. Map the powers of the spectrum obtained above onto the mel scale, using triangular overlapping windows. The relation between mel scale and linear frequency scale is Mel(f) = 1125 × ln (1 + f/700)

3. The logs of the powers at each of the Mel frequencies are taken.
4. Take the discrete cosine transform of the list of mel log powers.
5. The MFCCs are the amplitudes of the resulting spectrum.

45.5 Word Recognition Stage Using HMM

HMMs are created for every basic sound unit; in our case, it is word. Further, all HMMs are linked together to represent the vocabulary under consideration. This linked representation is known as search space for given problem. During recognition phase, this graph is searched for finding occurrence of given word. Given an N number of observation sequences of a word ON = {o 1 o 2 o 3…. o T}. How is the training of that model done to best represent the word. This is done by adjusting the parameters for the model $\lambda = (\pi, A, B)$. The adjustment is an estimation of the parameters for the model $\lambda = (\pi, A, B)$ that maximizes $P(O|\lambda)$. The Fig. 45.1 shows speech recognition system using HMM.

The MFCC matrix is calculated according to speech signal to mel-frequency cepstral coefficients. This is also used when testing an utterance against model.

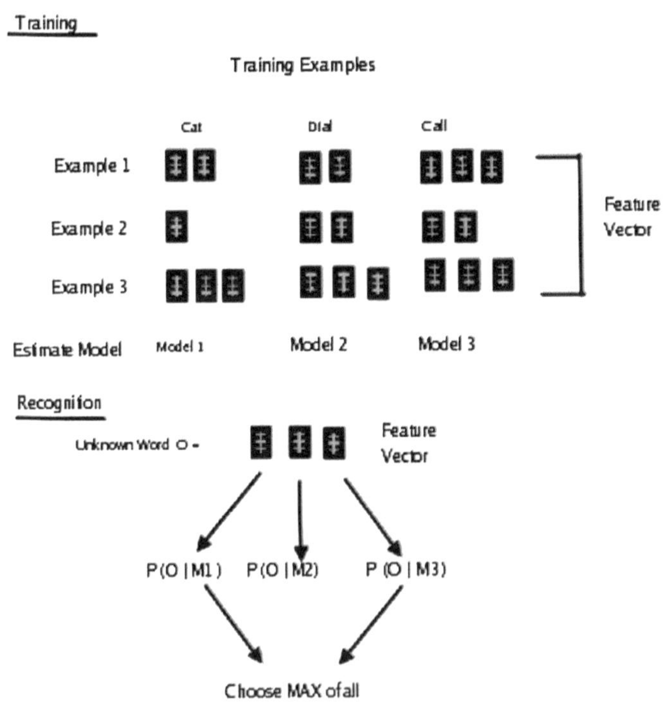

Fig. 45.1 Isolated word recognition system using HMM

When the MFCC is achieved, there is a need to normalize all the given training utterance. The matrix is divided into a number of coefficient times' number of states. Then, these are used for calculating the mean and variance of all the matrices. The mean is calculated using Eq. (45.1)

$$\underset{c}{\overline{x}} = \frac{1}{N}\sum_{n=0}^{N-1} x_c(n), \quad c = column \tag{45.1}$$

The variance is calculated using Eqs. (45.2) and (45.3).

$$\underset{c}{\overline{x^2}} = \frac{1}{N}\sum_{n=0}^{N-1} x_c^2(n), \quad c = column \tag{45.2}$$

$$\sigma_c^2 = \underset{c}{\overline{x^2}} - \left[\underset{c}{\overline{x}}\right]^2, \quad c = column \tag{45.3}$$

A more explicit example of calculating a certain index, e.g., the x_Σ(1, 1) is done according to the following equation (the grayed element in x_Σ(m, n)).

$$\mathrm{x}\_\sum(1,1) = \frac{sum(x(1:12).^2)}{12} - x\_\mu(1,1)^2 \tag{45.4}$$

The HMM model for a word is defined by $\lambda = (\pi, \mathrm{A}, \mathrm{B})$.

- π the initial state distribution vector
- A the state transition probability matrix
- B the continuous observation probability density function

As per [3], HMM model used for speech recognition is left-to-right model. There are actually three steps involved in creating HMM model:

1. Initialization (π, A)
2. Re-estimation of the initialized parameterize training
3. Recognition

45.5.1 Initialization

π_i, Initialize The Initial State Distribution Vector, Using The Left-To-Right Model:

As per speech recognition theory, left-to-right model is used for speech recognition. Here, we are using 5-state left-to-right model. Before the starting of training process, the initial distribution vector is taken as below:

$\pi_i = [1 \quad 0 \quad 0 \quad 0 \quad 0]$, $1 \le \mathrm{i} \le$ number of states, and in this case, i = 5.

It indicates the process of training initially assumed probability for state one rest, and other states are at zero probability

Initialization of A: The state transition probability matrix, which use the left-to-right model. The state transition probability matrix, A is initialized with the equal probability for each state.

$$A = \begin{bmatrix} 0.5 & 0.5 & 0 & 0 & 0 \\ 0 & 0.5 & 0.5 & 0 & 0 \\ 0 & 0 & 0.5 & 0.5 & 0 \\ 0 & 0 & 0 & 0.5 & 0.5 \\ 0 & 0 & 0 & 0 & 1 \end{bmatrix}$$

During the experimentation with the number of iterations within the re-estimation of A, the final estimated values of A were shown to deviate quite a lot from the beginning of the estimation. The final initialization values of A were initialized with the following values instead, which is more likely to the re-estimated values.

$$A = \begin{bmatrix} 0.85 & 0.15 & 0 & 0 & 0 \\ 0 & 0.85 & 0.15 & 0 & 0 \\ 0 & 0 & 0.85 & 0.15 & 0 \\ 0 & 0 & 0 & 0.85 & 0.15 \\ 0 & 0 & 0 & 0 & 1 \end{bmatrix}$$

The change in initialization values is not a critical event; thus, the re-estimation adjusts the values to the correct ones, according to the estimation procedure.

B, The Continuous Observation Probability Density Function Matrix:

For hidden Markov model, the direct observation of the state of the speech process is not possible, and some statistical calculation is needed. This is done by introducing the continuous observation probability density function matrix, B. The idea is that there is a probability of making a certain observation in the state, the probability that the model has produced the observed mel-frequency cepstral coefficients. The discrete observation probability can be used alternatively.

The advantage of using continuous observation probability density functions is that the probabilities can be calculated directly from the MFCC. The commonly used distribution to describe the observation densities is the Gaussian distribution. This is also used in this project. The continuous observation probability density function matrix, B, is represented or defined by the mean, μ, and variance, Σ

$$b_j(o_t) = \sum_{k=1}^{M} c_{jk} b_{jk}(o_t) \quad , \quad j = 1, 2, \ldots, N \tag{45.5}$$

And M is the number of mixture weights, c_{jk}. These are restricted due to

$$\sum_{k=1}^{M} c_{jk} = 1 \quad , \quad j = 1, 2, \ldots, N \tag{45.6}$$

$$c_{jk} \geq 0 \quad , \quad j = 1, 2, \ldots, N, \ k = 1, 2, \ldots, M \tag{45.7}$$

One x_mfcc feature vector is in the estimation versus each μ and Σ vector, i.e., each feature vector is calculated for all x_μ and x_Σ columns one by one. This results in state-dependent observation symbol probabilities matrix. The columns give the observation probabilities for each state.

Log (P(O|λ)), The Probability Of The Observation Sequence: The log(P(O| λ)) is saved in a matrix to see the adjustment of the re-estimation sequence. For every iteration, there is a summation of the sum (log(scale)), total probability. This summation is compared to the previous summation in previous iteration. If the difference between the measured values is less than a threshold, then an optimum can be assumed to have been reached. If necessary, a fixed number of iterations could be set to reduce calculations.

45.5.2 Re-estimation of the Parameters for the Model, $\lambda = (\pi, A, B)$

The most difficult task is to adjust the model parameter to accurately represent the word under consideration. In training mode, a large amount of voice data (from different speaker) are taken in the form of feature vectors, which in our case is MFCCs. Using this, HMMs adjust its probability distribution and transition matrix. Every HMM must be trained to maximize its (local optimum) recognition power. Initially, HMM for word (before learning) consists of 5 states and its adjacency matrix and output probability distribution are initialized randomly. It gets automatically updated once the training process starts.

The recommended algorithm used for this purpose is the iterative Baum–Welch algorithm that maximizes the likelihood function of a given model $\lambda = (\pi, \mathrm{A}, \mathrm{B})$ [3]. For every iteration, the algorithm re-estimates the HMM parameters to a closer value of the “global” (exist many local) maximum. The importance lies in that the first local maximum found is the global; otherwise, an erroneous maximum is found. The Baum–Welch algorithm is based on a combination of the forward algorithm and the backward algorithm.

45.5.2.1 A_Re-est, Re-estimate the State Transition Probability Matrix

The ξ variable is calculated for every word in the training session. This is used with the γ variable, which is also calculated for every word in the training session [3].

The re-estimation of the A matrix is quite extensive due to the use of multiple observation sequences. For the collection of words in the training session, there is an average estimation calculated with contribution from all utterances used in training session. The following equation is used.

$$x_{ij}(i) = \frac{\exp ected\, number\ of\ transitions\, from\ state\ i\ to\ state\, j}{\exp ected\, number\ of\ transitions\, from\ state\ i} = \frac{\sum_{t-1}^{T-1} \xi_t(i,j)}{\sum_{t-1}^{T-1} \gamma_t(i)} \quad (45.8)$$

45.5.2.2 μ_reest, Re-estimated Mean

A new x_μ(m,n) is calculated, which is then used for the next iteration of the process. Note that it is the concatenated $\gamma_t(j,k)$ that is used.

45.5.2.3 Σ_reest, Re-estimated Covariance

A new x_Σ (m,n) is calculated, which is then used for the next iteration. Note that it is the concatenated $\gamma_t(j,k)$ that is used. After the re-estimation is done, the model is saved to represent the specific observation sequences, i.e., an isolated word. The model is then used for recognition. The model is represented with the following denotation λ = (A, μ, Σ).

45.5.3 Recognition

Unknown word is given, and the output probability is calculated. The role of the recognizer is to do mapping between sequences of speech vectors and the wanted underlying symbol sequences. Two problems that arise during the recognition are as follows:

1. As there are large variations in the speech waveform, because of speaker variability, noise, and echoes, the mapping from symbols to speech is not one to one. There are also the chances that different languages may share same symbols.
2. The boundaries between symbols cannot be identified explicitly from the speech waveform. Hence, it is difficult to treat the speech waveform as a sequence of concatenated patterns.

45.6 Language Modeling

The basic aim of language modeling is to predict the language, given a phoneme sequence or small unit of speech or word. To make a good prediction of a sequence, a model is required. A model of a sequence describes the sequence behavior and allows decisions or guesses on what symbol comes next. This project uses bi-gram to model the behavior of the text sequence.

Bi-gram: This is the probability of a word being followed by a given word or pair of word in a given language. The sequential information related to word which is also specific to a language is captured. This method is a special case of so-called n-Gram approach, and we can go till any value of n, but as you increase value of n, complexity increases exponentially. Therefore, we have calculated till n = 2.

$$\text{Bi\_Prob}(w2 \mid X, w1) = \frac{\text{No. of time word w1 is followed byword w2 in the data}}{\text{No. of times word w1 occurs in the training data of language X}}$$

A bi-gram is simply a sequence of two symbols and can be thought of as an example or record of a transition between two symbols. By recording enough bi-gram in a sequence, the long-run behavior of transition between words in sentences is recorded. A record of every bi-gram contained in a sequence of words and the number of times they occurred are called bi-gram. To build bi-gram model of a sequence, all bi-gram of that sequence are extracted, counted, and stored in a table. For example, the sequence $\langle a, a, b, a, b, a, b, a, a, b, a, b, a, b, b, a\rangle$ has 15 bi-gram

bi-gram	count
⟨a,a⟩	2
⟨a,b⟩	6
⟨b,a⟩	6
⟨b,b⟩	1

To calculate the bi-gram probability, first the uni-gram probability is calculated. The Fig. 45.2 shows language modeling flow.

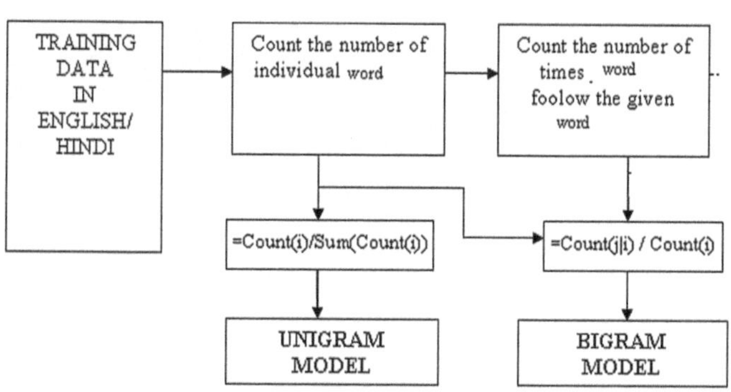

Fig. 45.2 Language modeling flow

45.7 Classifier

Maximum-likelihood classifier (MLC). In the system worked here, LID is performed by hypothesizing the language of the input utterance based on maximum-likelihood criterion. Probability evaluation log likelihood is done using the Alternative Viterbi Algorithm. The score according to the Viterbi algorithm is the highest probability that the compared model has produced for the given test utterance. It has been observed that the log-likelihood scores were biased in favor of one language over all other languages

45.8 Experiment and Results

Database used: English, French wave files from CSLU database and recorded Hindi wave files. We trained for three different languages, viz. English, Hindi, and French taking 20 wave files from each language.

1. Feature Extraction: Input Speech Signal (Continuous speech) (Figs. 45.3, 45.4 and 45.5).

Hidden Markov modeling training: We have used five-state left-to-right model for training.

Language modeling: We have used bi-gram language modeling.

Classifier: Maximum-likelihood classifier has been used (Tables 45.1 and 45.2).

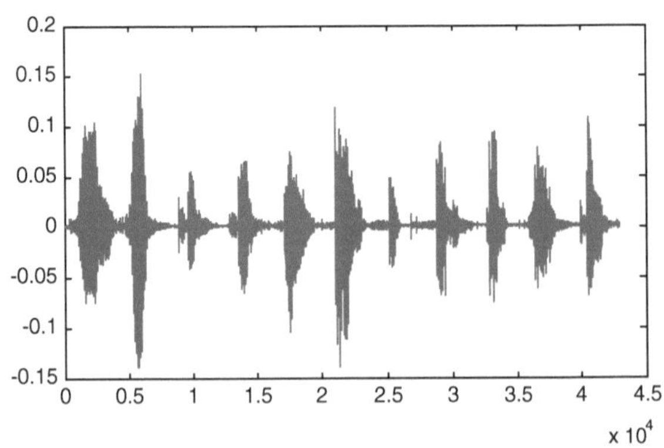

Fig. 45.3 Continuous speech signal

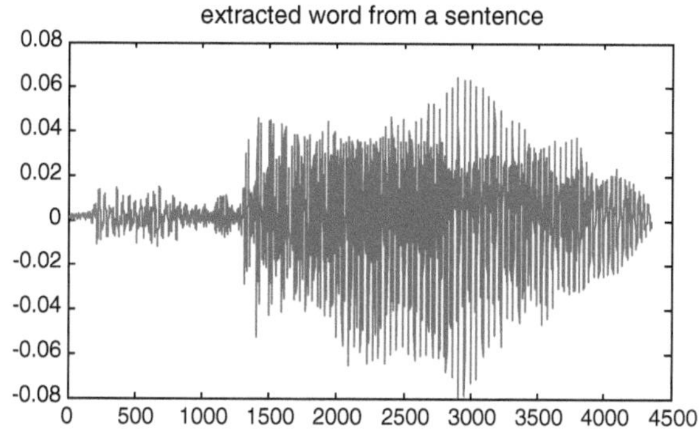

Fig. 45.4 Word from continuous speech

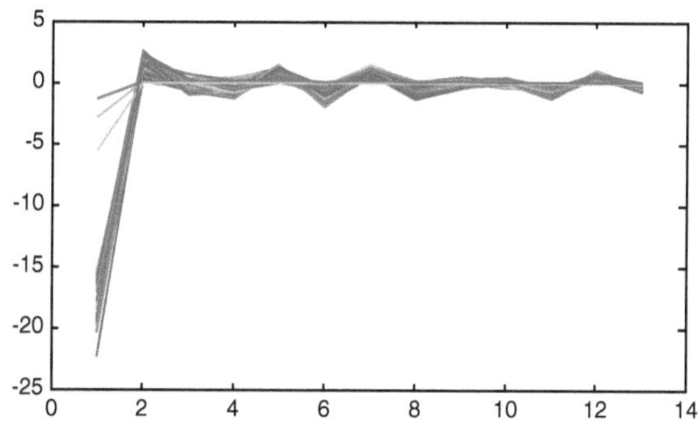

Fig. 45.5 MFCC of speech signal

Table 45.1 Test data belonging to training data

Languages	Number of HMM states	Language model	Result (%)
English	5	Bi-gram	85
French	5	Bi-gram	83
Hindi	5	Bi-gram	86

Table 45.2 Test data not belonging to training data

Languages	Number of HMM states	Language model	Result (%)
English	5	Bi-gram	75
French	5	Bi-gram	72
Hindi	5	Bi-gram	75

References

1. Dai Jayram AKV, Ramasubramanian VV, Sreenivas TV (2003) Language identification using parallel phone recognition. In: Proceedings of IEEE International Conference on Acoustics, Speech Signal Processing. pp 32–35
2. Muthusamy YK (1993) A segmental approach to automatic language identification. PhD thesis, Oregon Graduate Institute of Science and Technology, October 1993
3. Rabiner LR, Juang B-H (1993) Fundamentals of speech recognition. Prentice Hall, US

Chapter 46
Discrimination of Handwritten and Machine Printed Text in Scanned Document Images

Surabhi Narayan and Sahana D. Gowda

Abstract Discrimination of handwritten and machine printed text in a scanned document image is an important process as the Optical Character Recognizers (OCRs) available are domain specific. In this paper, a novel approach has been proposed to discriminate handwritten and machine printed word components based on the structure. In the binarized form of the word component, due to the informative foreground overlay on the null background, transitions from 0-1 and 1-0 occur at the contour of the component structure. The count and occurrence of these transitions are used to discriminate handwritten and machine printed word components. The proposed method is robust and simple. Extensive experimentation has been conducted over a wide range of data samples (English words).

Keywords Document image · Machine printed word · Handwritten word · Transition · Component structure · Critical lines

46.1 Introduction

Despite the fact that offices are moving into paperless, document existence is still necessary in many places such as reservation forms, bank cheques, application forms, tax returns, and legal documents. These document images contain handwritten and machine printed word components. Information extraction from these document images is cumbersome due to the presence of handwritten and machine printed components together. Discrimination of handwritten and machine printed

---

S. Narayan (✉) · S. D. Gowda
Department of CSE, BNM Institute of Technology, Bangalore, India
e-mail: surabhi.narayan@gmail.com

S. D. Gowda
e-mail: sahanagowda@rediffmail.com

V. Chakravarthi et al. (eds.), *Proceedings of International Conference on VLSI, Communication, Advanced Devices, Signals & Systems and Networking (VCASAN-2013)*, Lecture Notes in Electrical Engineering 258, DOI: 10.1007/978-81-322-1524-0_46,

Fig. 46.1 Printed word showing the four critical lines

text components is one of the major processing steps to facilitate the later OCR task. Based on the structure, the characters are classified as ascenders, descenders, and base-line characters in a word component [1–3].

In a machine printed word component, the characters occupy uniform space and the ascender and descender lines of the characters have the same height, whereas this is not true in case of handwritten components. In the proposed model, structure of the components is analyzed using transitions. Due to the overlay of informative foreground on the null background in the binarized document image, 1-0 and 0-1 transitions occur [4]. The transitions in the row and column are termed as row transition and column transition, respectively. Based on ascender, descender, and base-line structure of the character and the occurrence of transitions in the word components, critical lines can be drawn. Critical lines are top-line, x-line, base-line, and bottom-line as shown in Fig. 46.1.

For a machine printed word component, due to the uniform height and structure of the character, the top-line, x-line, base-line, and bottom-line can be clearly defined. Due to non-uniformity of the structure and height of characters in the handwritten components, top-line, x-line, base-line, and bottom-line do not fit the word structure. The detailed procedure is given in Sect. 46.3 with the rule-based engine to identify the handwritten and machine printed text.

The rest of the paper is organized as follows: Sect. 46.2 signifies the state of art, Sect. 46.3 focuses on the proposed model, Sect. 46.4 depicts the experimental results, and Sect. 46.5 gives the conclusion.

46.2 Literature Survey

Document images are processed for information extraction. The document images can be processed at line, word, or character level depending on the requirement. Discrimination of handwritten and machine printed components is also done at any of the above three levels [5]. Discrimination of handwritten and machine printed text is an offline processing problem, which needs many preprocessing stages. Assuming that the document image has undergone preprocessing stages, works have been reported only for the discrimination of handwritten and machine printed text components by many researchers.

Kavallieratou and Stamatatos [6] analyzed the horizontal profile of the text and found that the machine printed text lines have relatively stable height as compared to the handwritten text lines. Fan et al. [7] considered word block layout and applied Fisher classifier for discrimination. Guo J.K and Ma [8] used a hidden

Markov Model to classify the machine printed text and handwritten text. Tse et al. [5] classified the typewritten and handwritten portions using island grouping method. Angel Sanchez et al. [9] used classification rules mining by the WEKA tool. Farroq et al. [10] used Expectation Maximization-based probabilistic neural network for identifying the handwritten words in an Arabic document. Shirdhonkar et al. [11] used SVM to isolate signatures from the machine printed text.

The existing methods consider complex and large number of features and classifiers to discriminate the machine printed and handwritten components with fixed thresholds on the height and width of the components. In the proposed model, we seek to overcome the limitations of the existing methods by considering transitions and position of transitions as the discriminating features for handwritten and printed components which is invariant to size and width of the components.

46.3 Proposed Method

Every character component has a unique structure. The combination of characters and their position gives a new structure to the word. The word components extracted from the document image are binarized since the foreground information is always in contrast with the null background. Representation in the binary form makes the image matrix simple. The length and width of the component are the number of rows and columns in the image matrix. Due to the presence of 0's and 1's in the image matrix, transitions occur. There are two types of transitions: positive transitions and negative transitions. Positive transitions (t^+) are transitions from informative foreground (0) to null background (1) and negative transitions (t^-) are transitions from null background (1) to informative foreground (0). Based on the structure of the word component, transitions occur at the contour of the characters as depicted in Fig. 46.2b for the component in Fig. 46.2a.

Critical lines are defined based on the occurrence of transitions.

top-line: The line that marks the beginning of an ascender character.

x-line: The line that marks the beginning of the base-line and the descender characters.

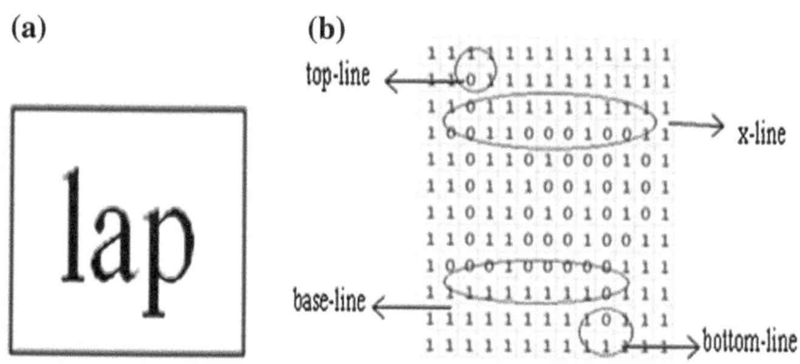

Fig. 46.2 **a** Printed word component. **b** Image matrix with transitions

base-line: The line that marks the end of base-line and the ascender character.
bottom-line: The line that marks the end of descender character.

Based on the occurrence of critical lines, word components are categorized into four types:

Type 2: The word component with top-line, x-line, base-line, and bottom-line.
Type 3: The word component with top-line, x-line, and base-line.
Type 4: The word component with x-line, base-line, and bottom-line.
Type 5: The word component with x-line and base-line.

46.3.1 Computation

For an image matrix of size m × n, where m is the number of rows and n is the number of columns,

Compute

$$r_j^+ = \sum_{i=1}^{n} ct^+ \text{ for all j} = 1 \text{ to m}$$

$$r_j^- = \sum_{i=1}^{n} ct^- \text{ for all j} = 1 \text{ to m}$$

where

ct^+ is the positive column transition
ct^- is the negative column transition
r_j^+ is the number of positive column transitions in the jth row of the image matrix
r_j^- is the number of negative column transitions in the jth row of the image matrix

Based on the number of column transitions in the corresponding rows, critical lines are marked.

top-line: Row with the first ct^- termed as r_t,
bottom-line: Row with the last ct^+ termed as r_{bt},
x-line: Row with $max(ct^-)$ termed as r_x and
base-line: Row with $max(ct^+)$ termed as r_b.

The critical lines are invariant to the size of the word component.

46.3.2 Rule Engine

Rule engine is developed for the discrimination of machine printed and handwritten word components. It is based on the following:

- Position of critical lines.
- Distance between the critical lines.
- Distance between the middle row of the component and the critical lines.

Rules are tabulated in Table 46.1.

r_t: Row number of the top-line.
r_x: Row number of the x-line.
r_b: Row number of the base-line.
r_{bt}: Row number of the bottom-line.

Every input word component is subjected to discrimination using the rules defined in Table 46.1. If the input component is machine printed, then it satisfies one of the rules and is labeled as machine printed component. If the input component is handwritten, it does not satisfy any of the rules and is labeled as handwritten component. The detailed experimental results are discussed in the next section.

46.4 Experimental Analysis

Extensive experimentation has been conducted on different types of word components. Word images were generated by different manuscripts. The analysis is

Table 46.1 Rule-based engine for discrimination

Rule 1	1. If $r_t \neq r_x$ and $r_b \neq r_{bt}$ Calculate mid $= (r_{bt} - r_t)/2$ 2. If $r_t <$ mid and $r_x <$ mid and $r_b >$ mid and $r_{bt} >$ mid 3. If $(r_x - r_t) == (r_{bt} - r_b)$	If all three conditions are satisfied identify the word as printed word component of type 1
Rule 2	1. If $r_t == r_x$ and $r_b \neq r_{bt}$ Calculate mid $= (r_{bt} - r_t)/2$ 2. If $r_t <$ mid and $r_b >$ mid and $r_{bt} >$ mid	If both conditions are satisfied, identify the word as printed word component of type 2
Rule 3	1. If $r_t \neq r_x$ and $r_b == r_{bt}$ calculate mid $= (r_{bt} - r_t)/2$ 2. If $r_t <$ mid and $r_x <$ mid and $r_{bt} >$ mid	If all four conditions are satisfied, identify the word as printed word component of type 3
Rule 4	1. If $r_t == r_x$ and $r_b = = r_{bt}$	If the condition is satisfied, identify the word as printed word component of type 4

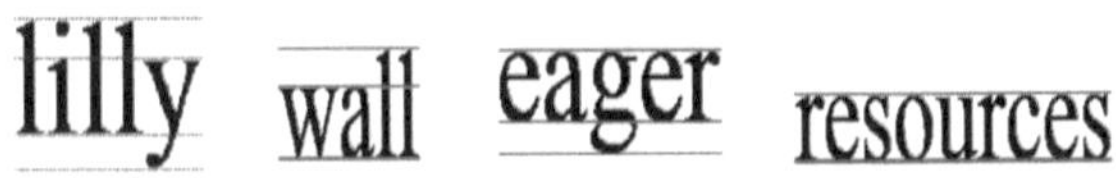

Fig. 46.3 Printed word components

Fig. 46.4 Handwritten word components

shown on the word which is machine printed and handwritten by different people (Figs. 46.3, 46.4).

The results evidently show that the structure of the machine printed components at character level within a word exhibit uniform space occupancy, whereas in handwritten components, the uniformity does not exist due to the individual style of writing. In handwritten word component, the top-line, x-line, base-line, and bottom-line do not indicate uniformity in width and height. The proposed model was tested on 200 machine printed and 200 handwritten words. Some miss-identification was recorded in case of printed word components. Word components which contain only character 't' as the ascender are miss-identified as handwritten since height of the ascender for any font size is smaller than any other ascender or descender characters. It does not satisfy the rules.

46.5 Conclusion

Discrimination of machine printed and handwritten text is a prerequisite for document processing. In this paper, a new method is proposed which considers the structure, occupancy, and uniformity of characters in the word component. Critical lines are drawn on the word component based on the count and occurrence of the transitions. The rule engine is developed for discrimination of handwritten and machine printed word components. The proposed method is simple and less time consuming.

Acknowledgments This work has been supported by Visveswaraya Technological University (VTU) under the Research Grant Scheme 2010-11 (Ref. No VTU/Aca./2011-12/A-9/13097). The authors acknowledge the support provided by VTU.

References

1. Doermann D, Li H, Kia O, Kilic K (1997) Detection of duplicates in document image databases. Technical report, February 1997
2. Hull JJ (1994) Document image matching and retrieval with multiple distortion-invariant descriptors. In: Proceedings of the international workshop on document analysis systems, pp 383–400
3. Spitz AL (1995) Using character shape codes for word spotting in document images. In: Proceedings of shape, structure and pattern recognition, World Scientific, Singapore, pp 382–389
4. Gowda SD, Nagabhushan P (2007) Equivalence between two document images in terms of geometry and entropy. In: Proceedings of advanced computer vision and information technology, Aurangabad, INDIA, pp 210–216
5. Tse J, Curtis D, Bunch J, Jones C, Yfantis EA, Thomas A (2007) Handwritten and typewritten word and character separation in unconstrained document images. Proceedings of IPCV
6. Kavallieratou E, Stamatatos S (2004) Discrimination of machine-printed from handwritten text using simple structural characteristics. In: Proceedings of 17th international conference on pattern recognition, vol I, 2004
7. Fan KC, Wang LS, Tu YT (1998) Classification of machine—printed and handwritten texts using character block layout variance. Pattern Recognit 31(9):1275–1284
8. Guo JK, Ma MY (2001) Separating handwritten material from machine printed text using hidden Markov models. In: Proceedings of 6th international conference in document analysis and recognition
9. da Silva LF, Conci A, Sanchez A (2009) Automatic discrimination between printed and handwritten text in documents. In: Proceedings of computer graphics and image processing, 2009
10. Farooq F, Sridharan K, Govindaraju V (2006) Identifying handwritten text in mixed documents. In: Proceedings of IEEE 18th international conference on pattern recognition (ICPR)
11. Shirdhonkar MS, Kokare MB (2010) Discrimination between printed and handwritten text in documents. In: Proceedings of IJCA special issue on "recent trends in image processing and pattern recognition

Chapter 47
Harmonic Analysis and its Elimination in Nonlinear Loads Using Wavelet Transform

H. A. Vidya, S. Priyashree and K. Venkatesha

Abstract The objective of the electric utility is to deliver sinusoidal voltage at fairly constant magnitude throughout their system. As the load is more nonlinear, complexities also increase since these nonlinear loads produce harmonic currents. These harmonic currents result in voltage and current distortions that can adversely impact the system performance in different ways. With the increase in harmonic producing loads at the installation, severity to address the issue also increases. As an impact of these phenomena, two important concepts are to be borne in mind with regard to power system harmonics. They are harmonic current generated due to nonlinear loads such as computers and printers, and the second is the effect of harmonic currents [1, 2]. In this work, the harmonics generated due to computer and printer is analyzed using power quality analyzer and harmonics are eliminated using wavelet transform.

Keywords Daubechies 19 · Harmonics · Nonlinear loads · Power quality · Power quality analyzer · Wavelet transform

47.1 Introduction

Power quality (PQ) is an important issue in concern with today's energy scenario. The PQ problem causes the deterioration of performance of various sensitive electrical and electronic equipments. Wide use of nonlinear devices such as

H. A. Vidya · S. Priyashree (✉) · K. Venkatesha
Department of EEE, BNM Institute of Technology, Banashankari II stage, 7087 Bangalore 560070, India
e-mail: priyashree_suresh@yahoo.com

H. A. Vidya
e-mail: vidyakrishna_ag@yahoo.co.in

V. Chakravarthi et al. (eds.), *Proceedings of International Conference on VLSI, Communication, Advanced Devices, Signals & Systems and Networking (VCASAN-2013)*, Lecture Notes in Electrical Engineering 258, DOI: 10.1007/978-81-322-1524-0_47,

rectifiers (power supplies, UPS units, discharge lighting), adjustable speed motor drives, ferromagnetic devices, DC motor drives and arcing equipment, battery chargers, electronic ballasts, and switching mode power supplies inject harmonics into the power supply. Voltage harmonic frequencies produce additional heat in the motor that increases eddy current and hysteresis losses. Potential sources of harmonic are as follows: switched mode power supplies, current regulators, frequency converter, voltage source inverters with pulse width modulated converters, low-power consumption lamps, electrical arc furnaces, and equipments with built-in switching devices or with internal loads with nonlinear voltage/current characteristics. In order to get good quality of power, demand-side management, analysis of harmonics generated, and elimination of harmonics are necessary [1].

Kamal Al-Haddad et al. proposed a new control scheme of a three-phase active power filter (APF) to eliminate harmonics, to compensate reactive power and neutral currents, and to remedy system unbalance, in a three-phase four-wire electric power distribution system, with unbalanced nonlinear loads. The APF is found effective to compensate reactive power and neutral current while eliminating harmonics with load balancing [2].

These PQ disturbances are detected, classified, and analyzed using power quality analyzer (PQA). Such PQ problems can be mitigated using suitable filters [3–5]. The analysis of different PQ problems can also be done simultaneously in both time and frequency domains. Wavelet transform (WT) can satisfy this type of analysis where multiresolution signal decomposition provides valuable information about location and type of PQ problems [6, 7].

Basu et al. presented the application of continuous wavelet transform to identify power system transients and disturbances. A mother wavelet basis function with dyadic nonoverlapping frequency bands has been used. A few power signals with PQ disturbances have been analyzed to show the effectiveness of the proposed technique. The paper highlights an alternative way of PQ signal analysis, which enables to identify the duration of disturbance accurately. Classification of various power quality disturbances is also possible through this time-frequency-based wavelet analysis [8].

47.2 Harmonics and its Effects

The term harmonic component of a waveform is defined as that component which occurs at an integer multiple of the fundamental frequency. According to Fourier theory, any repetitive waveform can be defined in terms of summing sinusoidal waveforms which are integer multiples (or harmonics) of the fundamental frequency that can be expressed as given in Eq. 47.1:

$$f(t) = \sum_{n=0}^{\infty} A_n \times \sin\left(\frac{n\pi t}{T}\right) \tag{47.1}$$

where f (t) is the time-domain function, n is the harmonic number (only odd values of n are required), A_n is the amplitude of the nth harmonic component, and T is the length of one cycle in seconds.

Total harmonic distortion (THD) [3] is used to describe voltage or current distortion and is calculated by Eq. 47.2.

$$\mathrm{THD} = \frac{\sqrt{(V_2^2 + V_3^2 + V_4^2 + \ldots + V_n^2)}}{V_1} * 100\,\% \tag{47.2}$$

47.3 Wavelet Transform

A wavelet is a transient waveform that has a zero average value and decays quickly to zero. Due to its transient nature, the analysis of transient signals with a wavelet is naturally more descriptive than that of periodic functions such as sine and cosine bases. The wavelet coefficients work as weights of the wavelets to represent the signal at these locations and scales. Wavelet transform can be performed on continuous and discrete functions, respectively, referred to as the continuous wavelet transform (CWT) and discrete wavelet transform (DWT).

47.3.1 Wavelet-Based Denoising

Wavelet decomposition theory is used to concentrate the signal energy to a small number of large coefficients. The coefficients, which fall into a rejection band, are dropped and those that are not located within the band are reserved or shrunk. The signal is usually decomposed into a few large coefficients, whereas the noise component gives rise to small coefficients only. This property of the wavelet transformation helps to suppress the noise part of the signal.

The general wavelet-based denoising method proceeds in three steps [9]: Step (1) decomposing the noisy signal, Step (2) thresholding the wavelet coefficients, and Step (3) reconstructing the signal as shown in Fig. 47.1

For efficient noise cancelation, wavelet denoising methods can be carried out using either hard or soft thresholding. Hard thresholding processes data in such a way that those wavelet coefficients whose absolute values (x) are greater than the threshold (λ) are kept and those less than the threshold are set to zero (Eq. 47.3):

$$\delta_\lambda^H = \begin{cases} x & if|x| > \lambda \\ 0 & if|x| \le \lambda \end{cases} \tag{47.3}$$

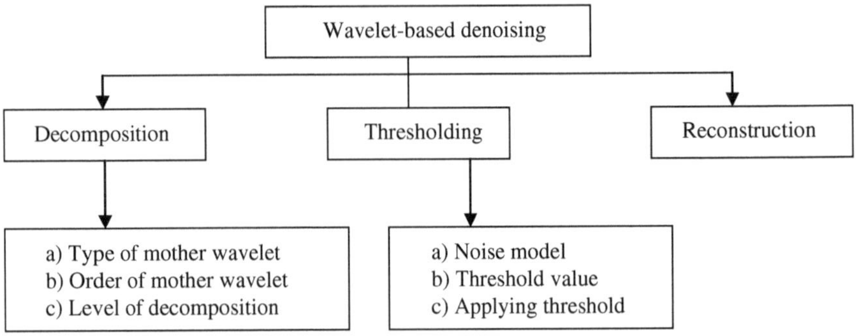

Fig. 47.1 Wavelet-based denoising procedure

Soft thresholding sets the wavelet coefficients below the threshold to zero. The coefficients greater than threshold are kept and then shrunk toward zero (Eq. 47.4)

$$\delta_{\lambda}^{S}(x) = \begin{cases} x - \lambda & if|x| > \lambda \\ 0 & if|x| \leq \lambda \\ x + \lambda & if|x| > -\lambda \end{cases} \tag{47.4}$$

The success of the noise cancellation can be assessed using signal-to-noise ratio (SNR) defined by Eq. 47.5. The concept of denoising is used to improve the SNR. If the value of SNR is positive, then greater is the power of the signal as compared to the noise, and for a negative value of SNR, greater is the power of noise as compared to the signal. Another measure evaluates the mean squared error (MSE) between the input signal and denoised signal as defined in Eq. 47.6 and minimizes it. The denoising action is guided by a reasonable minimum MSE and reduction in pulse amplitude as defined in Eq. 47.7. In simulation study, since a pulse sequence is used where the percentage reduction in the peak amplitude of the pulse recovered after denoising with respect to the corresponding pulse in the input sequence is calculated.

$$\mathrm{SNR} = 10 * \log \frac{\sum_{i=1}^{N} Y(i)^2}{\sum_{i=1}^{N} (X(i) - Y(i))^2} \tag{47.5}$$

$$\mathrm{MSE} = \sum_{i=1}^{N} \frac{(Z(i) - Y(i))^2}{N} \tag{47.6}$$

$$\%\ \text{Reduction} = \frac{X - Y}{X} * 100 \tag{47.7}$$

where X(i) is the original signal, Y(i) is the denoised signal, Z(i) is the noisy signal acquired, N is the number of pulses, X is the peak amplitude (positive going peak

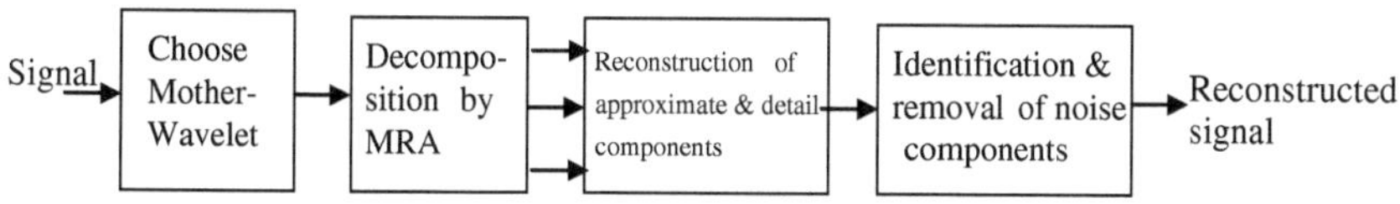

Fig. 47.2 Denosing by MRA

pulse in this case) of the original (reference) signal, and Y obtained from Eq. 47.7 is the peak amplitude of the recovered denoised pulse.

47.3.2 Proposed Technique

This type of denoising consists of two phases. In the first phase, input signal is decomposed into approximate and detail components up to a desired number of levels with the help of multiresolution analysis (MRA). This can be done by first choosing a mother wavelet according to the signal characteristics, and decomposition can be carried out by scaling and dilating the mother wavelet. Once the approximate and detail components are computed, the second phase starts. This phase involves identifying the Filtered Signal. Block diagram representation of denoising by MRA is shown in Fig. 47.2.

47.4 Experimental Setup

Experimental setup for analysis of harmonics generated by the nonlinear loads such as computer and printer is shown in Fig. 47.3. PQA PW3198 is used to analyze harmonics, and it is eliminated using wavelet transforms.

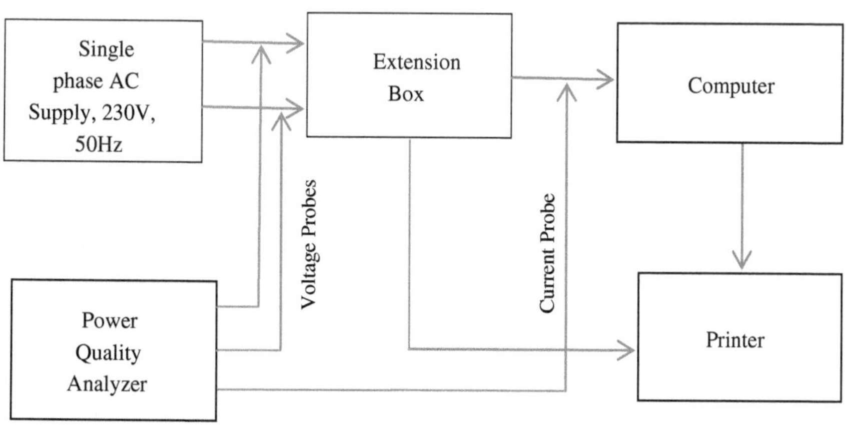

Fig. 47.3 Experimental setup with computer and printer as load

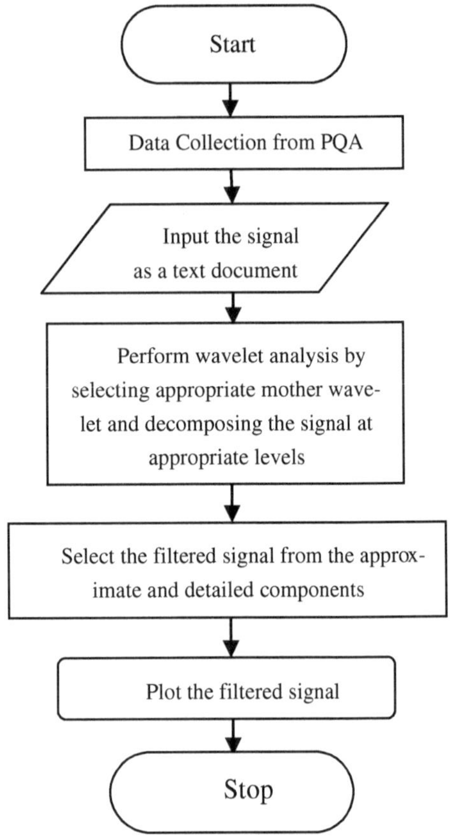

Fig. 47.4 Harmonic analysis using wavelet transform

47.5 Flow Diagram

Data collected from PQA are further analyzed to obtain a filtered signal using wavelet transform as mentioned in Fig. 47.4.

47.6 Results and Discussion

Set 1: Computer load combination of Internet, MS PowerPoint, MS Excel applications

Current drawn: 0.698A, Power consumed: 111 W, Current THD: 84.32 %, RPA = 7.9 %

Approximate component at 7th level is taken as the filtered signal as it is a fully denoised signal and pure sine waveform as shown in Fig. 47.5 (Table 47.1).

Set 2: Printer load with print operation

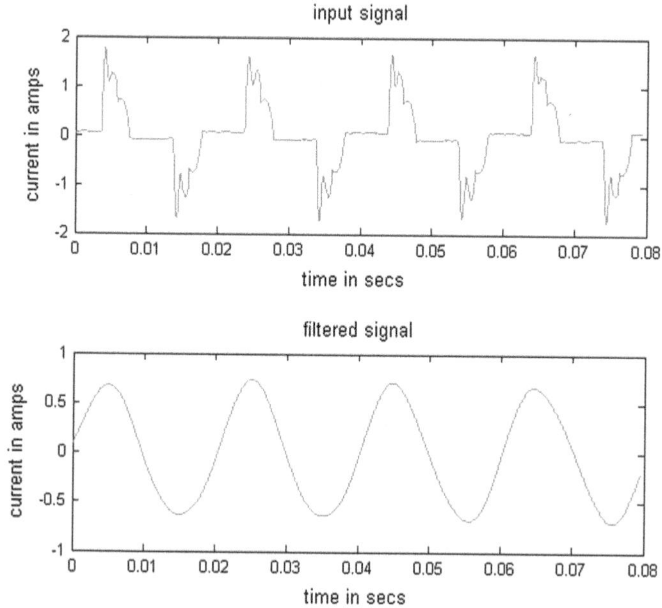

Fig. 47.5 Waveforms of input signal and filtered signal

Table 1 Harmonic analysis using power quality analyzer

Application	I_{fund} (amps)	V_{fund} (volts)	V_{THD} (%)	I_{THD} (%)	Voltage harmonics list (order)			Current harmonics list (order)			
					5	7	11	3	5	7	9
Desktop	0.465	228.5	3.25	87.8	5.38	2.2	3.87	0.33	0.154	0.097	0.105
MS Excel	0.711	228.0	3.34	86.3	5.49	2.28	3.91	0.44	0.166	0.12	0.125
Working on MS Excel	0.56	229.9	3.19	85.41	5.93	2.6	2.8	0.342	0.156	0.117	0.128
PowerPoint presentation	0.483	229.2	3.18	82.21	5.83	2.5	2.72	0.327	0.136	0.109	0.107
Slideshow	0.492	228.5	3.22	79.17	6.01	2.84	2.66	0.356	0.135	0.107	0.103
MATLAB	0.492	229.3	3.25	80.87	6.29	2.78	2.52	0.325	0.128	0.111	0.103
Gmail	0.504	226.4	3.42	82.19	6.32	2.69	2.77	0.334	0.113	0.113	0.122
Net, ppt, MS Excel	0.488	228.6	3.4	84.32	5.93	2.46	2.53	0.333	0.145	0.103	0.115
Multiple tabs in net	0.488	227.3	2.63	83.74	6.36	2.7	2.61	0.389	0.144	0.124	0.127
Program run in C++	0.570	227.6	3.39	70.14	6.26	2.87	2.7	0.356	0.089	0.106	0.108
Printer (on state)	0.088	228.8	3.54	87.8	5.7	2.3	2.6	0.347	0.123	0.09	0.128
Printer (working state)	0.760	226.6	2.94	98.49	6.53	2.47	2.82	0.367	0.105	0.087	0.143

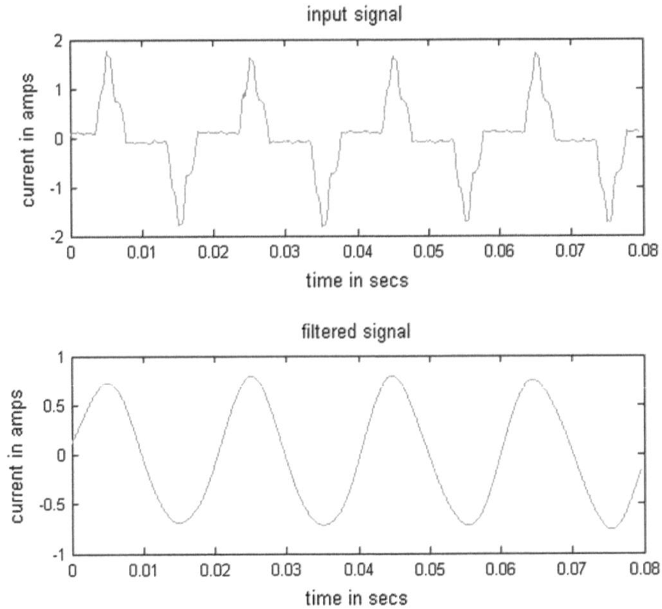

Fig. 47.6 Waveforms of input signal and filtered signal

Current drawn: 0.7607 A, Power consumed: 121.8 W, Current THD: 98.49 %, RPA = 2.9 %

Approximate component at 7th level is taken as the filtered signal as it is a fully denoised signal and pure sine waveform as shown in Fig. 47.6.

Observation:

(1) The 5th, 7th, and 11th order voltage harmonics were found to be dominant with voltage THD within IEEE standard ±3 %.
(2) The 3rd, 5th, 7th, and 9th order current harmonics were found to be dominant with current THD not within IEEE standard ±5 %.
(3) Power consumption depends on the type of the application. Ex: MS Excel—159 W, MATLAB—377 W.
(4) Power consumption increased with Internet application tool bar, and power consumption was found to be less with multiple tabs.

47.7 Conclusion

Importance of PQ, its problems, and need for harmonic elimination are discussed. Computer and printer are considered as nonlinear loads, and current harmonics are eliminated using wavelet tool in MATLAB. Daubechies 19 has been used as the mother wavelet by trial and error and successfully obtained filtered pure sine wave.

References

1. Ellis RG Power system harmonics, a reference guide to causes, effects and corrective measures. http://literature.rockwellautomation.com/idc/groups/literature/documents/wp/mvb-wp011_-en-p.pdf
2. Chandra A, Al-Haddad K, Singh B (1998) Harmonic elimination, reactive power compensation and load balancing in 3 phase 4 wire electric distribution system supplying nonlinear loads. Electr Power Syst Res 44(2):93–100
3. Blooming TM, Carnovale DJ (2007) Application of IEEE Std 519-1992 harmonic limits. IEEE IAS Atlanta section, September 17
4. Kim SS, Enjeti PN (2002) A new hybrid active power filter (APF) topology. IEEE Trans Power Electron 17(1):48–54
5. Attia HA, El-Metwally M, Fahmy OM (2010) Harmonic distortion effects and mitigation in distribution systems. J Am Sci 6:173–183
6. Baggini A (ed) (2002) Wavelet network based detection and classification of transients, Handbook of power quality. University of Bergamo, Italy 14PSCC, SEVILLA, 24–28 June 2002
7. Angrisani L, Daponte PD, Apuzzo M, Testa A (1998) A measurement method based on the wavelet transform for power quality analysis. IEEE Trans Power Deliv 13(4):990–998
8. Basu M, Basu B (2007) Analysis of power quality (PQ) signals by continuous wavelet transform 2007. IEEE power electronics specialists conference, June 2007
9. Vidya HA, Krishnan V (2009) Some investigations on denoising of partial discharge signals using wavelet transforms. PhD Thesis, VTU, Belgaum

Chapter 48
Lanczos Resampling for the Digital Processing of Remotely Sensed Images

B. N. Madhukar and R. Narendra

Abstract This paper presents theoretical and practical application of a relatively unknown and rare image resampling technique called Lanczos resampling. Application of this method on satellite remote sensing images is considered. Image resampling is the mathematical technique used to create a new version of the image with a different width and/or height in pixels. Interpolation is the process of determining the values of a function at positions lying between its samples. Sampling the interpolated image is equivalent to interpolating the image with a sampled interpolating function. Image registration is the process of overlaying two or more images of the same scene taken at different times, from different viewpoints, and/or by different sensors. It geometrically aligns two images: the reference and sensed images. In the interaction between interpolation and sampling processes, aliases occur on some occasions. Majority of the registration methods consist of the steps like feature detection, feature matching, transform model estimation and image resampling and transformation. The proprietary softwares that are commercially available for image processing that are capable of doing image registration do not provide us with performance metrics for assessing the resampling methods used. Lanczos resampling method has not been used in the digital processing of remotely sensed satellite images by any of the open source and the proprietary software packages that are available until now. In this paper, we have applied performance metrics (on satellite images) for analyzing the performance of Lanczos resampling method. Comparison of Lanczos resampling method with other resampling methods, such as nearest neighborhood resampling, and sinc resampling, is done based on the metrics pertaining to entropy, mean relative error, and time. We propose that Lanczos resampling method to be a good

B. N. Madhukar (✉)
ECE Department, Raja Rajeswari College of Engineering (RRCE), No. 14, Ramohalli Cross, Kumbalagodu, Mysore Road, Bangalore 560074, Karnataka, India
e-mail: madhukar_bn@rediffmail.com

R. Narendra
TCE Department, BNMIT, Bangalore 560070, Karnataka, India
e-mail: rnarendra29@gmail.com

V. Chakravarthi et al. (eds.), *Proceedings of International Conference on VLSI, Communication, Advanced Devices, Signals & Systems and Networking (VCASAN-2013)*, Lecture Notes in Electrical Engineering 258, DOI: 10.1007/978-81-322-1524-0_48,

method from qualitative and quantitative point of view when compared to the other two resampling methods. Also, it proves to be an optimal method for image resampling in the arena of remote sensing when compared to the other methods used. This, we hope, will enhance the understanding of the classified images' characteristics in a quantitative manner.

Keywords Resampling · Interpolation · Sinc · Lanczos · Window

48.1 Lanczos Resampling

Lanczos resampling, invented by Cornelius Lanczos, is an interpolation function that is used extensively in the arena of digital signal processing. It is basically a Fourier kernel. Its essentiality is for smoothly interpolating the value of a digital signal between its samples. Each of the given signal's samples is effectively mapped to give a translated and scaled copy of the Lanczos kernel. A Lanczos kernel is nothing but a sinc function apodized by the central hump of a dilated sinc function [1]. The sum of these shifted and scaled kernels is then evaluated at the requisite points. Lanczos resampling is also referred to as Lanczos filter. Lanczos resampling finds application for incrementing the sampling rate of a digital signal. It finds application in digital image processing for performing multidimensional interpolation, and it provides excellent results amongst several filters that exist in the literature. The baring effect that each input sample may cast on the interpolated values is defined and obtained by the filter's regeneration or reconstruction kernel *L*(*x*), which is renowned as the Lanczos kernel [2]. It is given by the normalized sinc function, apodized by the Lanczos window, or sinc window, which is the central lobe of a horizontally stretched sinc function *sinc*(*x*/*a*) for $-a \leq x \leq a$.

$$L(x) = \begin{cases} sinc(x).sinc\left(\frac{x}{a}\right), & -a < x < a \\ 0 & elsewhere \end{cases} \tag{48.1}$$

Note that $sinc(x) = \frac{\sin \pi x}{\pi x}$. This function has a value of 1 at $x = 0$ [3]. Simplifying the above equation, we get

$$L(x) = \begin{cases} 1, & x = 0 \\ \frac{a \sin(\pi x) \sin\left(\frac{\pi x}{a}\right)}{\pi^2 x^2}, & 0 < |x| < a \\ 0, & otherwise \end{cases} \tag{48.2}$$

The parameter *a* takes only positive integral values. A typical value of 2 or 3 determines the size of the kernel. The Lanczos kernel is having $(2a - 1)$ lobes: a positive lobe is located at the center, and $(a - 1)$ alternating negative and positive lobes are located on each side [4].

Consider a 1D signal which is having samples s_i for integral values of i. If $S(x)$ represents the interpolated value at an arbitrary real argument x, then it is obtained by the discrete-time linear convolution of the samples s_i with the Lanczos kernel. It is given by the following equation mentioned below:

$$S(x) = \sum_{i=\lfloor x\rfloor-a+1}^{\lfloor x\rfloor+a} s_i L(x-i) \tag{48.3}$$

Note that a represents the filter size parameter. The bounds of this sum are there which ensure that the kernel is zero outside of them. The function can also get negative values [5]. If the parameter a is a positive integer, then the Lanczos kernel is continuous everywhere. If this property is maintained, then its derivative also exists and is continuous everywhere even at $x = \pm a$, where both the sinc functions go to zero. Because of the above property, the regenerated or reconstructed signal $S(x)$ is also continuous with their differential coefficients also being continuous. The Lanczos kernel has a zero value at every integer argument x, excepting at the origin, i.e., at $x = 0$, where it has a value of unity. Hence, the regenerated signal interpolates the given samples accurately. Thus, we get $S(x) = s_i$ for every integer argument $x = i$ [6] (Fig. 48.1).

In 2D, the Lanczos kernel is defined by the product of two one-dimensional kernels

$$L(x, y) = L(x).L(y) \tag{48.4}$$

Consider a 2D signal s_{ij} which is defined at integer points (i, j) of the plane [7]. Then, reconstructed or regenerating function is given by the following definition

$$S(x, y) = \sum_{i=\lfloor x\rfloor-a+1}^{\lfloor x\rfloor+a} \sum_{j=\lfloor y\rfloor-a+1}^{\lfloor y\rfloor+a)} s_{ij} L(x-i).L(y-j) \tag{48.5}$$

If we resample a 2D signal at regularly spaced points (x, y), computation cost and computation time can be saved and high throughput can be accomplished by applying resampling technique on the full signal along a single axis and then

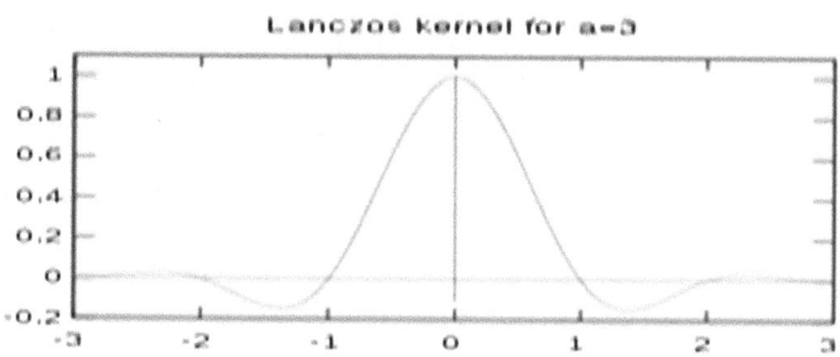

Fig. 48.1 Lanczos kernel for $a = 3$ [2]

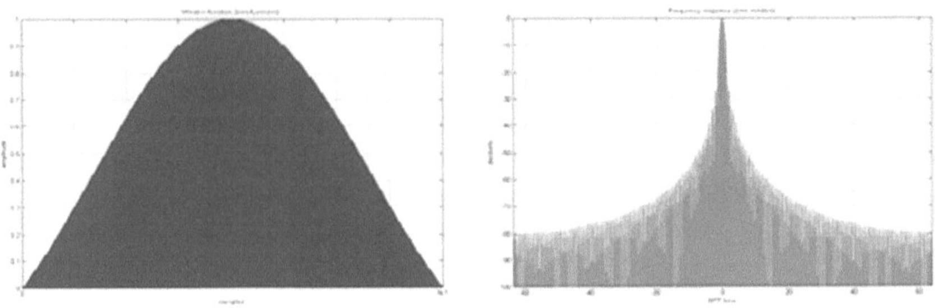

Fig. 48.2 A discrete Lanczos window and its frequency response [11]

resampling the 2D signal along the other axis. The same idea can be extended to N dimensions also [8] (Fig. 48.2).

Lanczos resampling is exclusively used in upscaling videos in various media utilities, such as AviSynth and FFmpeg [9].

48.2 Methodology

The different image resampling methods are applied on the same input image, and the output images are compared not only on visual perception but also on various performance metrics such as entropy, computation time, normalized least squares error (NLSE), and peak signal-to-noise ratio (PSNR). The image used is LISS IV of Bangalore. The entire procedure is implemented using MATLAB R2010b software. It is to be noted that although visual perception plays a pivotal role, analyses are carried out by applying relevant performance metrics on the methods used, as visual perception alone cannot help in comparing and contrasting the various methods applied. These are summarized in Table 48.1. The input (or original) satellite color image (of Bangalore) is a LISS IV image of size $282 \times 248 \times 3$, which is in "TIF" format. It is captured in the optical region of the electromagnetic spectrum. The satellite image corresponds to Bangalore city with the graticule (latitude and longitude) locations: 12°58′13″N, 77°33′37″E/ 2.970214, 77.56029. Entropy of input image in bits/symbol for 3 planes are R = 6. 850, G = 6.987, B = 6.89, respectively. These are summarized in Table 48.1.

The performance metrics that are used are computation time (in seconds), entropy (in bits/symbol), normalized least squares error (NLSE), and peak signal-to-noise ratio (PSNR) [8].

Entropy is defined as the average information generated by the source [10]. Entropy is denoted by H or H(S) [11].

$$H = -\sum_{k=1}^{L} p_k \log_2 p_k \tag{48.6}$$

Table 48.1 Result analysis of the image resampling techniques used

Method → Parameter ↓	Lanczos	Sinc (using Kaiser window ($\beta = 1.5$))	Sinc (using Hamming window)	Nearest neighborhood
Entropy of the output image in bits/symbol	R = 7.812 G = 7.654 B = 7.345	R = 6.475 G = 6.239 B = 6.416	R = 6.312 G = 6.218 B = 6.284	R = 5.121 G = 5.141 B = 5.431
PSNR (in dB)	R = 8.22 G = 7.31 B = 7.15	R = 7.23 G = 6.56 B = 6.11	R = 6.43 G = 5.85 B = 5.45	R = 5.28 G = 4.79 B = 5.12
NLSE	R = 0.1512 G = 0.2341 B = 0.0124	R = 0.4123 G = 0.4512 B = 0.3211	R = 0.5224 G = 0.4663 B = 0.4124	R = 0.6712 G = 0.5124 B = 0.6897
Time taken for execution (in seconds)	1.2378	3.8975	3.6514	1.1912

The entropy of a source gives the lower bound on the number of bits required to encode its output [12].

Normalized least squares error is given by the mathematical formula:

$$e_{NLSE} = \frac{\sum_{m=0}^{M-1}\sum_{n=0}^{N-1} |F(m,n) - \hat{F}(m,n)|^2}{\sum_{m=0}^{M-1}\sum_{n=0}^{N-1} |F(m,n)|^2} \tag{48.7}$$

NLSE should be minimum as far as possible between the input image and the output image [9].

The peak signal-to-noise ratio (PSNR) is defined as

$$PSNR = 10\log_{10}\left(\frac{MAX_I^2}{MSE}\right) = 20\log_{10}\left(\frac{MAX_I}{\sqrt{MSE}}\right), \tag{48.8}$$

Here, MAX_I is the maximum pixel value of the image [13]. When the pixels are represented using *8* bits per sample, this is 255. When normalized, $MAX_I = 1$ [13]. More generally, when samples are represented using linear PCM with *B* bits per sample, MAX_I is $2^B - 1$ [14]. MSE is the mean squares error and is defined for the input image, *I(m, n)*,and the output image, *K(m, n)*, by the formula

$$MSE = \frac{1}{MN}\sum_{i=0}^{M-1}\sum_{j=0}^{N-1} ||I(i,j) - K(i,j)||^2 \tag{48.9}$$

The PSNR is given in decibel units (dB), which measures the ratio of the peak signal and the difference between two images [15]. An increase of 20 dB corresponds to a tenfold decrease in the root mean square (RMS.) differences between two images [6] (Figs. 48.3, 48.4, 48.5, 48.6).

Fig. 48.3 Lanczos resampling. **a** Input image. **b** Output image

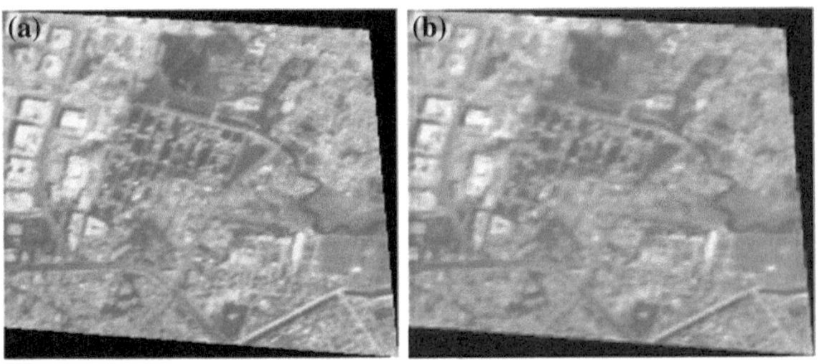

Fig. 48.4 Sinc resampling (by applying Kaiser window ($\beta = 0.5$). **a** Input image. **b** Output Image

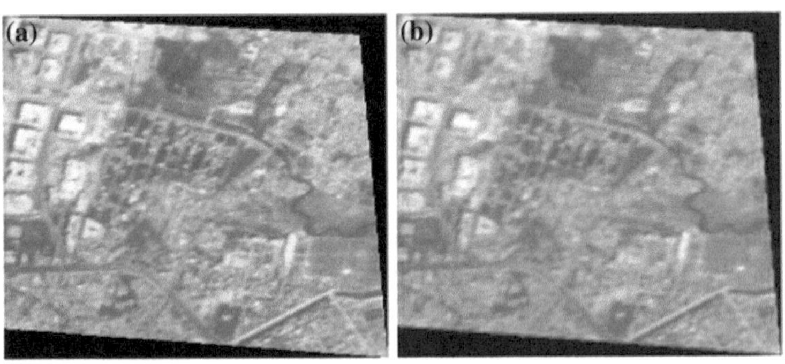

Fig. 48.5 Sinc resampling (by applying Hamming window. **a** Input image. **b** Output image

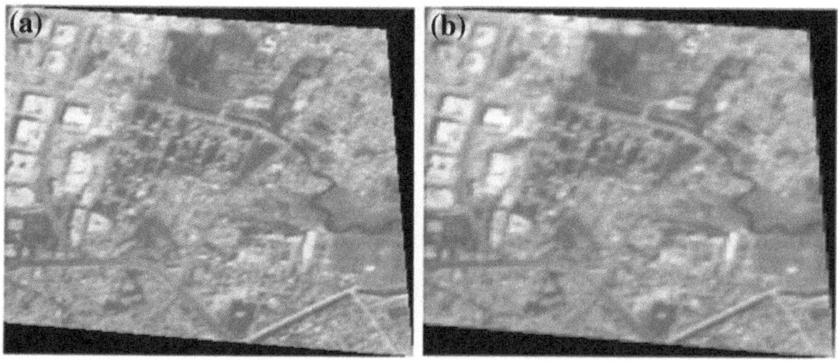

Fig. 48.6 Nearest neighborhood resampling (by applying Hamming window. **a** Input image. **b** Output image

48.3 Conclusions

From DSP theory, we know that the optimal reconstruction or regeneration filter for band-limited signals is the sinc filter; a sinc function has infinite support, i.e., it is having infinite duration. Lanczos filtering can be effectively used for approximating the sinc filtering technique that too with finite support. In Lanczos filtering, each interpolated value is obtained to be the weighted sum of $2a$ consecutive input samples. By doing so, we may have a trade-off between computation speed and improvised frequency response. The choice for having a smoother interpolation or preservation of sharper transients in the data can be easily done by suitably varying the appropriate parameter, viz., $2a$. For the processing of remotely sensed images, the trade-off is done between the reduction in aliasing artifacts and the preservation of sharp edges. One should exercise certain amount of caution in doing this procedure because the elimination of aliasing is necessary, and in doing so, there is a chance that preservation of sharp edges may go awry at times.

In this paper, the Lanczos resampling technique has been compared with other resampling techniques for discrete signals. The Lanczos filter with $a = 2$ gives the best result because it helps in the reduction in fold-over (aliasing), preserves sharpness, and decrements the ringing effect to a greater level, when compared with the truncated sinc resampling techniques of Bartlett-, Hamming-, Kaiser-, and Hanning-windowed sinc, for downsampling and expanding monochromatic and color images. The Lanczos kernel with $a = 3$ keeps low frequencies at bay and rejects high frequencies which is good when compared to the other filtering techniques considered.

Since the kernel assumes negative values for $a > 1$, the interpolated signal can be negative even if all samples are positive. More generally, the range of values of the interpolated signal may be wider than the range spanned by the discrete sample values. In particular, there may be ringing artifacts just before and after abrupt changes in the sample values, which may lead to clipping artifacts. However, these effects are reduced compared to the (non-windowed) sinc filter.

The ringing effect creates light and dark halos along the strong edges of the image. These bands at times will be visually erratic to look, but they also have the tenacity to help in increasing the perceived sharpness. This contributes toward providing edge enhancement. As a result, the overall subjective quality of the image is improvised.

The low-end clipping artifacts can be removed to a substantial level by transforming the data to a logarithmic domain before applying the filtering procedure. Thus, in this case, the interpolated values will be a weighted geometric mean of the input samples.

The partition of unity property is not existent in Lanczos kernel. The sum $U(x) = \sum_{i \in \mathbb{Z}} L(x - i)$ of all integer-shifted versions of the Lanczos kernel is unequal to unity. Because of this, the output of Lanczos resampling of a discrete-time signal having constant samples will not result in a constant function. This anomaly is seen when $a = 1$. At this point, the interpolated signal has zero derivative at each and every integer argument. A possible remedy for this anomaly is to choose proper value for a. The choice of $a = 3$ is a proper choice for compensating this problem.

It is to be noted that the usage of sinc interpolation using Kaiser window takes more time for computation because of the usage of modified Bessel function of zeros order in its definition than its realization using Hamming window. The entropies of R-, G-, and B-planes of Lanczos method are more than that of the other resampling methods considered. The PSNR of Lanczos method is also more when compared to the other methods. Also, the resampling procedure involving sinc interpolation produces an output image with a washed-out look, which is undesirable. The NLSE obtained using Lanczos method is very less when compared to the other methods used. Despite the nearest neighborhood technique being a simple one, its results are very poor when compared to the other methods. It displays ringing and blocking effects. The Lanczos resampling method proves to be the optimal method for resampling of remotely sensed images.

48.4 Future Scope

Until now, only fixed Fourier-based kernels have been used for the processing of remotely sensed images. The usage of variable kernels using wavelet transform needs to be exploited to the core in remote sensing image analysis. The above-discussed technique was applied after the input color image was decomposed into its 3 constituent monochromatic images, where in the method was applied individually on them, so the usage of these methods directly on the color image itself (and also on color images which are based on different color systems other than RGB) can also be exploited.

References

1. Gonzalez RC, Woods RE (2009) Digital image processing, 3rd edn. Pearson Education, Inc., New Delhi, India
2. Jain AK (1989) Fundamentals of digital image processing. Prentice Hall of India Private Limited, New Delhi, India
3. Joseph G (2009) Fundamentals of remote sensing, 2nd edn. Universities Press, Hyderabad, India
4. Lillesand T, Kiefer RW, Chipman J (2011) Remote sensing and image interpretation, 6th edn. Wiley India Private Limited, New Delhi, India
5. Rembold F, Atzberger C, Savin I, Rojas O (2013) using low resolution satellite imagery for yield prediction and yield anomaly detection. Remote Sens 5(4):1704–1733
6. Thévenaz P, Blu T, Unser M (2000) Interpolation revisited. IEEE Trans Med Imaging 19(7):739–758
7. Avcrbas I, Sankur B, Sayood K (2001) Statistical evaluation of quality measures. J Electron Imaging 11(2):206–223
8. Reichenbach RE (2003) Two-dimensional cubic convolution. IEEE Trans Image Process 12:857–865
9. Gotchev A, Vesma J, Saramaki T, Egiazarian K (2000) Digital image resampling by modified b-spline functions, IEEE nordic signal processing symposium, Sweden, pp 259–262
10. Hou HS, Andrew HC (1978) Cubic splines for image interpolation and digital filtering. IEEE Trans Acoust Speech Sig Process 26:508–517
11. Jensen JR (2007) Introductory digital image processing: a remote sensing perspective, 3rd edn. Pearson Publication, New Jersey, U.S.A
12. Schowengerdt RA (2007) Remote sensing models and methods for image processing, 3rd edn. Reed-Elsevier India Private Limited, New Delhi, India
13. Jähne B (2012) Digital image processing, 6th edn. Springer India Private Limited, New Delhi, India
14. Pratt WK (2009) Digital image processing, 4th edn. Wiley India Private Limited, New Delhi, India
15. Bose T (2009) Digital signal and image processing. Wiley India Private Limited, New Delhi, India

Chapter 49
Performance Analysis of Adaptive DOA Estimation Algorithms for Mobile Applications

A. M. Prasanna Kumar and K. Suresh

Abstract Spatial filtering for mobile communications has attracted a lot of attention over the last decade and is currently considered a very promising technique that will help future cellular networks achieve their ambitious goals. One way to accomplish this is via array signal processing with algorithms which estimate the direction-of-arrival (DOA) of the received waves from the mobile users. This paper evaluates the performance of a number of DOA estimation algorithms. In all cases, a linear antenna array at the base station is assumed to be operating typical cellular environment.

Keywords Direction-of-arrival · MUSIC · ESPRIT · Quaternion MUSIC

49.1 Introduction

Adaptive antennas have attracted a lot of attention over the past few years as possible solutions to some of the main problems associated with current mobile systems [1]. For this purpose, a number of *research* activities are being carried out in order to properly design, analyse, and implement adaptive antennas for mobile communications applications, *e.g.*, TSUNAMI Project (Technology in Smart antenna for the UNiversal Advanced Mobile Infrastructure) [2]. DOA estimation algorithms are used to improve the performance of an antenna by controlling the directivity of the antenna to reduce effects like interference, delay spread, and

A. M. Prasanna Kumar (✉)
ACS Engineering College, Bangalore, India
e-mail: pravijll@rediffmail.com

K. Suresh
DSCE, Bangalore, India
e-mail: ksece12@gmail.com

V. Chakravarthi et al. (eds.), *Proceedings of International Conference on VLSI, Communication, Advanced Devices, Signals & Systems and Networking (VCASAN-2013)*, Lecture Notes in Electrical Engineering 258, DOI: 10.1007/978-81-322-1524-0_49,

multipath fading. Several DOA estimation algorithms are available which are categorised by Krim and Viberg [3] into three categories: spectral estimation, parametric subspace-based estimation (PSBE), and deterministic parametric estimation (DPE).

Algorithms of the first category are MUSIC (MUltiple SIgnal Classification) algorithm [4], ESPRIT (Estimation of Signal Parameters via Rotational Invariance Techniques) [5], and all its variants such as Least Square (LS), Total Least Square (TLS) [6], and Unitary-ESPRIT [7] belong to the PSBE techniques. The DPE techniques include Maximum Entropy (ME) [8], Maximum Likelihood (ML) [9], Space-Alternating Generalised Expectation-maximisation (SAGE) [10], and Weighted Subspace Fitting (WSF) [11] methods.

The aim is to provide a comprehensive comparison of DOA estimation algorithms. The results from such a comparison can then be used to indicate solutions for different levels of applications, *e.g.*, for measurement systems with the capability to provide spatial information, for cellular base stations with the capability to improve range-capacity-service quality, etc, for user positioning systems.

49.2 Basic Channel Model

Consider a uniformly spaced, linear array consisting of N sensors on which plane waves from M $(M < N)$ narrow-band sources impinge from directions 01. Taking the first element in the array as the phase reference and assuming that the signal sources are in the far field, the complex vector received by the array can be expressed as [12]:

$$y(t) = \sum_{m=1}^{M} a(\theta \mathrm{m}) Sm(t) + w(t) \tag{49.1}$$

where s(t) is the signal vector, w(t) is the additive noise vector, and $a(\theta)$ is referred to as the *array response* or *array steering vector* for the direction θ, *i.e.*

$$\begin{aligned} \mathrm{a}(\theta) &= \left[1\mathrm{e}^{-\mathrm{j}\emptyset} \ldots \mathrm{e}^{-\mathrm{j}(\mathrm{N}-1)\emptyset}\right]^{\mathrm{T}}, \\ \phi &= 2\pi(\mathrm{d}/\lambda) \sin\theta \end{aligned} \tag{49.2}$$

where **T** is the transposition operator, d in the spacing between sensors, and λ is the wavelength of the received signal.

Representing (49.1) in a more compact matrix form of size N X 1:

$$\mathrm{y(t)} = \mathrm{As(t)} + \mathrm{w(t)} \tag{49.3}$$

where $s(t)$ is Mxl vector, and A = [a(θ1),. a(θM)] is the NxM matrix of steering vectors.

We assume that the w(t) is modelled as temporally white and zero-mean complex Gaussian process. Now, the *NxN* spatial correlation matrix of the observed signal vector y(t) can be defined as

$$R = E\left[y(t)y^{H}(t)\right] \tag{49.4}$$

where E[•] and **H** denote expectation operator and conjugate transpose, respectively. R is sometimes also referred to as the *array covariance matrix.* From our assumptions, this can also be represented as follows recognising also that *s(t)* and w(t) are statistically independent:

$$R = ASA^{H} + \sigma_{w}^{2}I \tag{49.5}$$

where S is the signals covariance matrix.

The subspace methods utilise the special eigenstructure of R which is expressed in terms of its eigenvalues and their corresponding eigenvectors e_n (n = 1, 2, … , N). We assume that $\lambda_1 \geq \lambda_2 \geq \lambda_3 \ldots \geq \lambda_N$. The first *M* eigenvalues will correspond to the directional sources, their values are larger than σ^2, and the remaining (N-M) eigenvalues are equal to σ^2. The eigenvectors corresponding to the signal eigenvalues can be used to describe the signal subspace: $Es = [e_1 \ldots C_M]$. E_n is the matrix containing the remaining N-M noise eigenvectors that describe the noise subspace, which is the orthogonal complement to the signal subspace: $E_n = [e_{M+1} \ldots e_N]$

The estimation of the number of sources M is carried by using information theoretic criteria [13], *e.g.*, Akaike's or the Minimum Description Length (MDL) criterion [14].

49.2.1 Music

The MUSIC algorithm developed by Schmidt [4] tends to exploit the orthogonality between the signal subspace and noise subspace. This is done by searching for peaks in the MUSIC spectrum which are a function of the look direction, 0, given by:

$$P_{MU}(\theta) = \frac{1}{a^{H}(\theta)E_{n}E_{n}^{H}a(\theta)} \tag{49.6}$$

The peaks in the spectrum occur at the points where the steering vector is orthogonal to the noise subspace, *i.e.*, the denominator in (49.6) goes to zero and therefore $P_{MC}^{®}$ peaks. The accuracy of the signal arrival directions estimated by MUSIC depends on the accuracy of the correlation matrix R. This can be improved by using more snapshots of data and/or having high Signal-to-Noise Ratio (SNR) [15]. Also, the accuracy of MUSIC estimate degrades when the incident signals are coherent, which can be improved by using spatial smoothing preprocessing before the MUSIC algorithm [16]. Root-MUSIC is a modified version of the MUSIC algorithm in which the DOAs are determined from the roots of a polynomial formed from the noise subspace [17]. Unlike MUSIC which is applicable to general array configurations, Root-MUSIC is restricted to uniform linear arrays. If we define polynomials using the noise eigenvectors [18]:

$$Dk(z) = \sum_{n=1}^{N} \mathrm{e}_{\mathrm{nk}} \mathrm{z}^{-(\mathrm{n}-1)}, \quad \mathrm{k} = \mathrm{M}+1, \ldots, \mathrm{N} \tag{49.7}$$

where e_{nk} are components of E_n. The above polynomials have roots at $\mathrm{Z} = \mathrm{e}^{\mathrm{j}2\pi(\mathrm{d}/\lambda)\sin\theta}$, $\mathrm{i} = 1, \ldots, M$. Now define the polynomial:

$$\mathrm{Q(z)} = \sum_{(k=M+1)}^{N} D_k(z) D_k(1/z) \tag{49.8}$$

It can be seen that Q(z) will have the same roots as D_k(z) and there will be *M* double roots lying on the unit circle in the z-plane. These roots will correspond to the actual incident signals and the other roots which do not lie on the unit circle will not correspond to the signals and they are called spurious roots [8]. It has been reported in many studies that Root-MUSIC shows better performance than MUSIC, especially in environments where the signals are located closer and/or they have low SNR [15, 18] and [19]

49.2.2 ESPRIT

ESPRIT [5] is a computationally efficient and robust method for estimating DOA which was developed in order to overcome the disadvantages of MUSIC. Other versions of ESPRIT have been developed to improve the technique, *e.g.*, Least Squares (LS-ESPRIT), Total Least Squares (TLS-ESPRIT) [6], and Unitary-ESPRIT [7]. Unitary-ESPRIT further reduces the computational complexity of the standard ESPRIT algorithm by using real-valued computations from start to finish. It not only estimates the DOA, but it can be used to estimate the number of sources present. It also it incorporates forward–backward averaging which overcomes the problem of coherent signal sources. In this paper, the standard version of ESPRIT and Unitary-ESPRIT is tested.

49.3 Test Scenarios

The propagation scenarios used in testing the DOA estimation algorithms are divided into two main categories: *Macro-* and *Microcells.* This categorisation is based on the size of the cells and the height of the base station antenna. Macrocells are further divided, based on the environmental characteristics, into four sub-categories: *urban, bad urban, sub-urban, and rural.* Results for all these environments were produced according to the parameters described in [20] and the values suggested therein. The different DOA estimation algorithms *were* tested for the uplink of a W-CDMA (wideband-code division multiple access) signal

Table 49.1 The values of some of the parameters employed in the test scenarios

Parameter	Value
Frequency (MHz)	1,800
No. of antenna elements (N)	8
Inter-element spacing (d)	2/2
Azimuth angle of MS (degrees)	20
W-CDMA chip rate (k chips/sec)	4,096
W-CDMA data rate (kbps)	16
SNR range (dB)	−5–20
Power window (dB)	1, 5, 15

transmitted by a single static mobile station (MS) and received at the base station (BS) by an 8-element linear antenna array. The W-CDMA signal was generated following the ETSI UMTS Terrestrial Radio Access **(UTRA) RTT** (Table 49.1).

DOA estimation algorithm when applied to the different simulated channel types with different values of power window for the impulse responses of the radio channel.

49.4 Simulation Results

The channel impulse responses (CIR) for 100 snapshots were produced by the propagation model, for different channel types. The CIRs were then filtered with different power windows, applying the DOA estimation algorithms on the filtered CIRs for a range of SNR values and in order to estimate the number of sources present the MDL criterion was employed [14]. The mean (over 100 snapshots) of the DOA error at each SNR level, for the different algorithms, for the different environments, with a power window for the CIRs of 1 dB. The figure also includes the mean estimate of M found by the MDL algorithm for the same SNR range. It is evident from the figure that as the SNR level increases the DOA estimate error decreases with a different rate for the different environments, with the highest values for the Microcellular Urban scenario (Fig. 49.le). It can be seen that the tested DOA estimation algorithms have the smallest error in the Macrocell Sub-Urban and Rural scenarios, Fig. 49.lc and d, respectively. This is due to that in these environments the BS receives small number of multipaths with relatively small angular spread. The Macrocell Urban and Bad Urban scenarios, Fig. 49.la and b, respectively, have a slightly higher level of DOA estimation error than the Macrocell Sub-Urban and Rural scenarios. This is due to the increase in the mean number of received multipaths that arrive at the BS with a higher angular spread. The DOA estimation algorithms produced the highest error in the Microcellular Urban environment, Fig. 49.le. This is due to that in this environment most of the time the BS receives at least two multipath with high angular spread. From the figures, it can be seen that Root-MUSIC produced the best results for all environments except the Bad Urban case. Also, the figures show that the DOA estimation

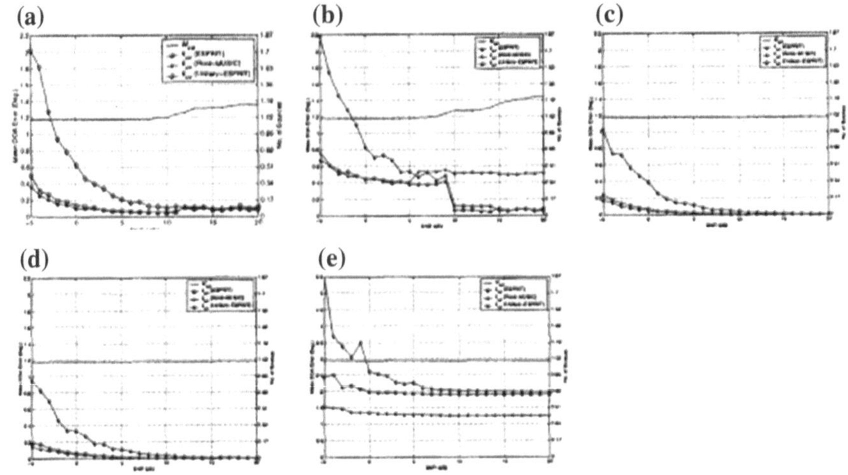

Fig. 49.1 Mean DOA error of the DOA estimation algorithms for the different propagation environments with 1 dB power window. **a** Macrocell urban. **b** Macrocell bad urban. **c** Microcell sub-urban. **d** Macrocell rural. **e** Microcell urban

error of Unitary-ESPRIT is the most sensitive to the change in the SNR levels as it produced high DOA error levels for low values of SNR that *decrease* by the increase in SNR. When using a power window of 50, the performance of the DOA estimation methods degrades except from the Sub-Urban and Rural scenarios .

For the Macrocell Urban scenario (Fig. 49.2a), the DOA estimate error reaches its lowest point at SNR = 7 dB, for all the DOA estimation algorithms, and then increases slightly at the higher SNR levels. This is due to the effect of the increase in the number of sources estimated by the MDL algorithm. The effect of the additional cluster in the Bad Urban scenario is clear on the performance of the tested DOA estimation algorithms and on the Root-MUSIC in particular. The performance of the DOA estimation algorithms still exhibits the best performance in the Macrocell Sub-Urban and Rural scenarios, Fig. 49.2c and e, respectively. The DOA estimation error increased in the Microcell Urban scenario compared to the 1 dB power window case, which shows that more multipaths with higher angular spread are present in the scenario compared to the other scenarios or to the same scenario with smaller power window.

DOA estimation algorithms. The DOA estimate error is higher in all the propagation scenarios compared to the cases when using smaller power windows. This is mostly evident in the Macrocellular Bad Urban and Microcellular Urban scenarios, Fig. 49.3b and e, respectively. This is due to the fact that when larger power windows are used more multipaths from different directions other than the required mobile direction might be also present. This will cause the MDL algorithm to estimate more sources in the environment. Since the DOA estimation algorithms use the estimate of M from the MDL algorithm, some of the distant scatterers will be considered as the actual required mobile.

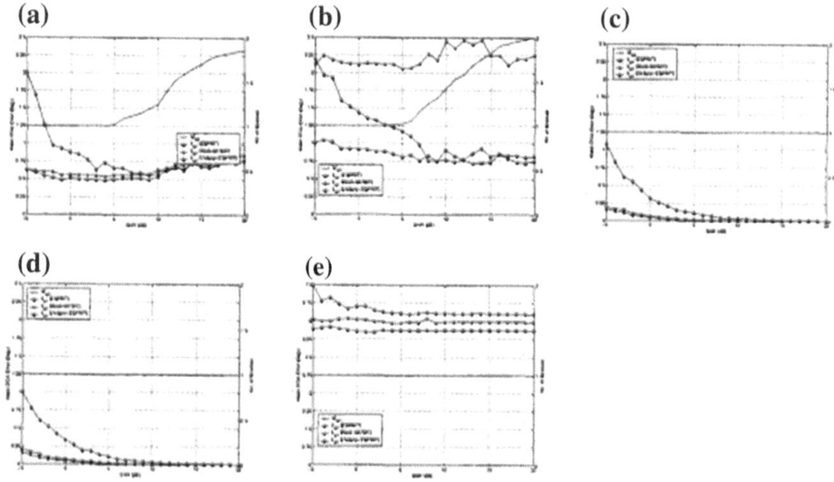

Fig. 49.2 Mean DOA error of the DOA estimation algorithms for the different propagation environments with 5 dB power window. **a** Macrocell urban. **b** Macrocell bad urban. **c** Microcell sub-urban. **d** Macrocell rural. **e** Microcell urban

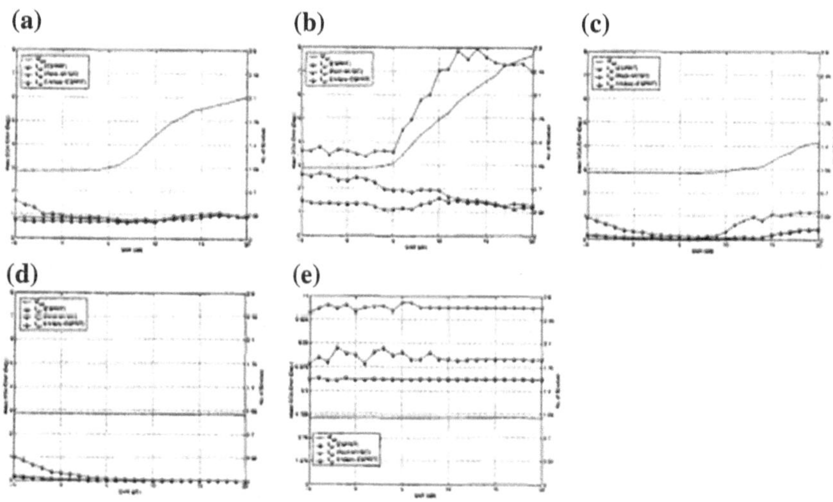

Fig. 49.3 Mean DOA error of the DOA estimation algorithms for the different propagation environments with 15 dB power window. **a** Macrocell urban. **b** Macrocell bad urban. **c** Microcell sub-urban. **d** Macrocell rural. **e** Microcell urban

We have tried to solve this problem by using the estimate of number of sources (M) calculated by the Unitary-ESPRIT from the most dominant eigenvectors [7] and tested it on the worst-case scenarios with worst-case power window. DOA estimate error in the Macrocell Urban, Macrocell Bad Urban and Microcell Urban scenarios when using the estimate of M calculated by the Unitary-ESPRIT

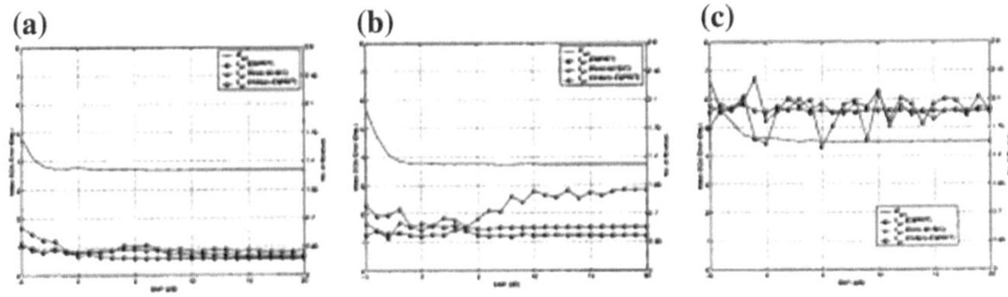

Fig. 49.4 Mean DOA error of the DOA estimation using estimate of M from Unitary-ESPRIT with a 15 dB power window. **a** Macrocell urban. **b** Macrocell bad urban. **c** Microcell urban

algorithm with a power window of 15 dB. The figure shows how the performance of the DOA estimation algorithms improved by using the better estimate of the number of sources. Root-MUSIC is more susceptible to the multipath cluster from the remote scatterers present in the Bad Urban scenarios. DOA estimation error increases when the power window size for the CIRs increases, especially in scenarios with more than one clusters of rays, *e.g.*, Macrocell Bad Urban and Microcell Urban. Figure 49.4 shows how the performance of the DOA estimation techniques could be improved by employing more accurate estimate of the number of sources (M) present in the environment.

Macrocell Bad Urban, Macrocell Sub-Urban, Macrocell Rural, and Microcell Urban. The algorithms have been applied at the BS to find the DOA of a W-CDMA signal transmitted by a single static MS. The received complex impulse responses (CIR) have been filtered using power window sizes of 1, 5 and 15 dB for an SNR range from −5 to 20 dB. The results showed that Root-MUSIC outperformed the other algorithms in the Macrocell Sub-Urban and Rural scenarios, and for the Macrocell Urban scenario when the power window was small. However, Root-MUSIC produced high DOA estimation errors in scenarios with more than one clusters of rays, e.g., Macrocell Bad Urban and Microcell Urban. The DOA estimation error produced by the Unitary-ESPRIT algorithm showed that this algorithm is more sensitive to the SNR changes than the other algorithms, in all the tested scenarios. Comparing the results when using the MDL algorithm to estimate the number of sources present, using the estimate calculated by Unitary-ESPRIT, it was seen that Unitary-ESPRIT produces more accurate estimates. Generally, the results showed the importance of power windows and good estimates for the number of sources in improving the accuracy of the DOA estimation algorithms.

49.5 Conclusion

In this paper, the performance of Root-MUSIC, standard version of ESPRIT and Unitary-ESPRIT has been tested in terms of the mean **DOA** error. This is carried out in five different scenarios: Macrocell Urban, Macrocell Bad Urban, Macrocell

Sub-Urban, Macrocell Rural, and Microcell Urban. The algorithms have been applied at the BS to find the DOA of a W-CDMA signal transmitted by a single static MS. The received complex impulse responses (CIR) have been filtered using power window sizes of 1.5 and 15 dB for an SNR range from −5 dB to 20 dB. The results showed that Root-MUSIC outperformed the other algorithms in the Macrocell Sub-Urban and Rural scenarios, and for the Macrocell Urban scenario when the power window was small. The DOA estimation error produced by the Unitary-ESPRIT algorithm showed that this algorithm is more sensitive to the SNR changes than the other algorithms, in all the tested scenarios. Comparing the results when using the MDL algorithm to estimate the number of sources present and when using the estimate calculated by Unitary-ESPRIT, it was seen that Unitary-ESPRIT produces more accurate estimates. Generally, the results showed the importance of power windows and good estimates for the number of sources in improving the accuracy of the DOA estimation algorithms.

References

1. Tsoulos G (1999) Smart antennas for mobile communication systems: benefits and challenges. IEE Electron Commun Eng J 11(2):84–94
2. Tsoulos G et al (1998) Space division multiple access (SDMA) field trials. Part 1: tracking and BER performance & part 2: calibration and linearity issues. IEE Proc Radar Sonar Nav 145(1):73–84
3. Krim, H, Viberg M (1996) Two decades of array signal processing research. IEEE SP Mag 13:67–94
4. Schmidt R (1986) Multiple emitter location and signal parameter estimation. IEEE Trans AP 34(3):276–280
5. Paulraj A et. al (1986) A subspace rotation approach to signal parameter estimation. Proc IEEE 74(7):1044–1045
6. Roy R, Kailath T (1989) ESPRIT-estimation of signal parameters via rotational invariance techniques. IEEE Trans ASSP 37(7):984–995
7. Haardt M, Nossek JA (1995) Unitary ESPRIT: how to obtain increased estimation accuracy with a reduced computational burden. IEEE Trans SP 43(5):1232–1242
8. Therrien C (1992) Discrete random signals and statistical signal processing. Prentice Hall, NJ
9. Capon J et al (1967) multidimensional maximum likelihood processing of a large aperture seismic array· Proc IEEE 55(2):192–211
10. Fleury B et al (1996) Wideband angle of arrival estimation using the SAGE algorithm. In: Proceedings of the ISSSTA '96, Mainz, Germany, Sept 1996, pp 79–85
11. Viberg M et al (1991) Detection and estimation in sensor arrays using weighted subspace fitting. IEEE Trans SP 39(11):2436–2449
12. Haykin S (ed) (1985) Army signal processing. Prentice-Hall, NJ
13. Wax M, Kailath T (1985) Detection of signals by information theoretic criteria. IEEE Trans ASSP 33(2):387–392
14. Wax M, Ziskind I (1989) Detection of the number of coherent signals by the MDL principle. IEEE Trans ASSP 37(8):1190–1196
15. Ogawa Y, Kikuma N (1995) High-resolution techniques in signal processing antennas. IEICE Trans Commun E78-B(11):1435–1441
16. Shan T-J et al (1985) On spatial smoothing for direction-of-arrival estimation of coherent signals. IEEE Trans ASSP 33(4):806–811

17. Barabell A (1983) Improving the resolution performance of eigenstructure-based direction-finding algorithms. In: Proceedings of ICASSP '83, Boston, MA, pp. 336–339
18. Rao B, Had K (1989) Performance analysis of root-MUSIC. IEEE Trans ASSP 37(12): 1939–1949
19. Zoltowski M et al (1993) Beamspace root-MUSIC. IEEE Trans SP 41(1):344
20. Piechocki R, Tsoulos G (1999) A macrocellular radio channel model for smart antenna tracking algorithms. In: Proceedings of VTC'99, Houston, TX, May 1999, pp 1754–1758

Part VI
Networking

Chapter 50
Design of Security Schemes for Wireless Sensor Networks Based on Attack Behavior: Proactive Approach

Anusha Anigol, Roshan Ahmed and R. C. Biradar

Abstract Wireless sensor networks (WSNs) are one of the emerging network fields, with a tremendous prospect to research and contributing itself in a vast arena of applications. In any sort of network, the utmost priority is given to its security feature, since any means of communication demands the need of securing and safe guarding the sensitive data from the adversary. Even WSNs are also susceptible to any offenders attacks. In this paper, an attempt is made to design an perpetual proactive security scheme at the sink node using sniffing mechanism of intrusion detection system by studying the behavioral aspect of the sensor nodes due to attack and to monitor the system with an rescue mechanism.

Keywords Wireless sensor network · Security · Proactive approach

50.1 Introduction

WSN are rapidly emerging as an important new area in wireless and mobile computing research. Wireless sensors facilitate many applications in a wide distance of areas such as traffic data collection in transportation, earthquake monitoring for

A. Anigol (✉) · R. Ahmed
Department of Electronics and Communication Engineering, Reva Institute of Technology and Management, Kattigenhalli, Yelhanka, Bangalore, Karnataka 560064, India
e-mail: anu.anigol@gmail.com

R. Ahmed
e-mail: roshanahmed31@gmail.com

R. C. Biradar
Department of Information Science and Engineering, Reva Institute of Technology and Management, Kattigenhalli, Yelhanka, Bangalore, Karnataka 560064, India
e-mail: raj.biradar@revainstitution.org
URL: http://www.revainstitution.org

V. Chakravarthi et al. (eds.), *Proceedings of International Conference on VLSI, Communication, Advanced Devices, Signals & Systems and Networking (VCASAN-2013)*, Lecture Notes in Electrical Engineering 258, DOI: 10.1007/978-81-322-1524-0_50,

emergency response, combat zone surveillance, and disease diagnosis in medical environments [1]. Sensing mechanisms will reveal previously unobserved phenomena. The various areas where major research activities are going on in the field of WSNs are deployment, localization, synchronization, data aggregation, dissemination, database querying, architecture, middleware, security, designing less power consuming devices, abstractions, and higher level algorithms for sensor specific issues [2].

50.1.1 Security and Intrusion Detection System in WSN

Security in a WSN is very challenging. Every WSN application should adopt basic security requirements like availability, data freshness, data confidentiality, robustness, data authenticity, data integrity, and time synchronization and some of them given in [2–4]. Some of the active research areas of security in WSN are designing secured routing protocols, symmetric key cryptography for WSNs, designing secure data aggregation protocols, designing IDS, and security systems for sensors nodes. An IDS is an interesting, underdeveloped service, useful for scenarios where there is a possibility for a node being subverted and controlled by an adversary. The major task of IDS is to monitor networks and systems to detect eventual intrusions in the network, alert users after specific intrusions that have been detected and if possible reconfigure the network and mark the root of the problem as malicious [5]. IDS is similar to an immune system that identifies and eliminates anomalies by measuring deviations from normal process using distributed identifiers over the system with an identifiable and adaptable relationship [6]. Fully functional IDS, there is a need for filtering the information provided by the system to detect malicious nodes and distinguish between possible errors and attacks launched against the network [7]. Wireless sensor nodes are usually small and have many physical constraints, making them vulnerable to insider or outsider attacks [8, 9]. Security can be achieved by means of secure key management against outsider attacks except for jamming which can be circumvented by spread spectrum on frequency hopping communication, locating the jamming area, and rerouting traffic [10].Types of IDS are generally classified into two types: (1) Signature-based detection uses characteristics of attacks to form signatures or patterns and detect attacks by matching activities against the signatures [11]. (2) Anomaly-based detection is based on the assumption that the attacker's behavior deviates from the normal network behavior [12].

50.1.2 Related Works

To protect sensors against physical attacks which completely destroy sensors functionalities, Xuan et al. [13] proposed that some nodes behave as sacrificial nodes by performing the function even with the knowledge of intruder lurking around, while other neighbor nodes can switch themselves off, thereby going undetected. This proposed approach protects sensors against attacks, which reprogram sensors through radio signals for malicious purpose. Instead of causing a complete out of service, the malicious nodes try to disguise themselves by functioning as normal while attacking the entire network sneakily. To identify sensors that pretend to function as normal while having malicious intent, Srivastava et al. [14] introduced reputation-based approach in RFSN. Similar to keynote [15] and Ebay [19], RFSN uses a trust-based system in which sensor nodes maintain reputation for each other. By maintaining this reputation, they are able to identify incorrect information and discard it. To identify disguised malicious nodes, semantic analysis and pattern-matching approach were proposed in [16–18, 20]. Michardi and Molva [21] use passive eavesdropping to isolate the misbehaving nodes and assign reputation to the nodes such that only good nodes in the network are trusted. Marti et al. [22] proposed the popular watchdog mechanism to monitor the neighbor nodes and detect misbehavior in order to reliably route packets around the good nodes in the network. Kachirski and Guha [7] propose the idea of a distributed IDS where cluster heads are elected and the IDS functionality is distributed between them.

Our Contributions: An attempt is made to design an efficient proactive security scheme using sniffers which continuously monitors the network for any sort of security attacks. Sniffers act as the security guard system as a watchdog to the sink node, protecting the network from any sort of external and internal attacks. Sniffers identify an intruder with the help of direction of the signal received from a node using proactive algorithm. The algorithm performs distance checking mechanism, direction, angle calculation, and node ID verification for the purpose of authentication. Sink node takes necessary steps to protect the network once it identifies the intruder using counter-based mechanisms.

50.2 Sniffer-Based Intrusion Detection

As the sensor network is a self-organized network by nature, we deploy these sniffing sensors around the sink node at the network organization state. Once the sensor nodes are randomly deployed in the network, they self-organize themselves into groups among their nearest nodes and form a cluster and a cluster head to aggregate the data and send to sink node. The sink node being the source of information is susceptible to attacks. To protect the sink node from any type of physical attacks, IDS is employed.

50.2.1 Proposed Scheme

To protect the sink node from any sort of intruders attack, some of the sniffing nodes (sniffers) around the sink node are deployed which acts as fence to the sink node. The sniffers are equipped with the entire network range, the packet format employed in the network communication, the direction of data transfer between each cluster head involved in the network. The cluster head deployed in the network signifies the efficient data communication with the sink. Every cluster head aggregates the data collected from each senor node and sends the aggregated data along with the subnet range which ensures that the signal received is from within the range. Once the attacker tries to attack the sink node, these sniffers utilize this information as their tool to verify that the data being sent to sink is from the attacker or an authenticated source. Since, the authentication is achieved by saving node IDS in the sink node, the deployed sniffers are programmed with the proactive algorithm. This algorithm helps the sniffer nodes to detect the intruder and protect the sink from attack by giving an alarm warning to sink to deactivate itself as the attacker is approaching the network. Since the sink appears to be deactivated to the attacker, the targeted attack fails hence the data are safeguarded. Figure 50.1 shows the deployment of sniffer nodes around the sink node.

50.2.2 Algorithm

The input to this algorithm is the total number of clusters involved in the network system n and currently the incident signal targeted from the *i*th cluster. First distance of each cluster head is calculated. On event e occurrence, sniffers validate the node ID of the received packet by the agreed packet format between each cluster and the sink. The packet consists of the node ID of sending source and the cluster information to which it belongs to along with range, i.e., x, y, z coordinates of the node. If the node is authenticated, it checks for the range. If the received signal range is outside the network grid, the packet is directly discarded from reaching it to sink. Signal range calculation at the sniffer is employed using the Eq. (50.1)

$$\text{Distance D} = \sqrt{\left(X_1^2 - X_2^2\right) + \left(X_1^2 + X_2^2\right)} \text{ and Range} = \sum\nolimits_i^n \text{Dn}_\text{i} + \text{Db} - \text{Dc}. \tag{50.1}$$

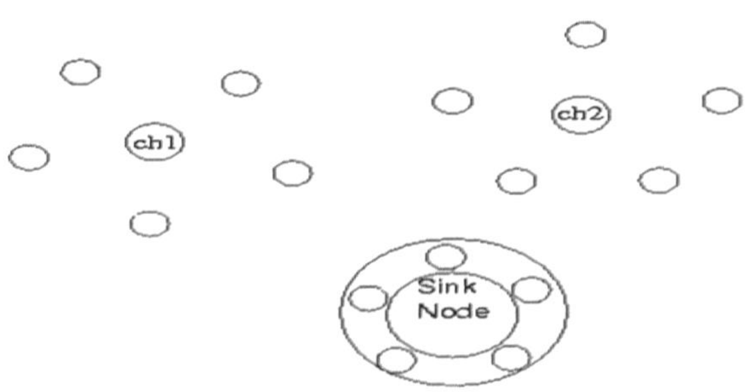

Fig. 50.1 Deployment of sniffer nodes and creation of proactive security shield

The distance is between the farthest nodes in the cluster to the sink. The distance from one cluster to another participating in the network which is denoted by Dc is ignored. The Db being the actual distance of the incident point to sink. The summation as Dpi is the distance from each sensing node to sink in the *i*th cluster. Now after authentication and range checking, there are possibilities that an intruder might be in a position in a very close proximity to a particular cluster and try to send the information to sink. Hence, direction checking is used as node center is the point from which the angle is calculated at the sink. The sink node is considered as the origin to calculate all the angles, and thus attacker node is clearly identified. The angle is calculated using the Eq. (50.2). The reception angle at the sink from all the nodes participating in the network is $90° - \theta°$ with respect to x–y plane.

$$\theta = \cos^{-1}\left(\sqrt{\mathrm{z}^2/(\mathrm{x}^2 + \mathrm{y}^2 + \mathrm{z}^2)}\right) \tag{50.2}$$

The number of clusters participating along with the sensing nodes in each cluster is fixed at the network organization stage, so the legitimate nodes are deployed at a predefined position from the sink at the time of deployment. If attacker trying to attack with same position, there will be angle difference which is not possible due to angle resolution with respect to the sink node position is +50 to +10. This resolution is fixed as threshold value to differentiate from the legitimate nodes and attacker nodes. Any angle beyond this threshold value is considered as the attacker node.

50.3 Simulation Results

Proactive approach–based sniffing mechanism is simulated using QualNet 5.2 Network Simulator to assess the performance and effectiveness of the approach. Event driven simulation is used in which the execution of various functions takes place at discrete events in a chronological sequence. Simulation environment consists of three models: network model, propagation model, and traffic model. *Network model:* A sensor network is placed in an area of $l * b$ square meters. It consists of *N number of nodes* that are assumed to be connected to a sink node at the boundary of a network.

Traffic model: Constant bit rate model is used to transmit fixed size packets, Trpkts. Coverage area around each node has a bandwidth which includes the channel frequency along with the noise factor shared among its neighbors.

50.3.1 Simulation Environment

The inputs are l = 100 m, b = 100 m, N = [5–25], Trpkts = multiples of (50), channel frequency = 2.4 GHz, noise factor = 10.

Steps involved in simulation:

1. Sensor network environment is developed by deploying sensor nodes along with sniffer nodes according to the discussed algorithm in a fixed network arena with the predefined Cartesian coordinates and the topology is static in nature. For every instant of simulation inputs, the performance is evaluated.
2. Performance parameters computed with different data inputs at various instants and plotted.

1. *Packet Delivery Ratio* (PDR)—The percentage of packets received correctly at the destination with total packets sent from the source as shown in Fig. 50.2. PDR decreases with increase in number of nodes due to bandwidth limitation, some of the packets may be lost. Also, PDR for 100 packets transmission is better than 50 packets transmission.
2. *Attacker Node*—Attacker node is classified with the help of reception angle at the sink. During the attack, variations is observed in the node position and hence in the reception angle. Table 50.1 clearly classifies the attacker node which is having the reception angle difference beyond the threshold limit.
3. *Packet Loss*—Loss in the received number of packets at the sink node ensures the presence of adversary node in the communicating subnet as shown in Fig. 50.3. The loss is increasing as the number of nodes increases. Presence of attacker node which sends extra packets in the network leads to loss of original packets.

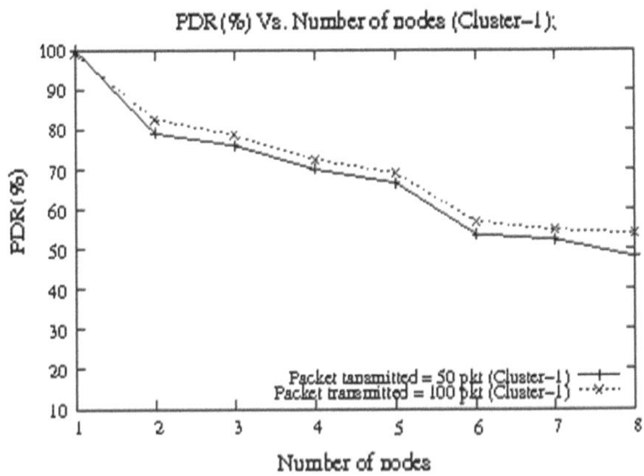

Fig. 50.2 PDR % versus number of nodes for single cluster

Table 50.1 Node reception angle

Node ID	Angle of reception at the sink in degrees	Angle difference with respect to sink node position
1	10.02498	3.72325
2	8.0494	1.74767
3	6.721369	0.149642
4	10.5197348	4.2180084
5	8.478713147	2.176986773
6	7.07055117	0.768824796
7	6.050746016	−0.250980358
8	17.54840061	11.24667373
9	35.264	28.96227363
10	19.47122	13.16949363

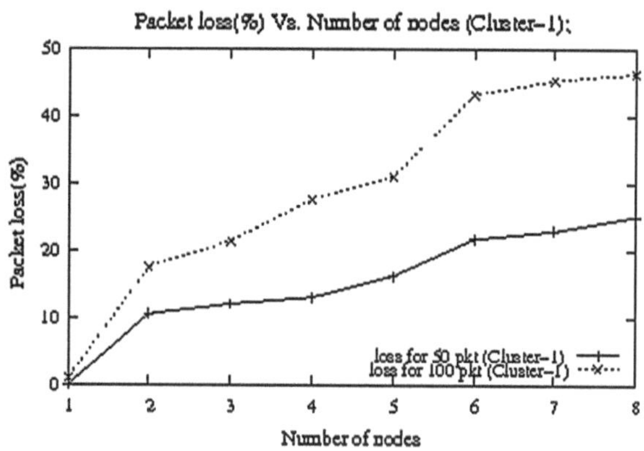

Fig. 50.3 Packet loss versus number of nodes

50.4 Conclusion

The need of security is alarming issue in any sort of network, especially with sensitive network such as wireless sensor network, which are highly vulnerable to security attacks, due to the highly obdurate constraints. These susceptible networks pose a great challenge for security provisioning. Every approach provides a great extent of security stipulation within its own arena.

This scheme provides a precautionary measure before an attack. This scheme helps in knowing the attacker before he attacks the sink node, which is our major focus to provide security. Since the sniffing nodes are deployed and are continuously active to protect the sink node, this is a proactive approach to monitor the entire network continuously providing an anti-attack protection shield to the sink node. This helps the sink node to save the data from the attacker.

Acknowledgments The authors wish to thank Visvesvaraya Technological University (VTU), Karnataka, India, for funding the part of the project under VTU Research Scheme (Grant No.VTU/Aca./2011-12/A-9/753, Dated: May 5, 2012.

References

1. Shetty S, Song, M, Ancel RAE, Bone K (2007) Wireless sensor payload design for sounding rocket. In: 22nd international conference on computers and their applications, March 2007
2. Culler D, Estrin D, Srivastava M (2004) Overview of sensor networks. IEEE Computer Society, Washington DC
3. Akylidiz I, Su, W, Sankarasubramaniam Y, Cayrici E (2002) A survey on sensor networks. IEEE Commun Mag 40(8):102–114
4. Singh R, Singh DK, Kumar L (2010) A review on security issues in wireless sensor network. J Inf Syst Commun 1(1):01–07
5. Ramen R, Lopez J, Gritzalis S (2008) Situation awareness mechanisms for wireless sensor networks. IEEE Commun Mag 46(4):102–107
6. Khanna R, Lin H (2008) Control theoretic approach to intrusion detection using distributed hidden Markov model. IEEE Wirel Commun 15(4):24–33
7. Kachirski O, Guha RK (2003) Effective intrusion detection using multiple sensors in wireless ad hoc networks. In: Proceeding of the 36th annual Hawaii international conference on system sciences, Hawaii, USA, Jan 2003, pp 57–64
8. Chan SP, Poovendran R, Sun MT (2005) A key Management Scheme in: Distributed Sensor Networks Using Attack Probabilities. Proc IEEE GLOBECOM 2:1007–1011
9. Rajasegarar S, Leckie C, Palansiwami M (2008) Anomaly detection in wireless sensor networks. IEEE Wirel Commun 15(4):34–40
10. Kakasageri MS, Manvi SS, Sorgavi GD (2006) Agent-based information access in wireless sensor networks. WSEAS Trans Commun 3(7):1369–1374
11. Cardei M, Wu J (2006) Energy-efficient coverage problems in wireless Ad-Hoc sensor networks. Comput Commun 29(4):413–420
12. Du X, Chen H-H (2008) Security in wireless sensor networks. IEEE Wirel Commun 15(4):60–66
13. Wang X, Gu W, Schosek K, Chellappan S, Xuan D (2005) Lifetime optimization of sensor networks under physical attacks. In: Proceedings of IEEE international conference on communications (ICC), May 2005, pp 32953301
14. Abrach H, Bhatti S, et.al (2003) MANTIS: System support for multimodal networks of in-situ sensors. In: 2nd ACM international workshop on wireless sensor networks and applications, pp 50–59
15. Blaze M, Feigenbaum J, Ioannidis J, Keromytis AD (1999) The key note trust management system version 2. Internet RFC 2704, Sept 1999
16. Christodorescu M, Jha S, Seshia S, Song D, Bryant R (2005) Semantics aware malware detection. In Proceedings of IEEE symposium on security and privacy, Oakland, May 2005
17. Christodorescu M, Jha S (2003) Static analysis of executables to detect malicious patterns. In: Proceedings of the 12th USENIX security symposium (Security 03), USENIX Association, Aug 2003, pp 169–186
18. Kruegel C, Robertson W, Vigna G (2004) Detecting kernel level root kits though binary analysis. In: Proceedings of the 20th annual computer security applications conference (ACSAC 04), Tucson, AZ, Dec 2004
19. Resnick P, Zeckhauser R (2000) Trust among strangers in internet transactions: empirical analysis of eBays reputation system. In: Proceedings of NBER workshop on empirical studies of electronic commerce, 2000

20. Chen H, Wagner D (2002) MOPS: an infrastructure for examining security properties of software. In: Proceedings of the 9th ACM conference on computer and communications security, Nov 2002, pp 235–244
21. Michiardi JP, Molva R (2002) Core: a collaborative reputation mechanism to enforce node cooperation in mobile Ad Hoc networks. In: IFIP communication and multimedia security conference
22. Marti S, Giuli TJ, Lai K, Baker M (2000) Mitigating routing misbehavior in mobile ad hoc networks. In: 6th ACM international conference on mobile computing and networking, Aug 2000, pp 255–265

Chapter 51
Ingress Flow Based Triple Token Bucket Traffic Control System for Distributed Networks

Veena S. Chakravarthi and M. Shilpa

Abstract Network processors in applications like cloud-based services call for sophisticated algorithms of traffic rate control mechanisms with different flexibility levels at different sites of the network. This paper presents the design, implementation, and analysis of distributed rate limiters, which work together to enforce a global rate limit across traffic aggregates at multiple sites, enabling the coordinated policing of a cloud-based service's network traffic. The implementation enforces global limit and ensures that the congestion/responsiveness appears as though it has passed through a single shared rate limiter.

Keywords Rate limiter · Bridge · Network processor · QOS · Average rate · Peak rate · Network management · Token bucket

51.1 Introduction

Network processors are the basic building blocks of today's high speed, high-demand, quality-oriented communication. Designing network processors requires a new programming paradigm and an in-depth understanding of network processing requirements. Smart traffic monitoring and controlling algorithms are getting implemented to achieve desired quality of service. The network processors are assessed by such sophisticated traffic management algorithms for a set of QOS and to achieve fairness in bandwidth. It is even truer for applications like cloud-based services. In such distributed networks, traffic has to be monitored and controlled based on different requirements at different network sites. These

V. S. Chakravarthi (✉) · M. Shilpa
Department of ECE, BNM Institute of Technology, Bangalore, India
e-mail: scveena@bnmit.org

M. Shilpa
e-mail: shilpachoti1486@gmail.com

V. Chakravarthi et al. (eds.), *Proceedings of International Conference on VLSI, Communication, Advanced Devices, Signals & Systems and Networking (VCASAN-2013)*, Lecture Notes in Electrical Engineering 258, DOI: 10.1007/978-81-322-1524-0_51,

demand different levels of flexibility at different nodes of the network connecting the cloud. The flexibility is achieved by implementing rate limiter's function based on the flow at the network site using triple token bucket algorithm which can be integrated to any network processor block.

51.1.1 The Network Traffic

The network traffic is characterized by different parameters like average rate, peak rate, burst rate, and maximum rate. It is required to configure these control parameters correctly to suit requirements of different nodes on distributed network. This work emphasizes generic rate limiter using triple token bucket algorithm which can be configured to suit the requirement of rate limiting to achieve traffic monitoring and control. The work demonstrates decision process of arriving at the right parameters through empirical formulae to achieve fair traffic throughput through a node in a distributed network.

51.1.2 Distributed Networks

Distributed network is a computing network system where the programming and data to be worked on are spread over more than one computer. It is a concept in which each station is connected to all adjacent stations rather than to a few switching points, as in a centralized system [1]. Cloud-based networking has recently revolutionized the distributed networks.

In distributed networks, it is essential to have a set of rate limiters together which subject the network traffic to a single aggregate global limit. While traffic policing is common in data centers and widespread in today's networks, such limiters typically enforce policy independently at each location [2]. For example, a resource provider with 10 hosting centers may wish to limit the total amount of traffic it carries for a particular service to 100 Mbps. Its current options are either to limit the service to 100 Mbps at each hosting center (running the risk that they may all use this limit simultaneously, resulting in 1 Gbps of total traffic) or to limit each center to a fixed portion (i.e., 10 Mbps), which overconstrains the service traffic aggregate and is unlikely to allow the service to consume its allocated budget unless traffic is perfectly balanced across the cloud [3].

51.2 Traditional Implementation

Traditional implementation of the scenario described in Fig. 51.1 would use one token bucket for limiting the overall rate of the network and another for limiting the rate of the individual nodes. The rate limit of the nodes is hard–configured, and any traffic exceeding the configured rate will be dropped as a policing action. This type of implementation does not assure high efficiency of bandwidth usage where some flows are inactive while the others are greedy [4]. The inactive flows are causing under utilization of the network bandwidth and greedy flows are being starved.

51.3 Flow-Based Rate Monitoring and Limiting

In this work, a novel approach is suggested to ensure fairness to all nodes in the network. The fairness is achieved through flow-based rate limiting. The incoming flow at each node is assessed by means of a traffic policer, and a weight is calculated based on the flow. These weights are used to change the rate of the rate limiter at each node.

Therefore, the rate limiter's parameters are soft-configured based on the incoming flow at each node. In Fig. 51.1, a distributed network is illustrated with 10 nodes as explained earlier. The overall network is rate-limited to 1 Gbps, and the individual nodes are limited to accept the traffic at a rate of 100 Mbps each.

The proposed approach monitors the incoming traffic at each node for a pre-defined time interval and calculates the weight of each node by an empirical formula,

$$Wi = \frac{flow[i]}{config\_rate[i]} \quad \forall i, \ where\ i \in network\ nodes$$

where flow[i] is the incoming flow rate and config_rate[i] is the configured rate limit of node i.

These weights are used to ensure fairness in the network. If the weight of a given node is equal to 1, then the node is understood to be using its configured rate. When the weight of a given node x is greater than 1, then the extra demand is

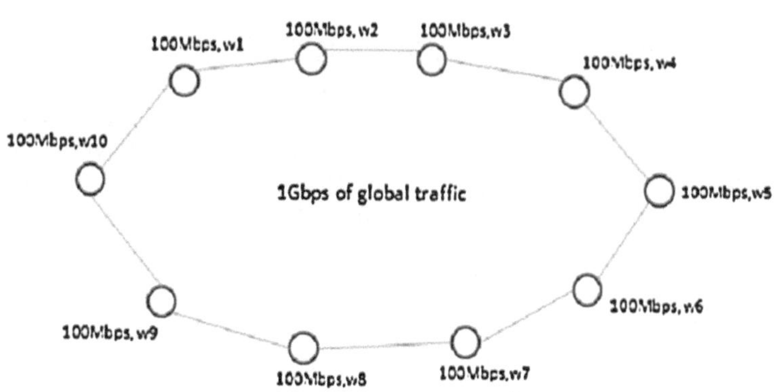

Fig. 51.1 Distributed network with 10 networking sites/nodes

catered to by providing the extra demand in proportion to the weight. The pseudocode for the weight calculation is shown below which is self-explanatory.

WEIGHT ()

Begin

For each flow of each node i, i = 1 to N.
r[i] ← ESTIMATE (f) //estimates the incoming flow rate
flow[i] ← r[i]
W[i] ← flow[i]/config_rate[i]
PROPAGATE (W[i]) //weights are propagated to all nodes
If (W[i] <= 1) **then**
Max_rate[i] ← config_rate[i]
Else
BURST_HANDLE_BLOCK (W[i])

End

Propagate () function takes care of broadcasting the weights of the nodes to all other nodes. This part is done in the upper layers of OSI as software configuration.

The work done in [3] estimates the demand with a different approach, and the weight calculation for each node based on the demand is done with a different empirical formulae.

51.4 Traffic Policing

The proposed approach enables the user to decide upon the services offered to the nodes based on the incoming flow which is calculated as demonstrated in [5] by using the formula,

$$R_{new} = (1 - e^{-Ti/K}) * \frac{Li}{Ti} + e^{-\frac{Ti}{K}} * R_{old}$$

where Li is the length of the packet at Ti and Ti is the interpacket arrival interval. Exponential averaging is used to estimate the rate of the flow. K is assigned a value between 100 and 500 ms [5]. The weights are calculated as mentioned in the previous section. Since token bucket algorithm is used for achieving rate limit function at each node, a lookup table is generated which consists of token bucket parameters customized for different weights of the network nodes. Based on the weights, a decision is taken whether to grant the extra demand for bandwidth or to deny the request.

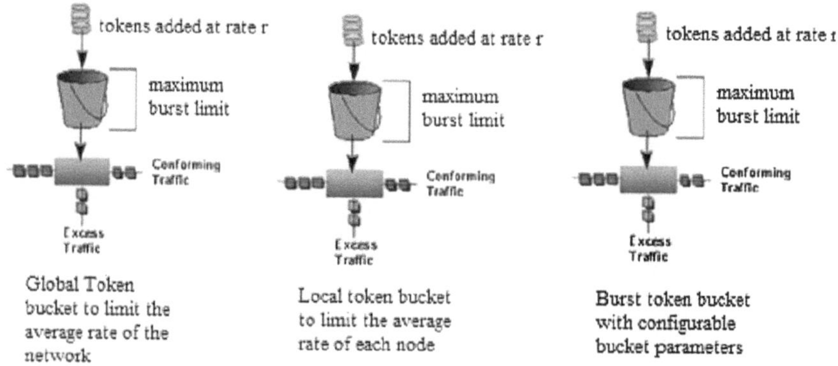

Fig. 51.2 Conceptual diagram of the triple token bucket (courtesy://vsaecomp.blogspot.in)

51.4.1 Triple Token Bucket Algorithm

Implementation of the flow-based traffic control for the distributed networks can be accomplished by using a triple token bucket algorithm (Fig. 51.2). A token bucket algorithm is a simple traffic shaping approach that permits burstiness but bounds it [6]. Token bucket is the most commonly used traffic policing algorithm because of its simplicity and ability to accommodate limited burstiness which is typical of data traffic [6].

Three token buckets are used to achieve the purpose of flow-based rate limit.

A global token bucket (GTB) is used for the rate limit function of the complete network, which discards any traffic that exceeds the average rate limit of the network.

A local token bucket (LTB) is used to limit the rate of the individual nodes, which is hard-configured to limit the average rate to 100 Mbps as discussed earlier.

A burst token bucket (BTB) is maintained for each node whose parameters are determined, depending on the weight of that particular node. These parameters are generated and saved a priory in the memory as a lookup table referenced by the weights of the nodes.

TOKEN_BUCKET (b, r)
Begin

Buck_max = FULL
GETTOKENS (r, timestamp, now) CONSUME (pkt_length) **End**

GETOKENS (r, timestamp)
Begin

When (pkt_arrived)
{now() ← time ()

delta ← now () - timestamp;
buck_occ = MIN (Buck_max, buck_occ + delta)

}

timestamp = now ()
end
CONSUME (pkt_length)
Begin

If (buck_occ > = pkt_length) **then**

Buck_occ ← buck_oc c - pkt_length

Else if (buck_occ < pkt_length)
Goto action block

End

Pseudocode for GETOKENS() and CONSUME() of TOKEN_BUCKET

BURST_HANDLE_BLOCK()
Begin

New_rate ← from LUT
New_buckmax ← from LUT
TOKEN_BUCKET (New_rate, New_buckmax)

End

Pseudocode for BURST_HANDLE_BLOCK

51.5 Results and Analysis

As in Fig. 51.3.a, the New_buckmax is the bucket depth configured for random traffic flow[i].

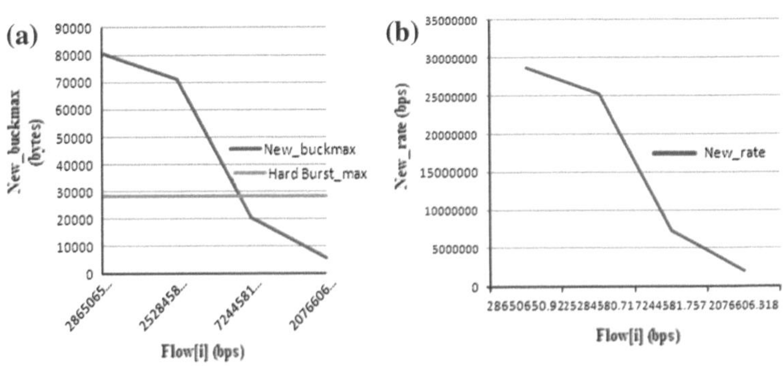

Fig. 51.3 **a** Plot of New_buckmax and hard burst_limit for different flow[i]. **b** Plot of New_rate parameter of the burst token bucket (BTB) for different flow[i]

It is evident that the proposed methodology provides flexibility in token bucket parameters based on the incoming flow. Figure 51.3.a shows the hard Burst_max which puts a hard limit on the bucket depth and hence not being able to use the bandwidth not used by other nodes.

51.6 Conclusion

The proposed approach provides fairness to network nodes by providing the share of bandwidth in proportion to their demands. Reconfigurability of the bucket parameters is the contribution of this work. The present work analyses random traffic for rate limiting. It is stated that in real scenario the traffic can be any of these distributions viz., poisson, bell curve and self-similar [7]. From this work, it is evident that fairness is achieved with comparable accuracy. The major challenge in this proposal would be the responsiveness of the system in reconfiguring the burst token bucket parameters to allow the demanded burst of traffic and the time interval for which the flow is monitored. Experimentation with different traffic profiles will help tackle this challenge. This methodology is critical in cloud-based services which operate their distributed networks with unprecedented levels of centralized control and where the services are subscription based.

References

1. Tang PP, Tai TY (2003) On Distributed Communications Networks. IEEE Transactions on Communications systems Vol. 1, pp 1–3
2. Packeteer. http://www.packeteer.com
3. Raghavan B, Vishwanath K, Ramabhadran S, Yocum K, Snoeren AC (2007) Cloud control with distributed rate limiting. In: Proceedings of the 2007 conference on applications, technologies, architectures, and protocols for computer communications, pp 337–348
4. Kim KS, Park E-C (2008) A token bucket based rate control algorithm with maximum and minimum rate constraints. In: IEICE Trans Commun E91-B(5):1623–1626
5. Stoicay I, Shenkerz S, Zhang H (2003) Core-stateless fair queuing: a scalable architecture to approximate fair bandwidth allocations in high speed networks. IEEE/ACM Trans Networking (TON) 11(1):33–46
6. Kidambi J, Ghosal D, Mukherjee B (1999) Dynamic token bucket (DTB): a fair bandwidth allocation algorithm for high-speed networks. In: Proceedings of the 8th international conference on computer communications and networks
7. Leland WE, Taqqu MS, Willinger W, Wilson DV (1994) On self similar nature of Ethernet traffic. IEEE/ACM Trans Networking (TON) 2(1):1–15

Chapter 52
An Algorithm to Find Minimal Cut-Sets of Complex Network

G. S. Prashanth and P. Manjunatha

Abstract Network reliability analysis is usually based on minimal path or cut enumeration from which the associated reliability expressions are deduced. The cut-set method is a popular approach in the reliability analysis of many systems from simple configurations to complex configurations. The computational requirements necessary to determine the minimal cut-sets of a network depend mainly on the complexity of the system. A new algorithm is presented in this paper to determine the minimal cut-sets. This algorithm can handle both simple and complex networks and considers both unidirectional and bidirectional branches. The applicability of the proposed technique is illustrated by application to a more complicated system.

Keywords Network reliability · Connection matrix · Minimal cut-sets · Minimal paths (tie-sets) · Incidence matrix

52.1 Introduction

Reliability evaluation is an important and integral element in the planning, design, and operation of engineering systems. The term 'reliability' is generally used to indicate the ability of a system to continue to perform its intended function. The cut-set method is a popular approach in the reliability analysis of many systems from simple configurations to complex configurations. Minimal cut-sets can be found from visual inspection, fault trees, and event trees. But these techniques

G. S. Prashanth (✉) · P. Manjunatha
Jawaharlal Nehru National College of Engineering, Shimoga, India
e-mail: gsp341@gmail.com

P. Manjunatha
e-mail: manjup.jnnce@gmail.com

V. Chakravarthi et al. (eds.), *Proceedings of International Conference on VLSI, Communication, Advanced Devices, Signals & Systems and Networking (VCASAN-2013)*, Lecture Notes in Electrical Engineering 258, DOI: 10.1007/978-81-322-1524-0_52,

become complicated for the complex configurations. Having the minimal paths, one way to determine the minimal cuts is to first find out the cut-sets of each path and then combine them with those of other paths in all possible ways [1, 2]. A new algorithm is presented in this paper to determine the minimal cut-sets. This algorithm can handle both simple and complex networks and considers both unidirectional and bidirectional branches. Computer software has been developed to deduce the minimal cut-sets of a network using this technique. The applicability of the proposed algorithm is illustrated by application to a more complicated system. Given the minimal cut-sets, the exact and approximate system reliability can be easily calculated using the existing techniques.

52.2 Proposed Algorithm

The proposed algorithm can handle both simple and complex networks and considers both unidirectional and bidirectional branches. The first step is to find the minimal tie-sets of the network using any of the existing techniques. Second step is to construct the incidence matrix from the minimal tie-sets. This algorithm checks all the possible combinations of the columns in the incidence matrix. By combining the elements of the columns of a combination using 'or' operation, the minimal cut-set is found. Two requisite to be a minimal cut-set are as follows:

1. It must be a cut-set.
2. Each component in a cut-set should contribute for its cut-set to become a cut-set.

The flowchart associated with the above steps is shown below (Fig 52.1).

52.2.1 Procedural Steps

1. Find the minimal tie-sets of the network using any of the existing techniques.
2. Establish the incidence matrix using the minimal tie-sets.
   Ex: If CEB, EAD, AB, and CD are the minimal tie-sets of a network. Its incidence matrix is shown in the Table 52.1.
3. There are five components in the incidence matrix('C', 'E', 'B', 'A', 'D'), and each component represents each column in the incidence matrix.
   Ex: In the incidence matrix Table 52.1, components 'C', 'E', 'B', 'A', and 'D' represent 2nd, 3rd 4th, 5th, and 6th columns, respectively.
4. As there are five components in the first row of the incidence matrix, there exists up to fifth order of cut-sets starting from the first order.

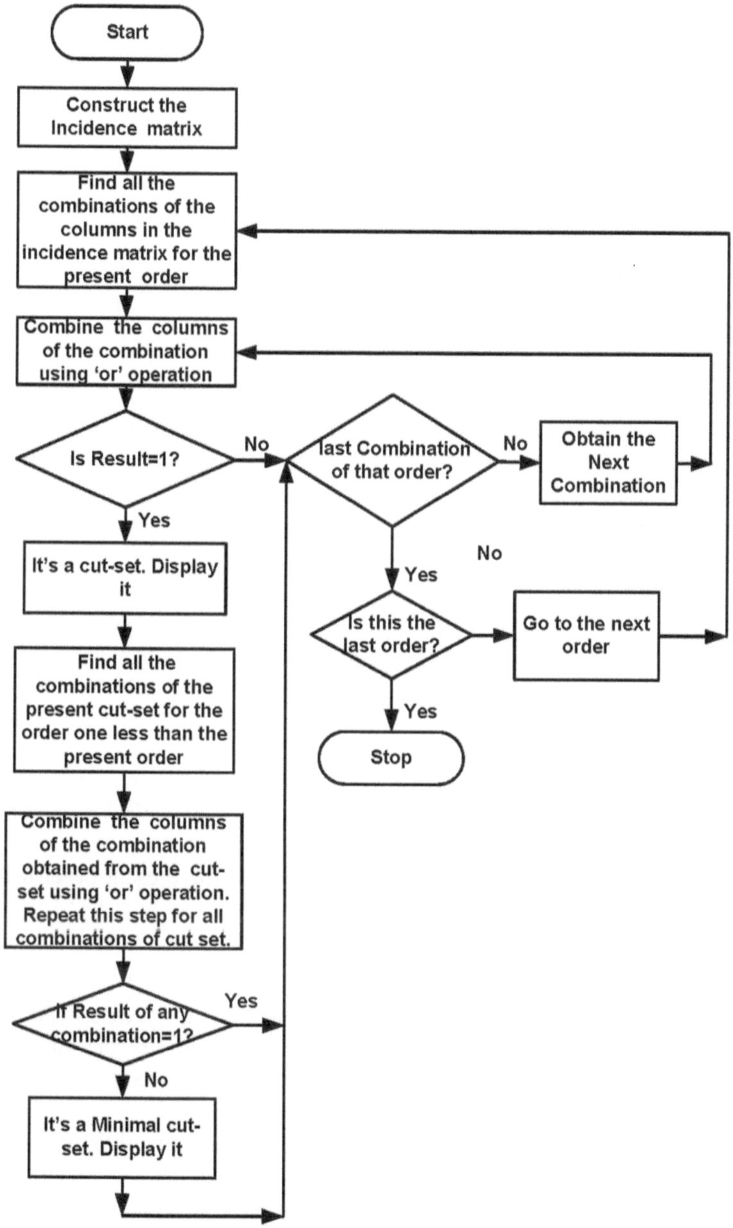

Fig. 52.1 Flowchart to determine the minimal cut-sets of complex network

5. If all the elements in the single column represented by any of the five components are one, then there exists a first order cut-set. As there are no columns with all ones in the above incidence matrix, first order cut-set does not exist.

Table 52.1 Incidence matrix

Minimal tie-sets	C	E	B	A	D
CEB	1	1	1	0	0
EAD	0	1	0	1	1
AB	0	0	1	1	0
CD	1	0	0	0	1

6. For the second order cut-sets, all the two possible ways of combinations of the five columns which are represented by five components are checked. The elements of two columns of each combination are combined using 'or' operation. If the output results in all ones, then that combination of components is a second order cut-set.
   Ex: Component 'C' represents the second column of the incidence matrix, and component 'A' represents fifth column of the incidence matrix. When second and fifth columns of the incidence matrix are combined using 'or' operation, it results in all ones. So, 'CA' is a second order cut-set. 'BD' is also a second order cut-set.
7. For the third order cut-sets, all the three possible combinations of the five columns represented by five components are checked. The elements of three columns of each combination are combined using 'or' operation. If the output results in all ones, then that combination of components is a third order cut-set.
   Ex: 'CEA' is the third order cut-set for the incidence matrix and is shown in Table 52.1.
8. In the similar way discussed above, higher order cut-sets are found. Procedure for determining the cut-sets and minimal cut-sets is explained in detail in the following steps.
9. Obtain the total number of combinations(count) of the present order using

$$n_{C_r} = \frac{n!}{r!(n-r)!} \tag{53.1}$$

where 'n' is the total number of individual components of the network appearing in the first row of the incidence matrix and 'r' is the order of the cut-set. Each individual component in the first row of the incidence matrix represents the each column of the incidence matrix.
   Ex: The total number of individual components of the network appearing in the first row of the above incidence matrix is 5. The individual components are 'C', 'E', 'B', 'A', and 'D'. If present order is one, then the count is $5_{C_1} = 5$.
10. Obtain all the combinations of the columns in the incidence matrix for the present order and store it.
    Ex: The five possible combinations of the first order are 'C', 'E', 'B', 'A', and 'D'. Each column represented by 'C', 'E', 'B', 'A', and 'D' is checked for all ones in their respective columns to find the first order cut-set. The ten possible

combinations of the second order are as follows: 'CE', 'CB', 'CA', 'CD', 'EB', EA, ED, BA, BD, and AD.

11. From the stored combinations, obtain the index of all the individual components of a particular combination.
    Ex: As the components 'C', 'E', 'B', 'A', and 'D' represent 2nd, 3rd, 4th, 5th, and 6th columns of the incidence matrix, the indexes are 2, 3, 4, 5, and 6 respectively.
12. Using the index of each individual component of the combination, combine the columns of the incidence matrix using the 'or' operation and store the result. If the result of all the elements is equal to one, it is a cut-set, else it is not a cut-set.
    Ex: If 'C', 'E', and 'B' components represent some columns in the incidence matrix, then CEB is a third order combination. Combine the columns represented by 'C', 'E', and 'B' using the 'or' operation. If the result of all the elements is equal to one, it is a cut-set.
13. Decrement the count. If it is a cut-set, move to the next step. Else, obtain the next combination of that order and repeat the steps from step 11.
14. If it is a cut-set, obtain all the combinations of that cut-set for the order one less than its present order and also find its number of combinations(Count2) using

$$r_{cp} = \frac{r!}{p!(r-p)!} \tag{53.2}$$

where 'r' is the present order and p = r-1. For third order cut-sets, r = 3 and p = 2. As $3_{c_2} = 3$, Count2 is 3. For the third order cut-set 'CEB', the three combinations are CE, CB, and EB.

15. Obtain the index of all the individual components of a combination for the present cut-set and combine their columns using 'or' operation. Decrement Count2 and repeat this step for all the combinations of a cut-set.
    Ex: CE, CB, and EB are the combinations of the cut-set CEB. For the combination CE, obtain the index of 'C' and 'E'. Combine the columns using 'or' operation and check whether CE is a cut-set or not. Use the same procedure for CB and EB. For the fourth order cut-set CEBA, the combinations are CEB, CEA, CBA, and EBA.
16. If none of the combinations of a cut-set after combining by 'or' operation is a cut-set, then the cut-set taken is a minimal cut-set. Else, it is just a cut-set, because each component should contribute for its cut-set to become a cut-set.
    Ex: CE, CB, and EB are the combinations of the cut-set CEB. If none of the combinations out of CE, CB, and EB are not the cut-set, then CEB is a minimal cut-set. If any one out of CE, CB, and EB is a cut-set, then CEB is not a minimal cut-set. For the fourth order cut-set CEBA, the combinations are CEB, CEA, CBA, and EBA are checked to find whether CEBA is a minimal cut-set or not.

17. Decrement the Count2. If Count2 is not equal to zero, repeat the steps from 15 to check whether the cut-sets are minimal cut-sets or not.
18. If Count2 is equal to zero, repeat the above steps from step 11 and check whether the next cut-set is a minimal cut-set or not.
19. If 'Count2' and 'count' both are zero, then increment the present order by one and get all the combinations of next order and repeat the above steps from step 9. Repeat the above steps for all the orders of the cut-set.

52.3 Implementation

We will explain the proposed algorithm step by step with bridge-type network.

1. Find the minimal tie-sets of the network using any of the existing techniques. Minimal tie-sets of the simple bridge-type network shown in the Fig. 52.2 are CEB, EAD, AB, and CD.
2. Establish the incidence matrix of the bridge-type network using minimal tie-sets. Incidence matrix is shown in Table 52.2.
3. Start with the order one. No single column in the incidence matrix contains all ones. So, first order minimal cut-set does not exist.
4. There are 5 individual components('C', 'E', 'B', 'A', 'D') in the incidence matrix, which represents the each column of the incidence matrix. So, there exists up to 5th order of cut-sets. For the second order cut-set, 'count' is $5_{c_2} = 10$. Combinations of second order are CE, CB, CA, CD, EB, EA, ED, BA, BD, and AD.
5. The component 'C' in the incidence matrix refers to the 2nd column, and 'E' refers to the 3rd column. The indexes of first combination 'CE' are 2 and 3. Combine the columns 2 and 3 using the 'or' operation and check the result of all the elements, it must produce all ones to be a cut-set.
6. As the columns 2 and 3 do not produce all ones when combined using 'or' operation, CE is not a cut-set. For the combinations AC and BD, 'or' operation produces all ones. So, AC and BD are the second order cut-sets. Once the cut-

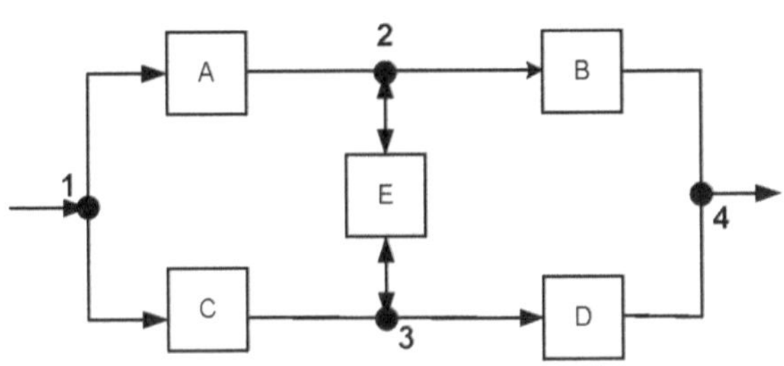

Fig. 52.2 Bridge-type network [1]

Table 52.2 Incidence matrix

Minimal tie-sets	C	E	B	A	D
CEB	1	1	1	0	0
EAD	0	1	0	1	1
AB	0	0	1	1	0
CD	1	0	0	0	1

sets are found, find whether it is a minimal cut-set or not using the below procedure.

7. For the second order cut-set, 'present order' is 2 and order one less than the present order is 1. So, $2_{c_1} = 2$. If it is a cut-set, obtain all the combinations of that cut-set for the order one less than its present order and check each combination of the cut-set. If none of its combination is a cut-set, then it is a minimal cut-set, else it is just a cut-set.
   Ex: For the cut-sets 'CA', the combinations are 'C' and 'A'. Both 'C' and 'A' are not the cut-sets individually. But when they are combined using 'or' operation, 'CA' becomes the cut-set. Here, 'C' and 'A' are both contributing for 'CA' to become a cut-set. So, 'CA' is a minimal cut-set. Same in case of 'BD'. So, 'CA' and 'BD' are the second order minimal cut-sets.
8. A number of 3rd order combinations are $5_{c_3} = 10$. Combinations are CEB, CEA, CED, CBA, CBD, CAD, EBA, EBD, EAD, and BAD. In these combinations, cut-sets are CEB, CEA, CBA, CBD, CAD, EBD, EAD, and BAD.
9. For the cut-set 'CEA', the combinations are CE, CA, and EA. As 'CA' is already a cut-set, 'CEA' is not a minimal cut-set. It is just a cut-set. For the cut-set 'CEB', the combinations are CE, EB, and CB. As none of the combinations of 'CEB' are cut-sets, CEB is a minimal cut-set. Same in case of EAD.
10. Third order minimal cut-sets are CEB and AED. There are no fourth and fifth order minimal cut-sets. As the number of components in the incidence matrix is five and we have checked up to the fifth order minimal cut-sets, further order cut-sets do not exist.

52.3.1 Sample Network

1. Minimal tie-sets are obtained from the 'Path Tracing Algorithm' [1]. For the sample network shown in the Fig. 52.3, minimal tie-sets are ABCJ, GCJ, ABCD, GHI, GCD, EHI, ABI, and EJ. The incidence matrix is constructed from minimal tie-sets as shown in the Table 52.3.
2. First order and second order minimal cut-sets do not exist for this network.
3. Third order minimal cut-sets are AGE, BGE, CJI, CIE, and JDI. Fourth order minimal cut-sets are BJDH, BJGI, BJGH, BCHE, BCJH, AJDH, AJGI, AJGH, ACHE, and ACJH. ← Further, order minimal cut-sets does not exist for the above sample network.

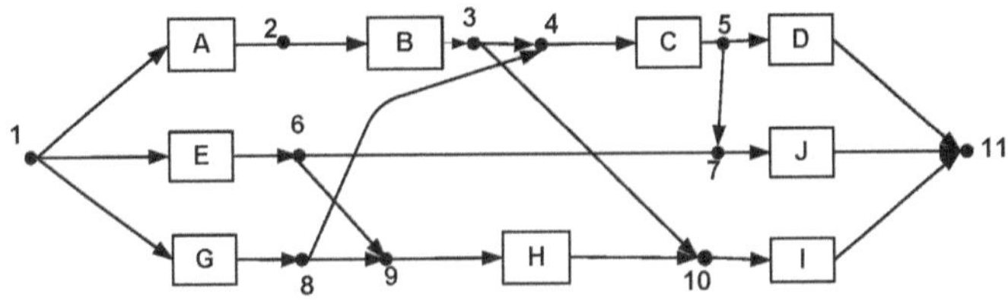

Fig. 52.3 Sample network [1]

Table 52.3 Incidence matrix

Minimal tie-sets	A	B	C	J	G	D	E	H	I
ABCJ	1	1	1	1	0	0	0	0	0
GCJ	0	0	1	1	1	0	0	0	0
ABCD	1	1	1	0	0	1	0	0	0
GHI	0	0	0	0	1	0	0	1	1
GCD	0	0	1	0	1	1	0	0	0
EHI	0	0	0	0	0	0	1	1	1
ABI	1	1	0	0	0	0	0	0	1
EJ	0	0	0	1	0	0	1	0	0

4. Having the minimal cut-sets, network reliability of the system can be found using the cut-set method. The network has 5 third order and 10 fourth order minimal cut-sets. If each component has a reliability of 0.9, its network reliability is 0.99459738.

52.4 Conclusion

As the complexity of the network increases, determining the minimal cut-sets becomes complicated. The proposed algorithm is easy to program and calculates the minimal cut-sets of simple networks as well as complex networks. Once the minimal cut-set of any network is found, its reliability can be found using the cut-set method.

References

1. Fotuhi-Firuzabad M, Billinton R, Munian TS, Vinayagam B (2004) A novel approach to determine minimal tie-sets of complex network. IEEE Trans Reliability 53(1):61
2. Billinton R, Allan R (1994) Reliability evaluation of engineering systems: concepts and techniques. Plenum Press, NY

Chapter 53
IPv6-Based Network Performance Metrics Using Active Measurements

N. Soumyalatha, Rakesh Kumar Ambhati and Manjunath R. Kounte

Abstract In the real-time network scenario, the network service providers need to ensure the quality of service (QoS) parameters. Network performance metrics (NPMs) are needed to measure the network performance and guarantee the QoS parameters like availability, delivery, latency, bandwidth which are also important for researchers and network equipment designers. One-way active measurement protocol (OWAMP) and two-way active measurement protocol (TWAMP) are the two active measurement approaches to measure the network performance. OWAMP measures one-way metrics and TWAMP measures two-way metrics. In this paper, currently prevalent active measurement methodologies and implementation of TWAMP approach are discussed. IPv6 TWAMP implementation for wireless networks is proposed, to obtain metrics, namely round-trip delay, two-way packet loss, jitter, packet reordering, packet duplication, and loss patterns.

Keywords Network performance metrics · Active measurement · Passive measurement · OWAMP · TWAMP

53.1 Introduction

IP networks enable users to transfer information in the form of voice, video, e-mail, and computer files. There are two Internet protocol versions: IP version 4 (IPv4) and IP version 6 (IPv6). IPv4 addresses are 32 bit long with 2^{32} addresses.

N. Soumyalatha (✉) · M. R. Kounte
Department of Electronics and Communication Engineering, REVA ITM, Bangalore, India
e-mail: soumyanaveen.u@gmail.com

M. R. Kounte
e-mail: manjunath.kounte@gmail.com

R. K. Ambhati
Central Research Laboratory, BEL, Bangalore, India
e-mail: rakesh.ambhati@gmail.com

V. Chakravarthi et al. (eds.), *Proceedings of International Conference on VLSI, Communication, Advanced Devices, Signals & Systems and Networking (VCASAN-2013)*, Lecture Notes in Electrical Engineering 258, DOI: 10.1007/978-81-322-1524-0_53,

IPv6 addresses are 128 bit long with 2^{128} address space. QoS is to be ensured by Internet Service Provider (ISP) to abide by the service level agreement (SLA) made between ISP and network users. NPMs are generally used by ISPs to measure the network performance and guarantee QoS parameters.

NPMs are useful [1] for network end users to cross-check QoS guaranteed by network provider and network equipment manufacturers for design and testing of newly developed networking equipments. It is also used by network researchers to experiment the quality of test networks deployed for research purposes.

53.1.1 Network Performance Metrics

As per RFC-2330, in an operational network, metric should have repeatable property (i.e., when same methodology is repeated under identical conditions, similar measurements must be obtained). Therefore, adhering to definition of metric, NPMs [2] are broadly categorized into four types: (1) availability, (2) loss, (3) delay, and (4) utilization. Each NPM is measured in terms of certain sub-metrics pertaining to it, by IP Performance Metrics Working Group (IPPMWG). These parameters mentioned below give insight into QoS of any fully operational network provided by network provider.

Brief explanations of NPMs are mentioned in the following sections:

Availability: Availability is measured in terms of connectivity and functionality in the network management layer. Parameters concerned to this metric are as follows: connectivity (RFC-2678) [3] and functionality.

Loss: Loss is the number of packet lost in transit from source to destination during a specific time interval. Parameters concerned to this metric are as follows: one-way loss (RFC-2680) [4], round-trip packet loss (RFC-6673) [5], and one-way loss patterns (RFC-3357) [6] (loss distance and loss period).

Loss distance: Difference in sequence numbers of two successively lost packets.

Loss period: The length of a packet loss event in successive lost packets.

Delay: Delay is the time taken for a packet to make the average round trip or one way from the sender to the distant destination and back. Parameters concerned to this metric are as follows: one-way delay (RFC-2679) [7], round-trip delay (RFC-2681) [8], and packet delay variation (RFC-3393) [9].

Utilization: Utilization is the throughput for the link expressed as a percentage of the access rate. Parameters concerned to this metric are as follows: link capacity, available bandwidth, and throughput.

53.1.2 Network Monitoring Methods

NPMs mentioned in previous section can be obtained by monitoring network passively or actively. Both passive monitoring and active monitoring approaches are mentioned below briefly.

(1) **Passive monitoring:**

Passive monitoring method needs additional hardware to be installed at the end points of network, currently being monitored, and all network traffic passing via installed hardware is logged. Logged network traffic is analyzed for the performance metrics required. This method does analysis on real-time traffic and is not intrusive in nature. Yet, this approach needs enormous storage space [10, 11]. Networking equipments such as gateways, routers, sniffers are required. One of the examples for passive monitoring is Wireshark.

Figure 53.1 depicts a typical passive monitoring metrics gathering scenario, where in all traffic passed via hub is logged into a database. This logged traffic is data mined later to obtain metrics of interest.

(2) **Active monitoring:**

Active monitoring method unlike passive monitoring injects additional traffic into network and thus consumes away the legitimate bandwidth of network that can be used for end-user applications. Therefore, any active monitoring protocol should test the network for concerned metrics with minimum consumption of available bandwidth.

Figure 53.2 depicts a typical active monitoring scenario wherein additional traffic is generated, i.e., time-stamped and injected into network from one network end point and same traffic is received at other end point. The packet received and sent time stamps are logged for deriving the metrics of interest.

Active monitoring approach unlike passive monitoring is intrusive in nature and does not require any special hardware installation. Both active monitoring and passive monitoring have advantages and disadvantages [12]. Focus of this paper is

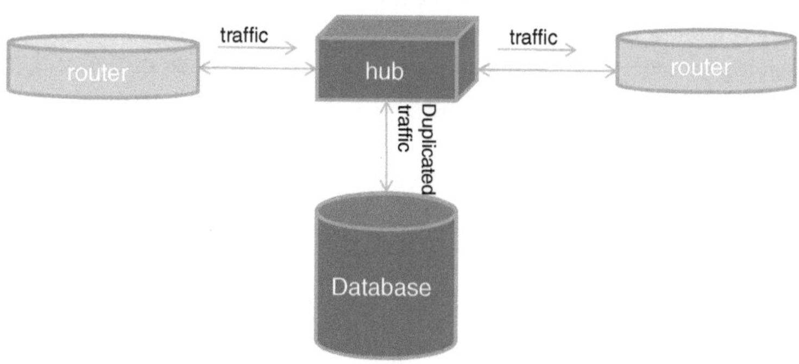

Fig. 53.1 Passive network monitoring

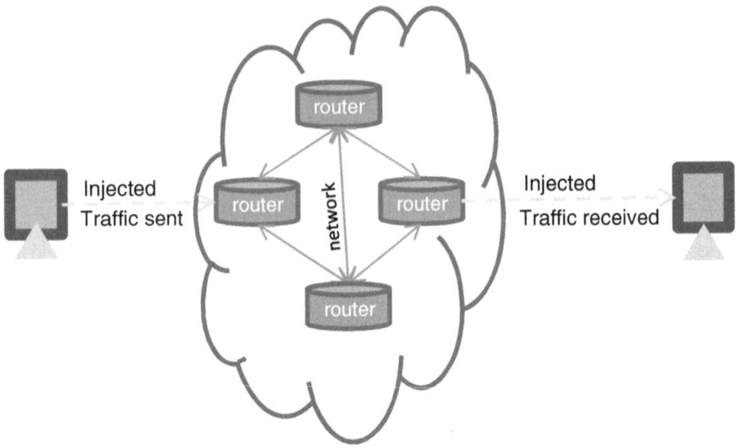

Fig. 53.2 Active network monitoring

on specific active monitoring strategies; therefore, two popular active measurement protocols and their working are briefly discussed in next section.

53.2 Active Measurement Protocols

Various active measurement tools [13], namely Pathchar, pchar, Cprobe, nettimer, Iperf, Ping, Owping, QoSMet, are available. Those are developed to obtain both one-way metrics and two-way metrics. Most of the tools not standardized and also based on Internet control message protocol (ICMP) are used; however, there are few limitations with these tools. Some routers reject the incoming ICMP packet because of security concerns.

OWAMP and TWAMP are active measurement protocols standardized by Internet Engineering Task Force (IFTF). Both generate UDP test traffic to obtain metrics. Both protocols use TCP connections to establish the initial control sessions between participating hosts. This TCP control session is used to negotiate the test session parameters like number of packets to be sent, size of the packet to be sent, UDP port to be used.

53.2.1 Owamp

One-way active measurement protocol defined in RFC-4656 measures one-way metrics such as one-way delay, one-way packet loss, one-way connectivity, one-way delay variation etc., across the network end points, by comparing the time stamps of the test packets on the sender's and receiver's end. Therefore, clocks of both the source and the destination should be synchronized.

OWAMP consists of two protocols, namely OWAMP-Control and OWAMP-Test.

OWAMP-Control: This protocol initiate, start, and stop test sessions and to fetch their results.

OWAMP-Test: This protocol exchanges test packets between two network nodes used to obtain metrics.

Figure 53.3 depicts the typical OWAMP architecture [14] and components involved in implementation of protocol.

Control-Client is a network node that starts and stops OWAMP-Test sessions.

Session-Sender is a network node, which sends test packets to the Session-Receiver during test sessions.

Session-Receiver receives the test packets to measure one-way metrics.

Server is a network node that facilitates one or more test sessions and publish metrics.

Fetch-Client is a network node that fetches results that are published by server.

The Control-Client starts a test session by sending an OWAMP-Control message to the server. Session-Sender sends test packets, which are time-stamped. Session-Receiver receives test packets, calculates relevant metrics, and sends results to server. Finally, the client fetches the result for analysis from the server.

(1) **Clock Synchronization.**

For measuring certain metrics, namely one-way delay, clocks must be synchronized in order to ensure that the metric is accurate. Clock synchronization [15] is achieved by using global positioning system (GPS), network time protocol (NTP), and precision time protocol (PTP). GPS gives most precise result, but it is more expensive. NTP gives accuracy of around few ms on a wide area network (WAN). PTP and GPS give nanosecond accuracy.

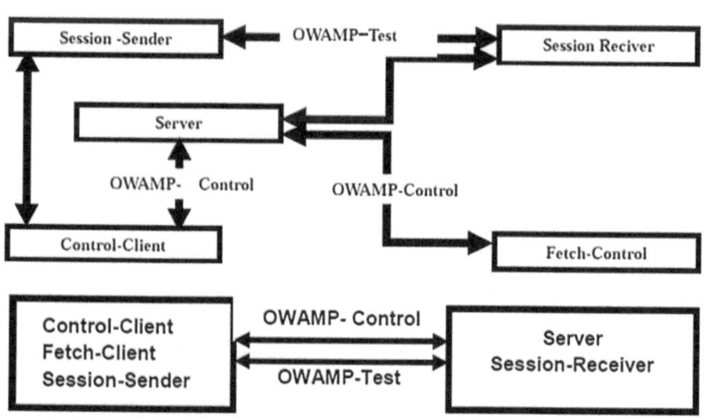

Fig. 53.3 OWAMP architecture

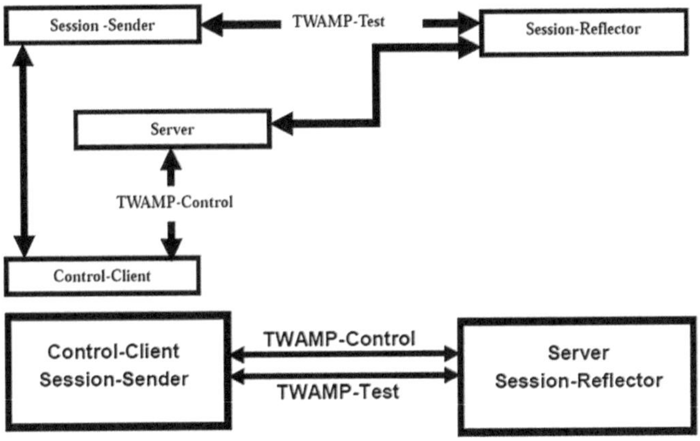

Fig. 53.4 TWAMP architecture

53.2.2 Twamp

Two-way active measurement protocol defined in RFC-5357, extension of OWAMP, is predominantly used to measure two-way metrics. Synchronization of clocks of hosts participating in protocol is not required to obtain two-way metrics, namely round-trip time, round-trip loss.

Figure 53.4 depicts the typical TWAMP architecture [16] and components involved in implementation of protocol.

TWAMP architecture is similar to OWAMP, except changes in the following modules:

Session-Receiver node is replaced by Session-Reflector node that is plainly reflecting test packets sent by Session-Sender, as part of test session.

Server component does not return the results of a test session as the Session-Reflector does not collect any results. Hence, Fetch-Client component is not required.

All the metrics are obtained, analyzed, and published by Session-Sender only.

53.3 Proposed System

The proposed system is targeted for IPv6-based wireless networks, especially to test 3G, Wi-Fi, Wi-Fi direct and developed on the Android Ice-cream Sandwich platform. The following NPMs are targeted, namely round-trip delay, round-trip loss, delay variation, packer reordering, packet duplication, loss distance, and loss period.

As part of the proposed system, two Android applications, namely Twamp-Client and Twamp-Server, are developed. Proposed system test setup is shown below in Fig. 53.5.

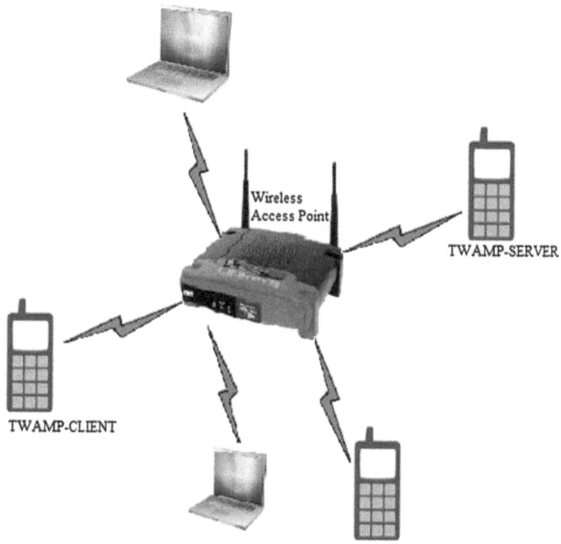

Fig. 53.5 Twamp-Client, Twamp-Server test setup

53.3.1 Wireless LAN

Wireless LAN is deployed using D-Link wireless routers. These wireless routers, which are IPv6-enabled, assign IPv6 addresses to mobile nodes that provide IPv6 wireless network.

53.3.2 Twamp-Server

Twamp-Server, Android application installed on mobile node, listens for possible Twamp-Client nodes on port 861 (as mentioned in RFC-5357). As Twamp-Client gets connected, Twamp protocol is initiated. Twamp-Server acts like "Session-Reflector" entity as mentioned in Twamp architecture.

53.3.3 Twamp-Client

Twamp-Client is another Android application installed on another mobile node and takes input from user, namely packet size, number of packets, payload pattern, delay between packets, and Twamp-Server IPv6 address. Twamp-Client sends control and session-initiation requests to Twamp-Server. Twamp-Client acts as Session-Sender as mentioned in Twamp architecture.

Twamp-Client initiates TCP connection with ***Twamp-Server*** that initiates ***TWAMP*** protocol control session. After successful establishment of control session, test session parameters are negotiated; final test results are cached and displayed at ***Twamp-Client*** application.

53.3.4 Test Results

Twamp-Client and Twamp-Server applications are run in congested wireless LAN, and following measurements pertaining to various metrics targeted are obtained and mentioned below.

Screenshot of the metrics obtained is shown in Fig. 53.6. Wherein the metrics, namely round-trip delay, packet lost, delay variation, packet duplication, reordering, loss distance, and loss period, are determined and displayed.

In the Table 53.1, first two columns show the metric under less network traffic scenario. Third and fourth columns show the metrics obtained under highly congested wireless network scenario. Nill indicates that no packet duplication and reordering.

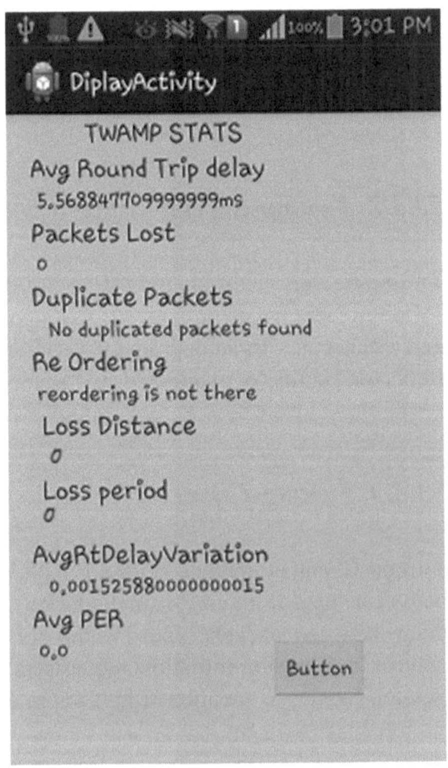

Fig. 53.6 Screenshot showing the results for 1,000 packets

Table 53.1 Test results

Packet size (in bytes) to number of packets	1000/1000	1200/2000	1200/5000	1300/10000
Average round-trip (RT) delay (ms)	5.5	6.0	7.5	9.0
RT loss	0	0	100	500
Jitter	0.00155	0.00165	0.00175	0.00195
Packet duplication	Nil	Nil	Nil	Nil
Packet reordering	Nil	Nil	Nil	Nil
Max loss distance	0	0	17	53
Max loss period	0	0	23	79

From the results in Table 53.1, it is clearly evident that under heavy traffic scenario, metrics such as average round-trip delay and loss distance and loss period undergo degradation.

53.3.5 Our Contributions

Currently, very less network performance measurement tools are available for measuring IPv6-based wireless network. Our contribution is developing a tool to measure the performance of IPv6-based wireless networks on Android platform by implementing TWAMP, which measures all two-way metrics such as round-trip time, round-trip packet loss, round-trip delay variation.

53.4 Conclusion

Current implementation of Twamp-Client and Twamp-Server Android application obtains metrics such as round-trip loss, round-trip delay, delay variation, packet duplication, packet reordering, loss distance, loss period for IPv6-based wireless network. In future work, Twamp-Client and Twamp-Server are intended to be implemented as Android service, which runs periodically to obtain more statistically accurate metrics.

References

1. Claffy K, Monk T (1997) What's next for internet data analysis? Status and challenges facing the community. In: Proceedings of the IEEE, California Univ, San Diego, La Jolla, CA, USA
2. Lee HJ, Kim MS, Hong JW, Lee GH (2002) QoS parameters to network performance metrics mapping for SLA monitoring. http://mail.apnoms.org/knom/knom-review/v5n2/4.pdf

3. Mahdavi J, Paxson V (1999) IPPM metrics for measuring connectivity. RFC-2498. http://www.ietf.org/rfc/rfc2678.txt
4. Almes G, Kaidindi S, Zekauskas M (1999) A one-way packet loss metrics for IPPM. RFC-2680. http://www.ietf.org/rfc/rfc2680.txt
5. Morton A (2012) Round-trip packet loss metrics. RFC- 6673. http://tools.ietf.org/html/rfc6673
6. Koodli R, Ravikanth R (2002) One-way loss pattern sample metrics. RFC-3357. http://tools.ietf.org/html/rfc3357
7. Almes G, Kalidindi S, Zekauskas M (1999) A one-way delay metrics for IPPM. RFC-2679. http://tools.ietf.org/html/rfc2679
8. Almes G, Kalidindi S, Zekauskas M (1999) A round-trip delay metric for IPPM. RFC-2681. http://tools.ietf.org/html/rfc2681
9. Demichelis C, Chimento P (2002) IP packet delay variation metric for IP performance metrics. RFC-3393. http://tools.ietf.org/html/rfc3393
10. Mohan V, Janardhan Reddy YR, Kalpana K (2011) Active and passive network measurements: a survey. Int J Comput Sci Inf Technol, ISSN: 0975-9646, 2(4):1372–1385
11. Calyam P, Krymskiy D, Sridharan M, Schopis P (2005) Active and passive measurements on campus, regional and national network backbone paths. Computer Communications and Networks, ICCCN 2005. In: Proceedings 14th International Conference on ISBN: 0-7803-9428-3, ISSN: 1095-2055, pp 537–542
12. Pezaros D, Hutchison D, Gardner RD, Garcia FJ, Sventek JS (2004) Inline measurements: a native measurement technique for IPv6. Networking and Communication Conference, INCC 2004. ISBN: 0-7803-8325-7, pp 105–110
13. Michaut F, Lepage F (2005) Application oriented network metrology: metrics and active measurement tools. Communications Surveys & Tutorials, IEEE ISSN: 1553-877X, 7(2):2–24
14. Shalunov S, Teitelbaum B, Karp A, Boote J, Zekauskas M (2006) A one-way active measurement protocol. RFC-4656. http://tools.ietf.org/html/rfc4656
15. Paxson V, Mahdavi J, Mathis M (1998) Framework for IP performance metrics. RFC 2330. http://tools.ietf.org/html/rfc2330
16. Hedayat H, Krzanowski R, Morton A, Yum K, Babiarz J (2008) A two-way active measurement protocol. RFC-5357. http://tools.ietf.org/html/rfc5357

Chapter 54
A Survey of Virtualization Techniques in Cloud Computing

Nivedita Manohar

Abstract The way of usage of computer is changed due to Internet and new technology called cloud computing. Cloud computing is the merging of another two new technologies like service-oriented architecture (SOA) and virtualization. The virtualization is gaining new goals and attracting those who need to perform the "miracle of multiplication," as it is to fit more information into less space. The present survey paper is addressing about the virtualization which is one of the key features of the cloud computing, types of virtualization and its approaches, types of hypervisor, and building private cloud with the virtualization. The security of cloud computing is also discussed with new idea of introducing optical network as an access network and its devices in the data centers for an energy efficient centers.

Keywords Cloud computing · Virtualization · Hypervisor

54.1 Introduction

Cloud computing is paradigm of distributed computing to provide the customers on demand and utility-based services. The new techniques like service -oriented architecture (SOA) and virtualization are combined for this paradigm. Cloud computing is partitioned into three parts as: providers, customers, and users. Providers provide services to end user with owing the physical resources as data centers, virtualization technology. This helps to customers to start their business without investing for physical hardware for which they can get rent from the providers. The services to be served by cloud are termed as "X" as a Service (XaaS). "X" may be infrastructure (IaaS), software (SaaS), platform (PaaS),

N. Manohar (✉)
Department of CSE, Alliance College of Engineering and Design,
Alliance University, Bangalore 562106, India
e-mail: nivedita.manohar@alliance.edu.in

V. Chakravarthi et al. (eds.), *Proceedings of International Conference on VLSI, Communication, Advanced Devices, Signals & Systems and Networking (VCASAN-2013)*, Lecture Notes in Electrical Engineering 258, DOI: 10.1007/978-81-322-1524-0_54,

storage (StaaS), security (SeaaS), etc. [1–4]. Google App Engine is an example for a Web platform as Service (PaaS) which enables to deploy and dynamically scale Python- and Java-based Web applications. Cloud computing is also the one of the IT resources and services that are abstracted from the underlying infrastructure and provided "On demand" and "at scale" in a "multitenant environment." Cloud computing is involved with these three key attributes. On demand which resources can be provisioned immediately when needed, released when no longer required and billed only when used. At Scale, it is a service to provide the illusion of infinite resources availability in order to meet whatever demands are made of it. Multitenant environment—the resources are provided to many consumers from a single implementation, saving the provider significant costs [5]. Cloud computing has its components as clients, the data center and distributed servers which have a specific role in delivering a functional cloud–based application. Clients are the devices that any end users use to interact as well as to manage their information on the cloud. There are three categories: mobile, thin client, and thick clients [6]. PDAs, smartphones, and iPhone are included in mobile. The thin client includes computers which do not have internal hard drives, the servers do all the work for it, and it displays the information. The thick clients are a regular computer uses a Web browser like Firefox or Internet Explorer to connect to the cloud. Second component of the cloud computing is data center which is the collection of servers where the application to which you subscribe is housed. It could be a large room in the basement of your building or a room full of servers on the other side of the world that you access via the Internet. The number of virtual servers that can exist on a physical server depends on the size and speed of the physical server and what applications will be running on the virtual server. Distributed servers are the other components of cloud computing. Reference model of cloud computing is as shown in Fig. 54.1. The reference model is stacked with Infrastructure as a Service (IaaS) is a base which delivers hardware on demand in the form of virtual hardware, storage, and networking. The provider creates virtual hardware based on the users' request. The users are also provided with tools as well as interfaces required to configure the software stack to be installed in the virtual machine. To execute the applications, Platform as a Service (PaaS) will support and it provide scalable and elastic runtime environments on demand to the applications of the users. Both applications and services are provided by the Software as a Service (SaaS). PaaS is preferable when for developing applications and are more appropriate when user requires developing new system [7].

IaaS is preferable when it is required to develop new system with services provided by SaaS. On the other hand, a large number of new companies also are spawned with competitive services relayed on those provided cloud computing systems. In terms of their provisions, these services are categorized into six as Data as a Service (Daas), Identity and Policy Management as a Service (IPMaaS), Network as a Service (NaaS), Infrastructure as a Service (IaaS) [8]. Every service is charged based on level of service, its characteristics, and quality.

The issues with cloud computing are multitenancy security issues, identity management issues, attacks by the insiders, virtualization issues, cryptography and

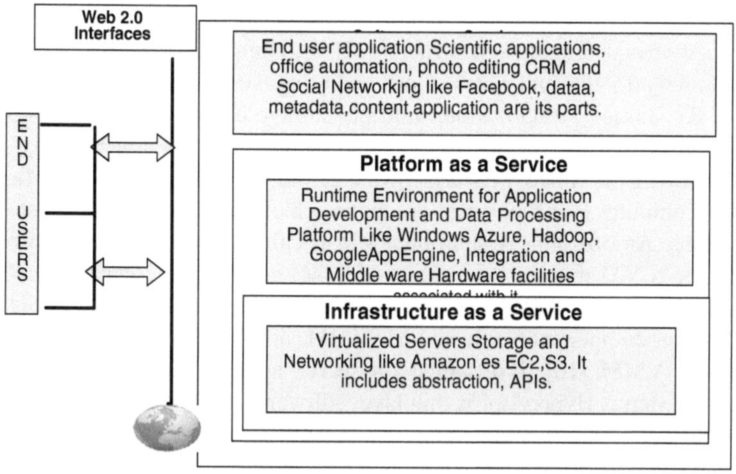

Fig. 54.1 Reference model of cloud computing

key management issues, software interface issues, migration of cloud and cloud providers, etc. This survey paper concentrates on virtualization-basic building block of cloud computing is the virtualization it allows different computing environments. Virtualization is a key element for cloud computing to achieve its objective. Virtualization helps to overcome the security challenges [9]. Virtualization is essential technology that allows creation of different computing environment which are named as virtual. Virtuals simulate the interfaces those are expected by the guest. Guest is the machine in cloud where the actual operating system runs. There are different types of virtualization like full virtualization and paravirtualization, based on the material to be virtualized, server virtualization, desktop virtualization, storage virtualization, network virtualization, etc. The next sections of this survey paper are focused on virtualization and its components, building a private cloud security of virtualized and non-virtualized cloud.

54.2 Virtualization and Its Components

Virtualization is a technology to helping organizations to optimize their application performance in a cost effective manner, but it can also present its share of application delivery challenges that cause some security risks. It hides the physical characteristics of computing resources, to simplify the way in which other systems, applications, or end users interact with those resources. There are different types of virtualization like server virtualization, desktop virtualization, storage virtualization, storage virtualization, and network virtualization. Server virtualization is also called as hardware virtualization.

It runs multiple operating system on a single machine. Desktop virtualization is also called virtual desktop infrastructure. Here, computing environment is delivered remotely to the end users. Host operating system and data are centrally located. By storage virtualization, multiple storage devices appear as a common shared media. Combining network resources and network functionality into a virtual network is called network virtualization. Virtual machine (VM) is the software computer like a physical computer which runs an operating system and applications. An operating system on a VM is called a guest operating system. A layer called a VM monitor or manager (VMM) creates and controls the VM are other subsystems. This VMM is referred as hypervisor [2]. As shown in Fig. 54.2, virtualization components include VM, VMM, and hardware. Hypervisor is also referred as VMM [10]. It installed on server and is responsible to run guest operating systems, Hypervisor is thin layer software. It allows multiple software to run concurrently on a host computer. It is capable to dynamically partitioning and sharing the physical resources such as CPU, storage, memory, and I/O devices [3, 4, 11]. In virtualization, may be full virtual or paravirtual depending upon requirement [12]. Full virtualization allows an unmodified Guest OS. Paravirtualization requires modification of guest OS to yield better performance but require an open source kernel. Xen is a hypervisor which allows paravirtualization. Virtualization is in no way mandatory for cloud computing, though it is certainly very common. There are several cloud providers that offer non-virtualized resources. This is commonly referred to as a bare-metal cloud. For example, SoftLayer offers bare-metal cloud. Bare-metal clouds are therefore “closer” to traditional data center hosting, though you typically still get an API that allows you to provision resources. Bare-metal clouds as shown in Fig. 54.4, will generally

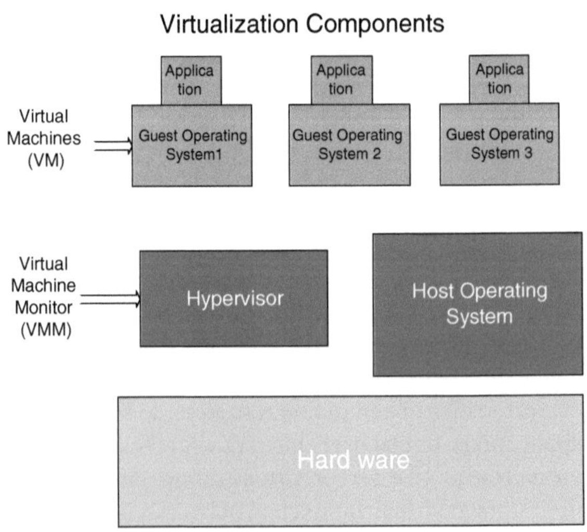

Fig. 54.2 Virtualization components

offer better performance when compared to a similar sized virtual resource as they do not carry the virtualization overhead theoretically. The removal of virtualization means just removes virtualization layer and retains the management capabilities needed by cloud. So each guest operating systems runs directly on hardware [11, 13].

The components of hypervisor are shown in the Fig. 54.3 [10]. Its main job is to control the sharing of system resources across multiple VMs. It is available at time of booting for above said function. The other function of hypervisor is to provide isolated environments for each virtual machine. The architecture in which privileged partitions have visibility and control over the VMs will establish the controllable environment and can perform additional security tool [11]. There are two types of hypervisors: Type 1 and Type 2 hypervisor as shown in Fig. 54.4 [10]. Type 1 hypervisor runs directly on bare metal instead of within an operating system environment. Such type of hypervisor provides the best performance and security. CitrxXen Server is an example of Type 1 hypervisor. Type 2 hypervisor runs with in an, operating system environment running on the host computer. Guest operating systems then run within virtual machine. This type of virtualization is called hosted virtualization. VMware server is an example for this type of hypervisor. Cloud middleware is another component of the cloud. A software which is called middleware is used to integrate services applications and content available on the cloud. The main key features of the cloud middleware are data management, identity, security, service hosting, mediation and management, user interface and portals, billing and metering and monitoring. All most all networking issues are handled by middleware. In the above section discussed about overview of components of cloud and in the following section, it is addressed about requirements to build own cloud as well as steps to it and issues with virtual cloud and non-virtual cloud. There are different approaches to the virtualization, with

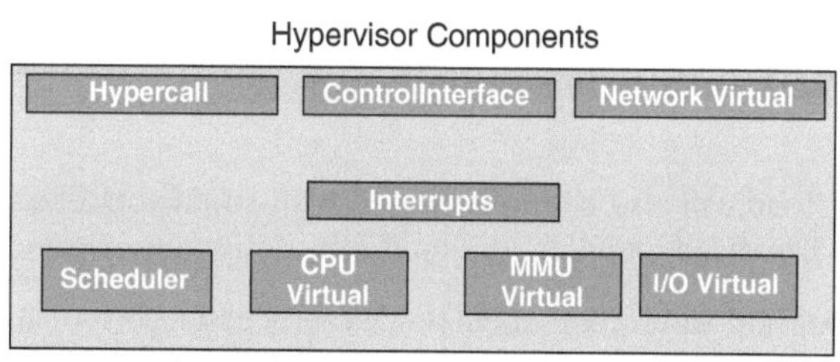

Fig. 54.3 Hypervisor components

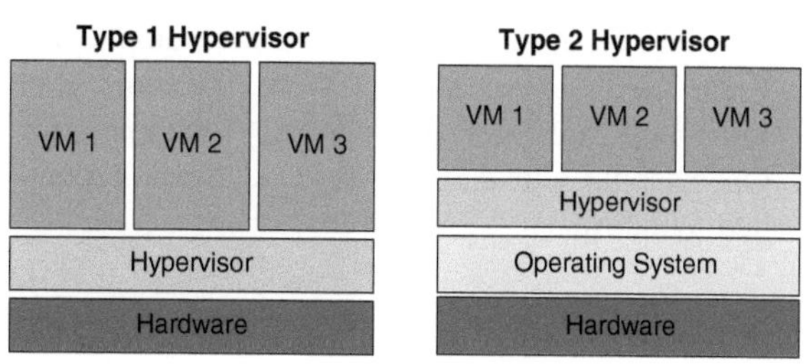

Fig. 54.4 Types of hypervisor

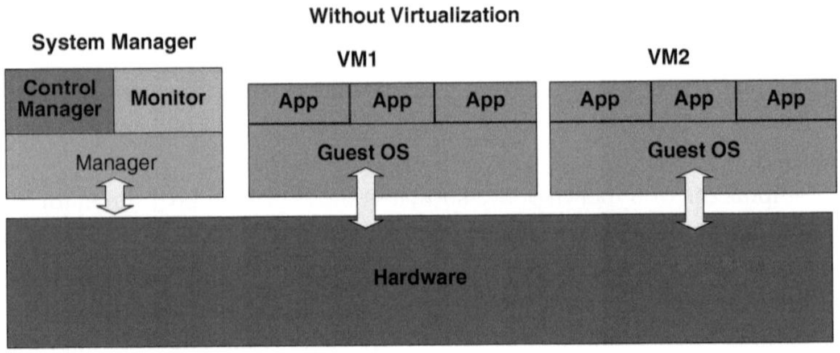

Fig. 54.5 Without virtualization

different pattern of control over VMs. These are classified as operating system–based virtualization, application-based virtualization, and hypervisor-based virtualization. In operating system-based virtualization, a single physical server is enabled by a hosting operating system and supports multiple virtualized guest operating systems. This supports to host operating system to have complete control on VMs. The other approach of virtualization is application-based virtualization, in which virtualized host is placed at top of the hosting operating system. Here, virtualization emulates VM. VM is having its own guest operating system as well as application required to run. Usually, this architecture is used in commercial environments. In the hypervisor-based virtualization, hypervisor is placed within the hardware infrastructure or the hosting operating system kernel [11]. In cloud computing, virtualization is main technique to create the cloud. The cloud can be built without hypervisor as shown in Fig. 54.5 [10].Next part of this section is about to create own private cloud with virtualization.

54.2.1 Building Own Private Cloud

Cloud can also be implemented with simulators like GreenCloud, VMcloud. Own virtualized cloud can be built with following requirements:

- Hardware requirements—One processor core or hyper thread for each virtualized CPU, RAM of minimum 2 GB, and hard disk minimum 6 GB free disk space per guest.
- Software requirements—hypervisor; examples for hypervisors are Xen, KVM, Citrix Xen Server, and VMWare. Xen and KVM are the open source, and remaining are the commercial. Examples for middleware are Nimbus, OpenNebula, Cloud Stock, and OpenStock. Examples for host are Linux, UNIX, and Windows.

54.2.2 Steps to Build A Private Cloud

- Install any Llinux/Unix OS on all machines.
- Middleware node and Hypervisor node(s).
  - Hypervisor Node(s): for one or more machine.
  - Middleware Node: One machine is sufficient.
- On Hypervisor node install co
  - Middleware Client Component
  - Hypervisor and Virtualization components.
  - Setup local DHCP server on hypervisor node to allocate IP address to VM running.
  - Configure Bridge Network.
- Middleware Node: Install M/W server. The inbuilt components are:
  - Scheduler
  - Network
  - Compute
  - Database
  - Storage Component that stores image of VM.

In this setup, there will be three nodes like Cloud client, Cloud Server, Hypervisor Node which have to interact with each other dynamically [2–4, 11]. VM creation is the next process, which is given below:

1. The user request is sent to middleware which act as a manager. At middleware, scheduler component in MWnode selects best physical node which acts like resource provider and provides.
2. CPUs, RAM, and IP to the VM from available nodes and transfers the image.
3. Now VM is ready to use. User name and password and IP are provided to the user through portal and user can access VM through SSH Connection.

Once the cloud is built, the deployment of cloud come to the picture which may be public cloud, private cloud, hybrid cloud which are not discussed in this paper. The following section is revealing about security of virtualized cloud.

54.3 Security of Virtualized Cloud

Security is one of the main issues with cloud which require in both virtualized and non-virtualized. In a virtualized cloud, hypervisor is responsible for the task like scheduling virtual machine, memory management, emulating, network packet processing, and starting virtual machine, stopping virtual machine, and migrating virtual machine. If the virtualization is based on the hypervisor, it is single point of failure because if attacker gets control over it, and then, he can get all VMs under his control. In operating system–based virtualized cloud, the attacker can inject his kernel script in hosting OS; by this, he can run all guest OS on this kernel [11]. The same threat is in case of application-based virtualization [11]. The other security risk of cloud computing is data integrity, abuse and nefarious, insecure interfaces and API malicious insiders, shared technology, data loss or leakage, account or service hijacking, and unknown risk profile. In data integrity, the stored data are going to be damaged during transition operations from or to the cloud storage provider. Cloud computing providers are actively being targeted partially because their relatively weak registration systems facilitate anonymity, and these cloud provider's fraud detection capabilities are limited. Insecure interfaces exposes to different security issues due to weak set of interfaces. Data centers are also consuming more power; so, vision must be there to have energy efficient data centers. For security purpose, it is essential to store the data in the service-level agreement (SLA) by the vendors and he has to gain the trust of customers for maintaining data center. There are many security benefits of non-virtualized cloud like making secured data available confidentially with integrity of data and software as well as side channels. [2, 13–16].

It is an idea to have a secure cloud computing with optical network. Optical network is one in which data transmission is in terms of light rather than bits and it is highly secured as well as consumes less power for transmission of information [16]. The data centers with fiber optic cables will be safe until an illegal fiber eavesdropping device in optical network placed for security [17, 18]. Data centers also consume more power which is another challenge to store data. Data centers can be built with redundant array of independent disk (RAID) with optical storage [19]. A secured virtualization can lead to a virtualized network which is prone to different types of security attacks that can be launched by a guest VM, an advanced cloud protection system, and it monitors the guest VM without being noticed, and hence any suspicious activity can be blocked and system's security system notified [8].

54.4 Conclusion and Future

The cloud computing is becoming one of the utilities of us. It will provide the basic level of computing service that is considered essential to meet the everyday needs of the general community with help of Internet. The new techniques like service-oriented architecture (SOA) and virtualization are combined for this paradigm. In this survey paper, virtualization, type of virtualization, and hypervisor components are studied, along with building a private cloud. Without virtualization, also a cloud can build, and it is seen that these clouds also secured in terms of availability, confidentiality, integrity, etc. From the study, non-virtual is suitable for the private cloud with all security solution. In future, a cloud can be build up with an optical network with or without virtualization so that security can be improved during transmission also. To build an energy efficient data centers, RAID technology can be implemented with optical storage. So, it is an idea to have optical network for access network in cloud computing to avoid malicious insiders, service hijacking, etc., with new devices to avoid eavesdropping.

Acknowledgments The encouragement given by all faculties of Alliance College of Engineering and Design, dean and stakeholders of this organization are highly acknowledged.

References

1. Tianfield H (2011) Cloud computing architectures, systems, man, and cybernetics (SMC). IEEE International Conference on ISBN 978-1-4577-0652-3, ISSN: 1062-922X, pp 1394–1399
2. Prahlad Rao BB, Saluja P, Sharma N, Mittal A, Sharma SV (2012) Cloud computing for IOT & sensing based applications. In: IEEE international conference on sensing technologies, ICST 2012, Kolkata, India, Dec 2012
3. Prahlad Rao BBP, Arackle VS, Tutorial HD (2012) Open source platforms for scientific clouds. CSI's annual conference on advances in cloud computing (ACC2012), Bangalore, 2012
4. Saluja P, Rao P, Mittal A, Ahmad R (2012) CDAC scientific cloud: on demand provisioning of resources for scientific application. In: Proceedings of 18th international conferences PDPTA 2012, Las vegas, Nevada, USA pp 16–19
5. Srinivasan MK, Sarukesi K, Rodrigues P, Manoj MS, Revathy P (2012) State-of-the-art cloud computing security taxonomies a classification of security challenges in the present cloud computing environment. ICACCI, CHENNAIACM 978-1-4503-1196-0/12/08, India
6. Ramaswami R, Sivarajan K, Sasaki G (2009) Optical networks: a practical perspective, 3rd Edition. Morgan Kaufmann Publication, Burlington. ISBN 978-0-12-374092-2
7. Troppen U, Erkens R, Muller W (2008) Storage networks explained : basic and applications of fibre channel SAN, NAS, ISCSI and Infiniband. Willey Publisher, New York, ISBN-13 9788126518326
8. Bhadauria R, Chaki R, Chaki N, Sanyal S (2011) A survey on security issues in cloud computing, arXiv:1109.5388v1[cs.CR]
9. AnthonyT, Velte T, Velte J, Elsenpeter R (2010) Cloud computing: a practical approach. Tata McGraw-Hill Edition, Noida, ISBN-13:978-0-07-068351-8
10. Prahlad Rao BBP, Arackle VS, Tutorial HD (2013) Cloud computing. PARACOMPTECH the national conference on parallel computing technologies, Bangalore, 2013
11. Sabahi F (2011) Virtualization level security in cloud computing, 3rd IEEE international conference on communication software and networks (ICCSN), pp 250–254, 978-1-61284-486-2
12. Buyya R, Vecchiola C, Selvi ST (2013) Mastering cloud computing. Mc-Graw Hill Education ISBN-13:978-25-902995-0
13. Keller E, Szefer J, Rexford J, Lee RB (2006) No-Hype: virtualization infrastructure without the virtualization. ISCA ACM 978-1-4-4503-0053
14. Kortchinsky K (2009) Hacking 3D (and Breaking out of VMWare). BlackHat, USA
15. Anonymous (2007) Xbox 360 hypervisor privilege escalation vulnerability. http://www.h-online.com/security/news/item/Xbox-360-hack-was-the-real-deal-732391.html
16. CVC-2007-4993 (2007) Xen guest root can escape to domain 0 through pygrub. http://cve.mitre,org/cgibin/cvename.cgi?name=CVE-2007-4993

17. Shaneman K, Gray S (2004) Optical network security: technical analysis of fiber tapping mechanisms and methods for detection & prevention. Military communications conference, MILCOM IEEE
18. Jessica T (2011) Connecting data centres over public networks. IPEXPO.ONLINE. http://online.ipexpo.co.uk/2011/04/20/connecting-data-centres-over-public-networks
19. Zhou M, Zhang R, Zeng D, Qian W (2010) Services in the cloud computing era: a survey. Universal communication symposium (IUCS), 4th international conference, pp 40–46, 2010
20. Yang CT, Huang KL, Liu JC, Chen WS, Hsg WH (2012) On construction of cloud IaaS using KVM and open nebula for video services. 41st international conference on parallel processing workshops, IEEE pp 212–220

Author Index

V. Chakravarthi et al. (eds.), *Proceedings of International Conference on VLSI, Communication, Advanced Devices, Signals & Systems and Networking (VCASAN-2013)*, Lecture Notes in Electrical Engineering 258, DOI: 10.1007/978-81-322-1524-0,